Springer Series in Statistics

Advisors:
P. Bickel, P. Diggle, S. Fienberg, U. Gather,
I. Olkin, S. Zeger

Springer Series in Statistics

Andersen/Borgan/Gill/Keiding: Statistical Models Based on Counting Processes.
Atkinson/Riani: Robust Diagnostic Regression Analysis.
Atkinson/Riani/Cerioli: Exploring Multivariate Data with the Forward Search.
Berger: Statistical Decision Theory and Bayesian Analysis, 2nd edition.
Borg/Groenen: Modern Multidimensional Scaling: Theory and Applications.
Brockwell/Davis: Time Series: Theory and Methods, 2nd edition.
Bucklew: Introduction to Rare Event Simulation.
Chan/Tong: Chaos: A Statistical Perspective.
Chen/Shao/Ibrahim: Monte Carlo Methods in Bayesian Computation.
Coles: An Introduction to Statistical Modeling of Extreme Values.
David/Edwards: Annotated Readings in the History of Statistics.
Devroye/Lugosi: Combinatorial Methods in Density Estimation.
Efromovich: Nonparametric Curve Estimation: Methods, Theory, and Applications.
Eggermont/LaRiccia: Maximum Penalized Likelihood Estimation, Volume I: Density Estimation.
Fahrmeir/Tutz: Multivariate Statistical Modelling Based on Generalized Linear Models, 2nd edition.
Fan/Yao: Nonlinear Time Series: Nonparametric and Parametric Methods.
Farebrother: Fitting Linear Relationships: A History of the Calculus of Observations 1750-1900.
Federer: Statistical Design and Analysis for Intercropping Experiments, Volume I: Two Crops.
Federer: Statistical Design and Analysis for Intercropping Experiments, Volume II: Three or More Crops.
Ghosh/Ramamoorthi: Bayesian Nonparametrics.
Glaz/Naus/Wallenstein: Scan Statistics.
Good: Permutation Tests: A Practical Guide to Resampling Methods for Testing Hypotheses, 2nd edition.
Good: Permutation Tests: Parametric and Bootstrap Tests of Hypotheses, 3rd edition.
Gouriéroux: ARCH Models and Financial Applications.
Gu: Smoothing Spline ANOVA Models.
Györfi/Kohler/Krzyżak/ Walk: A Distribution-Free Theory of Nonparametric Regression.
Haberman: Advanced Statistics, Volume I: Description of Populations.
Hall: The Bootstrap and Edgeworth Expansion.
Härdle: Smoothing Techniques: With Implementation in S.
Harrell: Regression Modeling Strategies: With Applications to Linear Models, Logistic Regression, and Survival Analysis.
Hart: Nonparametric Smoothing and Lack-of-Fit Tests.
Hastie/Tibshirani/Friedman: The Elements of Statistical Learning: Data Mining, Inference, and Prediction.
Hedayat/Sloane/Stufken: Orthogonal Arrays: Theory and Applications.
Heyde: Quasi-Likelihood and its Application: A General Approach to Optimal Parameter Estimation.

(continued after index)

Phillip Good

Permutation, Parametric and Bootstrap Tests of Hypotheses

Third Edition

With 22 Illustrations

Phillip Good
205 W. Utica Avenue
Huntington Beach, CA 92648
USA
pigood@verizon.net

Library of Congress Cataloging-in-Publication Data
Good, Phillip I.
   Permutation, Parametric and Bootstrap Tests of Hypotheses
Phillip Good.— 3rd ed.
     p. cm. — (Springer series in statistics)
   Includes bibliographical references and index.
   ISBN 0-387-20279-X (hardcover : alk. paper)
   1. Statistical hypothesis testing. 2. Resampling (Statistics) I. Title. II. Series.
QA277.G643 2004
519.5′6—dc22   2004050436

ISBN 0-387-20279-X

Printed on acid-free paper.

© 1994, 2000, 2005 Springer Science+Business Media, LLC
All rights reserved. This work may not be translated or copied in whole or in part without the written permission of the publisher (Springer Science+Business Media, LLC, 233 Spring Street, New York, NY 10013, USA), except for brief excerpts in connection with reviews or scholarly analysis. Use in connection with any form of information storage and retrieval, electronic adaptation, computer software, or by similar or dissimilar methodology now known or hereafter developed is forbidden.
The use in this publication of trade names, trademarks, service marks, and similar terms, even if they are not identified as such, is not to be taken as an expression of opinion as to whether or not they are subject to proprietary rights.

Printed in the United States of America.   (MVY)

9 8 7 6 5 4 3 2

springer.com

Preface to the Third Edition

This text is intended to provide a strong theoretical background in testing hypotheses and decision theory for those who will be practicing in the real world or who will be participating in the training of real-world statisticians and biostatisticians. In previous editions of this text, my rhetoric was somewhat tentative. I was saying, in effect, "Gee guys, permutation methods provide a practical real-world alternative to asymptotic parametric approximations. Why not give them a try?" But today, the theory, the software, and the hardware have come together. Distribution-free permutation procedures are the primary method for testing hypotheses. Parametric procedures and the bootstrap are to be reserved for the few situations in which they may be applicable. Four factors have forced this change:

1. Desire by workers in applied fields to use the most powerful statistic for their applications. Such workers may not be aware of the fundamental lemma of Neyman and Pearson, but they know that the statistic they want to use—a complex score or a ratio of scores, does not have an already well-tabulated distribution.
2. Pressure from regulatory agencies for the use of methods that yield exact significance levels, not approximations.
3. A growing recognition that most real-world data are drawn from mixtures of populations.
4. A growing recognition that missing data is inevitable, balanced designs the exception.

Thus, it seems natural that the theory of testing hypothesis and the more general decision theory in which it is embedded should be introduced via the permutation tests. On the other hand, certain relatively robust parametric tests such as Student's t continue to play an essential role in statistical practice.

As the present edition is intended to replace rather than supplement existing graduate level texts on testing hypotheses and decision theory, it includes material on parametric methods as well as the permutation tests and the bootstrap. The revised and expanded text includes many more real-world illustrations from economics, geology, law, and clinical trials. Also included in this new edition are a chapter on multifactor designs expounding on the theory of synchronous permutations developed by Fortunato Pesarin and his colleagues and sections on sequential analysis and adaptive treatment allocation.

Algebra and an understanding of discrete probability will take the reader through all but the appendix, which utilizes probability measures in its proofs. A one-semester graduate course would take the student through Chapters 1–3 and any portions of Chapters 4, 5, and Appendix that seem appropriate. The second semester would take the student through Chapters 6 and 7, and whatever portions of the remaining chapters seem germane to the instructor and students' interests.

An appendix utilizing measure theory has been provided for the benefit of the reader and instructor who may wish to have a mathematically rigorous foundation for the theory of testing hypotheses for continuous as well as discrete variables. For example, Section 2 of the appendix extends the proof of the Fundamental Lemma in Chapter 3 to the continuous case.

The number of exercises has been greatly increased from previous editions. Exercises range from the concept-driven, designed to develop the student's statistical intuition in practical settings, to the highly mathematical. Instructors are free to pick and choose in accordance with the mathematical and practical sophistication of their classes and the objectives of their courses.

To ensure greater comprehension of fundamental concepts, many essential results are now presented in the form of exercises. Although the primary motivation for this change came from instructors, feedback from the autodidact has persuaded us that full understanding can only be gained from actual usage.

Hopefully, this edition reflects the lessons I've learned from a series of interactive on-line courses offered through statistics.com. Immediate feedback from my students has forced me to revise the text again and again. The late Joseph Hodges once said, "The ideal mathematics lecture would be entirely free of symbols." The current text avoids symbols to the degree I am capable of doing so, their occasional use a weakness. Asymptotic results are avoided; the emphasis in this strongly theoretical work is on the practical.

If you find portions of this text particularly easy to understand the credit goes to Cliff Lunneborg for his insightful review of the entire text, to Fortunato Pesarin for his many contributions to Chapter 9, and to Norman Marshall for his comments on Chapter 8.

September 2004

Phillip Good
Huntington Beach, California
brother_unknown@yahoo.com

Preface to the Second Edition

In 1982, I published several issues of a *samdizat* scholarly journal called *Randomization* with the aid of an 8-bit, 1-MH personal computer with 48 K of memory (upgraded to 64 K later that year) and floppy disks that held 400 Kbytes. A decade later, working on the first edition of this text, I used a 16-bit, 33-MH computer with 1 Mb of memory and a 20-Mb hard disk. This preface to the second edition comes to you via a 32-bit, 300-MH computer with 64-Mb memory and a 4-Gb hard disk. And, yes, I paid a tenth of what I paid for my first computer.

This relationship between low-cost readily available computing power and the rising popularity of permutation tests is no coincidence. Simply put, it is faster today to compute an exact p-value than to look up an approximation in a table of the not-quite-appropriate statistic. As a result, more and more researchers are using Permutation Tests to analyze their data.

Of course, some of the increased usage has also come about through the increased availability of and improvements in off-the-shelf software, as can be seen in the revisions in this edition to Chapter 12 (Publishing Your Results) and Chapter 13 (Increasing Computation Efficiency).

These improvements helped persuade me it was the time to publish a first course in statistics based entirely on resampling methods (an idea first proposed by the late F.N. David). As a result, *Permutation Tests* has become two texts: one, *Resampling Methods*, designed as a first course, and this second edition aimed at upper division graduate students and practitioners who may already be familiar with the application of other statistical procedures. The popular question section at the end of each chapter now contains a number of thesis-level questions, which may or may not be solvable in their present form. While the wide applicability of permutation tests continues to be emphasized here, their limitations are also revealed. Examples include expanded sections on comparing variances (Chapter 3, Testing Hypotheses), testing interactions in balanced designs (Chapter 4, Experimental Design), and multiple regression (Chapter 7, Dependence).

Sections on sequential analysis (Chapter 4) and comparing spatial distributions (Chapter 8) are also new. Recent major advances in the analysis of multiple dependent tests are recorded in Chapter 5 on multivariate analysis.

My thanks to the many individuals who previewed chapters for this edition, including, in alphabetical order, Brian Cade, Mike Ernst, Barbara Heller, John Kimmel, Patrick Onghena, Fortunato Pesarin, and John Thaden.

April 2000
Phillip Good
Huntington Beach, California

Preface to the First Edition

Permutation tests permit us to choose the test statistic best suited to the task at hand. This freedom of choice opens up a thousand practical applications, including many which are beyond the reach of conventional parametric statistics. Flexible, robust in the face of missing data and violations of assumptions, the permutation test is among the most powerful of statistical procedures. Through sample size reduction, permutation tests can reduce the costs of experiments and surveys.

This text on the application of permutation tests in biology, medicine, science, and engineering may be used as a step-by-step self-guiding reference manual by research workers and as an intermediate text for undergraduates and graduates in statistics and the applied sciences with a first course in statistics and probability under their belts.

Research workers in the applied sciences are advised to read through Chapters 1 and 2 once quickly before proceeding to Chapters 3 through 8, which cover the principal applications they are likely to encounter in practice.

Chapter 9 is a must for the practitioner, with advice for coping with real-life emergencies such as missing or censored data, after-the-fact covariates, and outliers.

Chapter 10 uses practical applications in archeology, biology, climatology, education, and social science to show the research worker how to develop new permutation statistics to meet the needs of specific applications. The practitioner will find Chapter 10 a source of inspiration as well as a practical guide to the development of new and novel statistics.

The expert system in Chapter 11 will guide you to the correct statistic for your application. Chapter 12, more "must" reading, provides practical advice on experimental design and shows how to document the results of permutation tests for publication.

Chapter 13 describes techniques for reducing computation time; a guide to off-the-shelf statistical software is provided in an appendix.

The sequence of recommended readings is somewhat different for the student and will depend on whether he or she is studying the permutation tests by themselves or as part of a larger course on resampling methods encompassing both the permutation test and the bootstrap resampling method.

This book can replace a senior-level text on testing hypotheses. I have also found it of value in introducing students who are primarily mathematicians to the applications which make statistics a unique mathematical science. Chapters 1, 2, and 14 provide a comprehensive introduction to the theory. Despite its placement in the latter part of the text, Chapter 14, on the theory of permutation tests, is self-standing. Chapter 3 on applications also deserves a careful reading. Here in detail are the basic testing situations and the basic tests to be applied to them. Chapters 4, 5, and 6 may be used to supplement Chapter 3, time permitting (the first part of Chapter 6 describing the Fisher exact test is a must). Rather than skipping from section to section, it might be best for the student to consider one of these latter chapters in depth—supplementing his or her study with original research articles.

My own preference is to parallel discussions of permutation methods with discussion of a second resampling method, the bootstrap. Again, Chapters 1, 2, and 3—supplemented with portions of Chapter 14—are musts. Chapter 7, on tests of dependence, is a natural sequel. Students in statistical computing also are asked to program and test at least one of the advanced algorithms in Chapter 12.

For the reader's convenience the bibliography is divided into four parts: the first consists of 34 seminal articles; the second of two dozen background articles referred to in the text that are not directly concerned with permutation methods; the third of 111 articles on increasing computational efficiency; and a fourth, principal bibliography of 574 articles and books on the theory and application of permutation techniques.

Exercises are included at the end of each chapter to enhance and reinforce your understanding. But the best exercise of all is to substitute your own data for the examples in the text.

My thanks to Symantek, TSSI, and Perceptronics without whose GrandView® outliner, Exact® equation generator, and Einstein Writer® word processor this text would not have been possible.

I am deeply indebted to Mike Chernick for our frequent conversations and his many invaluable insights, to Mike Ernst, Alan Forsythe, Karim Hiriji, John Ludbrook, Reza Modarres, and William Schucany for reading and commenting on portions of this compuscript and to my instructors at Berkeley, including E. Fix, J. Hodges, E. Lehmann, and J. Neyman.

February 1994
Phillip Good
Huntington Beach, California

Contents

Preface to the Third Edition v

Preface to the Second Edition vii

Preface to the First Edition ix

1 **A Wide Range of Applications** 1
   1.1   Basic Concepts 1
       1.1.1   Stochastic Phenomena 1
       1.1.2   Distribution Functions 3
       1.1.3   Hypotheses 3
   1.2   Applications 5
   1.3   Testing a Hypothesis 7
       1.3.1   Five Steps to a Test 8
       1.3.2   Analyze the Experiment 8
       1.3.3   Choose a Test Statistic 8
       1.3.4   Compute the Test Statistic 9
       1.3.5   Determine the Frequency Distribution of
                the Test Statistic 9
       1.3.6   Make a Decision 10
       1.3.7   Variations on a Theme 10
   1.4   A Brief History of Statistics in Decision-Making 10
   1.5   Exercises 12

2 **Optimal Procedures** 13
   2.1   Defining Optimal 13
       2.1.1   Trustworthy 13
       2.1.2   Two Types of Error 14
       2.1.3   Losses and Risk 16
       2.1.4   Significance Level and Power 17

		2.1.4.1	Power and the Magnitude of the Effect	18
		2.1.4.2	Power and Sample Size	18
		2.1.4.3	Power and the Alternative	20
	2.1.5		Exact, Unbiased, Conservative	21
	2.1.6		Impartial	22
	2.1.7		Most Stringent Tests	23
2.2	Basic Assumptions			23
	2.2.1		Independent Observations	23
	2.2.2		Exchangeable Observations	24
2.3	Decision Theory			25
	2.3.1		Bayes' Risk	26
	2.3.2		Mini-Max	27
	2.3.3		Generalized Decisions	28
2.4	Exercises			29

3 Testing Hypotheses ... 33

3.1	Testing a Simple Hypothesis		33
3.2	One-Sample Tests for a Location Parameter		34
	3.2.1	A Permutation Test	34
	3.2.2	A Parametric Test	36
	3.2.3	Properties of the Parametric Test	37
	3.2.4	Student's t	38
	3.2.5	Properties of the Permutation Test	39
	3.2.6	Exact Significance Levels: A Digression	39
3.3	Confidence Intervals		40
	3.3.1	Confidence Intervals Based on Permutation Tests	41
	3.3.2	Confidence Intervals Based on Parametric Tests	42
	3.3.3	Confidence Intervals Based on the Bootstrap	43
	3.3.4	Parametric Bootstrap	45
	3.3.5	Better Confidence Intervals	46
3.4	Comparison Among the Test Procedures		46
3.5	One-Sample Tests for a Scale Parameter		48
	3.5.1	Semiparametric Tests	48
	3.5.2	Parametric Tests: Sufficiency	48
	3.5.3	Unbiased Tests	50
	3.5.4	Comparison Among the Test Procedures	50
3.6	Comparing the Location Parameters of Two Populations		51
	3.6.1	A UMPU Parametric Test: Student's t	51
	3.6.2	A UMPU Semiparametric Procedure	51
	3.6.3	An Example	54
	3.6.4	Comparison of the Tests: The Behrens–Fisher Problem	54

	3.7	Comparing the Dispersions of Two Populations	57
		3.7.1 The Parametric Approach	57
		3.7.2 The Permutation Approach	58
		3.7.3 The Bootstrap Approach	61
	3.8	Bivariate Correlation	62
	3.9	Which Test?	63
	3.10	Exercises	63
4	**Distributions**		67
	4.1	Properties of Independent Observations	67
	4.2	Binomial Distribution	67
	4.3	Poisson: Events Rare in Time and Space	68
		4.3.1 Applying the Poisson	69
		4.3.2 A Poisson Distribution of Poisson Distributions	70
		4.3.3 Comparing Two Poissons	70
	4.4	Time Between Events	71
	4.5	The Uniform Distribution	71
	4.6	The Exponential Family of Distributions	72
		4.6.1 Proofs of the Properties	73
		4.6.2 Normal Distribution	74
	4.7	Which Distribution?	75
	4.8	Exercises	75
5	**Multiple Tests**		79
	5.1	Controlling the Overall Error Rate	79
		5.1.1 Standardized Statistics	80
		5.1.2 Paired Sample Tests	81
	5.2	Combination of Independent Tests	81
		5.2.1 Omnibus Statistics	82
		5.2.2 Binomial Random Variables	82
		5.2.3 Bayes' Factor	83
	5.3	Exercises	84
6	**Experimental Designs**		85
	6.1	Invariance	85
		6.1.1 Some Examples	86
	6.2	k-Sample Comparisons—Least-Squares Loss Function	87
		6.2.1 Linear Hypotheses	87
		6.2.2 Large and Small Sample Properties of the F-ratio Test	89
		6.2.3 Discrete Data and Time-to-Event Data	90
	6.3	k-Sample Comparisons—Other Loss Functions	91
		6.3.1 F-ratio	91
		6.3.2 Pitman Correlation	92

xiv   Contents

    6.3.3 Effect of Ties 95
    6.3.4 Cochran–Armitage Test 96
    6.3.5 Linear Estimation 96
    6.3.6 A Unifying Theory 97
  6.4 Four Ways to Control Variation 97
    6.4.1 Control the Environment 98
    6.4.2 Block the Experiment 98
      6.4.2.1 Using Ranks 99
      6.4.2.2 Matched Pairs 100
    6.4.3 Measure Factors That Cannot Be Controlled 101
      6.4.3.1 Eliminate the Functional Relationship.... 101
      6.4.3.2 Selecting Variables..................... 102
      6.4.3.3 Restricted Randomization 102
    6.4.4 Randomize 103
  6.5 Latin Square.. 104
  6.6 Very Large Samples 106
  6.7 Sequential Analysis 107
    6.7.1 A Vaccine Trial 107
    6.7.2 Determining the Boundary Values 110
    6.7.3 Power of a Sequential Analysis 110
    6.7.4 Expected Sample Size 111
    6.7.5 Curtailed Inspection 112
    6.7.6 Restricted Sequential Sampling Schemes 112
  6.8 Sequentially Adaptive Treatment Allocation 113
    6.8.1 Group Sequential Trials 113
    6.8.2 Determining the Sampling Ratio 113
    6.8.3 Exact Random Allocation Tests 114
  6.9 Exercises 115

7 Multifactor Designs.. 119
  7.1 Multifactor Models 119
  7.2 Analysis of Variance 120
  7.3 Permutation Methods: Main Effects 124
    7.3.1 An Example 125
  7.4 Permutation Methods: Interactions 126
  7.5 Synchronized Rearrangements........................... 127
    7.5.1 Exchangeable and Weakly Exchangeable
      Variables....................................... 128
    7.5.2 Two Factors 129
    7.5.3 Three or More Factors 132
    7.5.4 Similarities 133
    7.5.5 Test for Interaction 135
  7.6 Unbalanced Designs 137
    7.6.1 Missing Combinations 138
    7.6.2 The Boot-Perm Test............................. 139

	7.7	Which Test Should You Use? 140
	7.8	Exercises 140

8 Categorical Data 143
    8.1 Fisher's Exact Test 143
        8.1.1 Hypergeometric Distribution.................... 145
        8.1.2 One-Tailed and Two-Tailed Tests 145
        8.1.3 The Two-Tailed Test 146
        8.1.4 Determining the p-Value 146
        8.1.5 What is the Alternative? 148
        8.1.6 Increasing the Power 148
        8.1.7 Ongoing Controversy 149
    8.2 Odds Ratio 150
        8.2.1 Stratified 2×2's 151
    8.3 Exact Significance Levels 152
    8.4 Unordered $r \times c$ Contingency Tables 154
        8.4.1 Agreement Between Observers 156
        8.4.2 What Should We Randomize? 157
        8.4.3 Underlying Assumptions 158
        8.4.4 Symmetric Contingency Tables 158
    8.5 Ordered Contingency Tables 160
        8.5.1 Ordered $2 \times c$ Tables 160
            8.5.1.1 Alternative Hypotheses................. 161
            8.5.1.2 Back-up Statistics 162
            8.5.1.3 Directed Chi-Square 162
        8.5.2 More Than Two Rows and Two Columns 163
            8.5.2.1 Singly Ordered Tables................. 163
            8.5.2.2 Doubly Ordered Tables................ 163
    8.6 Covariates 164
        8.6.1 Bross' Method 164
        8.6.2 Blocking 165
    8.7 Exercises 166

9 Multivariate Analysis 169
    9.1 Nonparametric Combination of Univariate Tests........... 169
    9.2 Parametric Approach 171
        9.2.1 Canonical Form 171
        9.2.2 Hotelling's T^2 172
        9.2.3 Multivariate Analysis of Variance
            (MANOVA) 173
    9.3 Permutation Methods.................................. 173
        9.3.1 Which Test—Parametric or Permutation? 175
        9.3.2 Interpreting the Results......................... 176

9.4	Alternative Statistics	177
	9.4.1 Maximum-t	177
	9.4.2 Block Effects	177
	9.4.3 Runs Test	178
	9.4.4 Which Statistic?	180
9.5	Repeated Measures	181
	9.5.1 An Example	181
	9.5.2 Matched Pairs	182
	9.5.3 Response Profiles	183
	9.5.4 Missing Data	183
	9.5.5 Bioequivalence	184
9.6	Exercises	185

10 Clustering in Time and Space ... 189
- 10.1 The Generalized Quadratic Form ... 189
  - 10.1.1 Mantel's U ... 189
  - 10.1.2 An Example ... 189
- 10.2 Applications ... 190
  - 10.2.1 The MRPP Statistic ... 190
  - 10.2.2 The BW Statistic of Cliff and Ord [1973] ... 191
  - 10.2.3 Equivalances ... 192
  - 10.2.4 Extensions ... 192
  - 10.2.5 Another Dimension ... 192
- 10.3 Alternate Approaches ... 193
  - 10.3.1 Quadrant Density ... 193
  - 10.3.2 Nearest-Neighbor Analysis ... 193
  - 10.3.3 Comparing Two Spatial Distributions ... 193
- 10.4 Exercises ... 194

11 Coping with Disaster ... 195
- 11.1 Missing Data ... 195
- 11.2 Covariates After the Fact ... 197
  - 11.2.1 Observational Studies ... 197
- 11.3 Outliers ... 198
  - 11.3.1 Original Data ... 199
  - 11.3.2 Ranks ... 199
  - 11.3.3 Scores ... 200
  - 11.3.4 Robust Transformations ... 201
  - 11.3.5 Use an L_1 Test ... 201
  - 11.3.6 Censoring ... 201
  - 11.3.7 Discarding ... 202
- 11.4 Censored Data ... 202
  - 11.4.1 GAMP Tests ... 202
  - 11.4.2 Fishery and Animal Counts ... 204

	11.5	Censored Match Pairs 204
		11.5.1 GAMP Test for Matched Pairs 205
		11.5.2 Ranks 206
		11.5.3 One-Sample: Bootstrap Estimates 206
	11.6	Adaptive Tests 207
	11.7	Exercises 208

12 Solving the Unsolved and the Insolvable 209

- 12.1 Key Criteria 209
  - 12.1.1 Sufficient Statistics 209
  - 12.1.2 Three Stratagems 210
  - 12.1.3 Restrict the Alternatives 210
  - 12.1.4 Consider the Loss Function 212
  - 12.1.5 Impartiality 213
- 12.2 The Permutation Distribution.............................. 213
  - 12.2.1 Ensuring Exchangeability 213
    - 12.2.1.1 Test for Parallelism 214
    - 12.2.1.2 Linear Transforms That Preserve Exchangeability 215
- 12.3 New Statistics 216
  - 12.3.1 Nonresponders 216
    - 12.3.1.1 Extension to K-samples 217
  - 12.3.2 Animal Movement 217
  - 12.3.3 The Building Blocks of Life..................... 217
  - 12.3.4 Structured Exploratory Data Analysis 218
  - 12.3.5 Comparing Multiple Methods of Assessment 219
- 12.4 Model Validation 221
  - 12.4.1 Regression Models.............................. 221
    - 12.4.1.1 Via the Bootstrap 221
    - 12.4.1.2 Via Permutation Tests 221
  - 12.4.2 Models With a Metric 222
- 12.5 Bootstrap Confidence Intervals 223
  - 12.5.1 Hall–Wilson Criteria............................ 224
  - 12.5.2 Bias-Corrected Percentile 225
- 12.6 Exercises 226

13 Publishing Your Results 229

- 13.1 Design Methodology 229
  - 13.1.1 Randomization in Assignment 229
  - 13.1.2 Choosing the Experimental Unit 230
  - 13.1.3 Determining Sample Size 231
  - 13.1.4 Power Comparisons 231
- 13.2 Preparing Manuscripts for Publication 231
  - 13.2.1 Reportable Elements 232
  - 13.2.2 Details of the Analysis 232

14 Increasing Computational Efficiency 233
- 14.1 Seven Techniques 233
- 14.2 Monte Carlo 233
  - 14.2.1 Stopping Rules 234
  - 14.2.2 Variance of the Result 235
  - 14.2.3 Cutting the Computation Time 235
- 14.3 Rapid Enumeration and Selection Algorithms 236
  - 14.3.1 Matched Pairs 236
- 14.4 Recursive Relationships 236
- 14.5 Focus on the Tails 237
  - 14.5.1 Contingency Tables 239
    - 14.5.1.1 Network Representation 239
    - 14.5.1.2 The Network Algorithm 241
  - 14.5.2 Play the Winner Allocation 242
  - 14.5.3 Directed Vertex Peeling 242
- 14.6 Gibbs Sampling 243
  - 14.6.1 Metropolis–Hastings Sampling Methods 244
- 14.7 Characteristic Functions 245
- 14.8 Asymptotic Approximations 246
  - 14.8.1 A Central Limit Theorem 246
  - 14.8.2 Edgeworth Expansions 246
  - 14.8.3 Generalized Correlation 247
- 14.9 Confidence Intervals 247
- 14.10 Sample Size and Power 248
  - 14.10.1 Simulations 248
  - 14.10.2 Network Algorithms 249
- 14.11 Some Conclusions 250
- 14.12 Software 251
  - 14.12.1 Do-It-Yourself 251
  - 14.12.2 Complete Packages 252
    - 14.12.2.1 Freeware 252
    - 14.12.2.2 Shareware 252
    - 14.12.2.3 $$$$ 252
- 14.13 Exercises 253

Appendix: Theory of Testing Hypotheses 255
- A.1 Probability 255
- A.2 The Fundamental Lemma 257
- A.3 Two-Sided Tests 258
  - A.3.1 One-Parameter Exponential Families 259
- A.4 Tests for Multiparameter Families 262
  - A.4.1 Basu's Theorem 262
  - A.4.2 Conditional Probability and Expectation 263
  - A.4.3 Multiparameter Exponential Families 263

A.5		Exchangeable Observations	268
	A.5.1	Order Statistics	269
	A.5.2	Transformably Exchangeable	270
	A.5.3	Exchangeability-Preserving Transforms	271
A.6		Confidence Intervals	272
A.7		Asymptotic Behavior	273
	A.7.1	A Theorem on Linear Forms	273
	A.7.2	Monte Carlo	274
	A.7.3	Asymptotic Efficiency	274
	A.7.4	Exchangeability	275
	A.7.5	Improved Bootstrap Confidence Intervals	276
A.8		Exercises	276

Bibliography 279

Author Index 303

Subject Index 309

1
A Wide Range of Applications

This is a book about testing hypotheses; more accurately, it is a book about making decisions. In this chapter we introduce some basic concepts in statistics related to decision theory, including events, random variables, samples, variation, and hypothesis. We consider a simple example of decision-making under uncertainty and review the history of statistics in decision-making.

1.1 Basic Concepts

1.1.1 Stochastic Phenomena

The two factors that distinguish the statistical from the deterministic approach are variation and the possibility of error. The effect of this variation is that a distribution of values takes the place of a single, unique outcome.

I found freshman physics extremely satisfying. Boyle's Law, $V = kT/P$, with its tidy relationship between the volume, temperature, and pressure of a perfect gas is just one example of the perfection I encountered there. The problem was I could never quite duplicate this (or any other) law in the freshman physics' laboratory. Maybe it was the measuring instruments, my lack of familiarity with the equipment, or simple measurement error, but I kept getting different values for the constant k.

By now I know that variation is the norm—particularly in the clinical and biological areas. Instead of getting a fixed, reproducible V to correspond to a specific T and P, one ends up, due to errors in measurement, with a distribution F of values instead. But I also know that with a large enough sample the mean and shape of this distribution are reproducible.

Figure 1.1a and 1.1b depict two such distributions. The first is a *normal* or *Gaussian distribution*. Examining the distribution curve, we see that the normally distributed variable can take all possible values between $-\infty$ and $+\infty$, but most of the time it takes values that are close to its median (and mean).

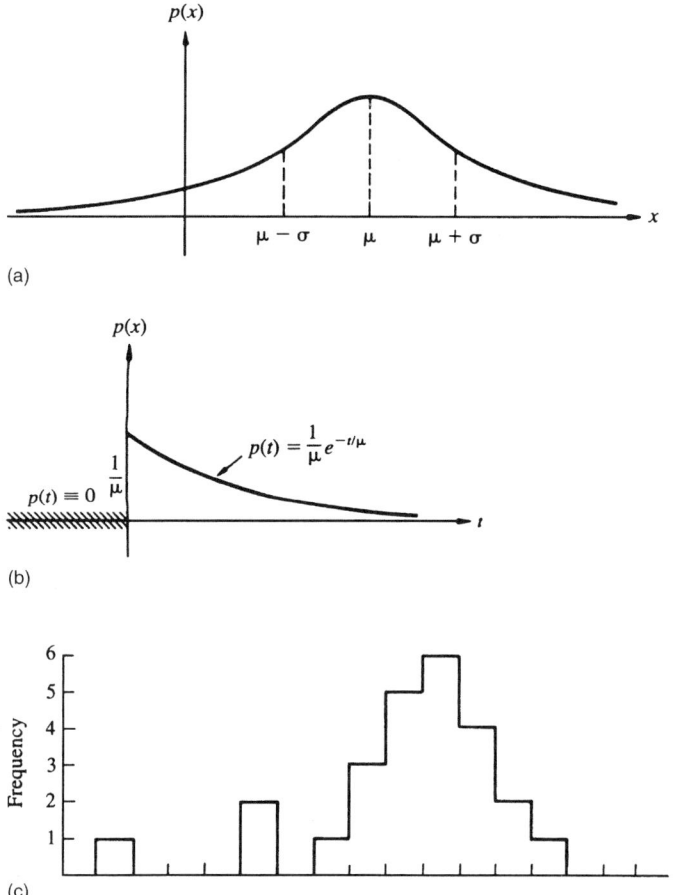

Fig. 1.1. Distributions: a) normal distribution, b) exponential distribution, c) distribution of values in a sample taken from a normal distribution.

The second is an *exponential distribution*: The exponentially distributed variable only takes positive values; half of the observations are small, crowded together in the left half of the distribution, but the balance is stretched out across a far wider range.

These distributions are both limiting cases: they represent the aggregate result of an infinite number of observations; thus, the distribution curves are smooth. The choppy histogram in Figure 1.1c is typical of what one sees with a small, finite sample of observations—in this case, a sample of 25 observations taken from a normal distribution. Still more typical of real-world data is the histogram of Figure 7.2, based on a sample taken from a mixture of two normal distributions.

The sample is not the population. To take an extreme, almost offensive example, suppose that every member of a Los Angeles, California jury were to be nonwhite. Could that jury really have been selected at random from the population as the statutes of the State of California requires? The answer is "yes"; there are court districts in Los Angeles in which less than 30% of the population is white; the probability of a jury of 12 individuals containing no whites in such a district is approximately 0.7 raised to the 12th power or about one in a hundred. With hundreds of juries being empanelled each week, nonwhite juries are not uncommon; nonetheless, they are not representative.

The good news is that as a sample grows larger, it will more and more closely resemble the population from which it is drawn. How large is large? The answer as we shall see in subsequent chapters depends both upon the underlying distribution—is it exponential? normal? a mixture of normals?— and the population parameters we wish to estimate.

1.1.2 Distribution Functions

We observe an event ω belonging to a set of events Ω. As often, we observe a real-valued random variable $X[\omega]$ or a vector $\boldsymbol{X}[\omega]$ of real-valued random variables whose values depend upon the event ω. The (cumulative) distribution function F of such a random variable, is $F[x] = \Pr\{X \leq x\}$. From the definition of a probability, we see that if $x < y$, then $0 \leq F[x] \leq F[y] \leq 1$.

If the observations are discrete, that is, if they can take at most a countably infinite number of values such as 1, 2, 2.5, 3, ..., then a graph of F would be a series of steps of increasing height and $F[x] = \sum_{x_j \leq x} \Pr\{X = x_j\}$. If the observations can be measured on a continuous scale, and a *probability density* $f(x)$ exists as described in the appendix, then we may write $F[y] = \int_{-\infty}^{y} f(x)dx$.

Accordingly, the *expected value* of such observations may be written either as $EX = \sum_{i=1}^{k} x_k \Pr\{X = x_k\}$ or $EX = \int_{-\infty}^{\infty} xf(x)dx$. If $EX^2 < \infty$, their variance may be written as $\sum_{i=1}^{k}(x_k - EX)^2 \Pr\{X = x_k\}$ or $\int_{-\infty}^{\infty}(x - EX)^2 f(x)dx$.

Let $F_n[x]$ denote the empirical distribution function of a sample of size n taken from a population. That is, if we have a set of observations $x_1, \ldots, x_n$, then $F_n[x]$ denotes the fraction of these observations that are less than x. The sample is not the population, so $F_n[x] \neq F[x]$. Nonetheless, if n observations are made independently and at random from the same population, then $F_n[x]$ converges to $F[x]$ as n grows larger.

Several distributions that arise in practice are found in Chapters 3 and 4.

1.1.3 Hypotheses

On the basis of our observations, we wish to choose an optimal decision d from a set D of possible decisions. Let us consider the sort of things we might observe and the type of decisions we might make.

Example 1: High blood pressure (hypertension) can be the source of many physical problems up to and including death of the individual unless it can be brought under control. Let us suppose our firm has developed a drug we think might be successful in lowering blood pressure in most of the patients who will receive it. (Note my use of the weasel word "most." Seldom can we predict "all" or "none" in advance of observing stochastic phenomena.)

We administer our drug to a patient and observe that his blood pressure drops 10 mb. We also observe that he has no other side effects or symptoms apart from the drop in blood pressure. In response, we make one of the following four decisions:

i) try an increased dose of the drug on the same patient;
ii) try the same dose of the drug on several other patients;
iii) put the drug into mass production and launch an advertising campaign;
iv) put this drug back on the shelf and start looking for a new and better drug.

Suppose we've opted to try the dose on other patients and must now decide how many to observe. A fundamental rule of decision-making under uncertainty, which we expand on in subsequent chapters, is that the *greater the number of observations, the more likely are our decisions to be correct.*

But clinical trials are not inexpensive ($10,000 per patient in the US is a good working number). Also, we want to keep the number of test subjects low until we can be reasonably confident the drug is harmless. On the other hand, we want to observe sufficient numbers so that we can justify a claim that the drug will prove safe and effective for the population at large.

Working against us is that *variation* appears to be the norm: Blood pressure can vary from hour to hour in the same patient. Even though the nurse may have just taken my blood pressure, my personal physician will usually take it again, and these two skilled observers frequently come up with quite different answers. If I'd been smoking before my blood pressure was taken—a practice I abandoned decades ago—my blood pressure would be elevated over what it would have been had I not been using tobacco.

A second fundamental rule of decision-making is that *the less the intrinsic variation, the fewer the observations necessary to ensure a high probability of making a correct decision.* (Note our continued use a weasel words such as "more likely" and "high probability." As long as there is intrinsic variation, there can never be guarantees, at least not for finite, realizable sample sizes.)

Out of some misguided sense of chivalry the first clinical trials were confined to men as the experimental subjects. But who is to say that a woman will necessarily respond to a given pharmaceutical in the same way as a man? Or that an antihypertensive drug will be as effective with a smoker as a nonsmoker? Or that individuals with a concurrent medical condition such as diabetes can be successfully treated with the same drug as those who are disease free? Either we must limit the scope of our conclusions or make sure individuals of all ages, races, sexes, smoking habits, and concurrent conditions are included in our observations.

When we formulate a hypothesis, we need to spell out in detail the population(s) to which it applies.

Example 2: Let us suppose our trials are now complete, the regulatory agency has granted approval for our drug, and it is up to our advertising department to see that sales match its prospects.

The advertising department has come up with two different circulars and needs to choose one to announce the availability of the new drug. Although there are many compelling theoretical arguments why one circular ought to generate more sales than another, the department head is reluctant to commit a million dollar advertising budget on the basis of speculation. To test the hypothesis that letter A is more effective than letter B, it is proposed to send out copies of the circulars to a sample of physicians offering them a trial supply of the new drug simply for writing back. The rationale behind this experiment is that the number of responses will be a direct indicator of the effectiveness of the circular and the advertising copy it contains.

Example 3: One of the company's executives feels these latter trials would be a waste of time. "Just flip a coin, and use it to decide which circular to use." Her proposal fails when it is noted that all previous coin flips have come out in her favor. "It's a fair coin," she insists.

Here, too, is a hypothesis that may be tested. Successive flips of her coin can be made, the numbers of heads and tails recorded and a decision reached as to whether her coin is fair or not.

The similar use of observation and statistics in decision-making can be found in virtually every field of endeavor, not merely advertising and pharmaceuticals. If you are new to research (or a "pure" mathematician), we suggest you read one or two of the articles in the following section to get a feel for the type of experimentation that motivates this text.

1.2 Applications

The theory of testing hypotheses has been applied in

- agriculture [Eden and Yates, 1933; Higgins and Noble, 1993; Kempthorne, 1952],
- aquatic science [Quinn, 1987; Ponton and Copp, 1997],
- anthropology [Fisher, 1936; Konigsberg, 1997; Gonzalez, Garcia-Moro, Dahinten, and Hernandez, 2002],
- archaeology [Klauber, 1971; Berry, Kvamme, and Mielke, 1980, 1983],
- astronomy [Zucker and Mazeh , 2003],
- atmospheric science [Adderley, 1961; Tukey, Brillinger, and Jones, 1978, Gabriel and Feder, 1969],
- biology [Daw et al., 1998; Howard, 1981],
- biotechnology [Vanlier, 1996; Park et al., 2003; Xu and Li, 2003],
- botany [Mitchell-Olds, 1987; Ritland and Ritland, 1989],

- cardiology [Chapelle et al., 1982],
- chemistry [van Keerberghen, Vandenbosch, Smeyers-Verbeke, and Massart, 1991],
- climatology [Hisdal et al., 2001; Robson, Jones, Reed, and Bayliss, 1998],
- clinical trials [Potthoff, Peterson, and George, 2001; Salsburg, 1992; Wei and Lachin, 1988],
- computer science [Yucesan, 1993; Laitenberger et al., 2000; Rosenberg, 2000],
- demographics [Jorde et al., 1997],
- dentistry [Mackert, Twiggs, Russell, and Williams, 2001],
- diagnostic imaging [Arndt et al., 1996; Raz et al., 2003],
- ecology [Cade 1997; Manly, 1983],
- econometrics [Kennedy, 1995; Kim, Nelson, and Startz, 1991; McQueen, 1992],
- education [Schultz and Hubert, 1976; Manly, 1988],
- endocrinology [O'Sullivan et al., 1989],
- entomology [Bryant, 1977; Mackay and Jones, 1989; Simmons and Weller, 2002],
- epidemiology [Glass et al., 1971; Wu et al., 1998],
- ergonomics [Valdesperez, 1995],
- forensics [Good 2002; Solomon, 1986],
- genetics [Levin, 1977; Karlin and Williams, 1984; North et al., 2003; Varga and Toth, 2003],
- geography [Royaltey, Astrachen, and Sokal, 1975; Hubert, 1978],
- geology [Clark, 1989; Orlowski et al., 1993],
- gerontology [Miller et al., 1997; Dey et al., 2001],
- immunology [Makinodan et al., 1976; Roper et al., 1998],
- linguistics [Romney et al., 2000],
- medicine [Bross, 1964; Feinstein, 1973; McKinney et al., 1989],
- molecular biology [Barker and Dayhoff, 1972; Karlin et al., 1983],
- neurobiology [Edgington and Bland, 1993; Weth, Nadler, and Korsching, 1996],
- neurology [Lee, 2002; Faris and Sainsbury, 1990],
- neuropsychopharmacology [Wu et al., 1997],
- neuropsychology [Stuart, Maruff, and Currie, 1997],
- oncology [Hoel and Walburg, 1972; Spitz et al., 1998],
- ornithology [Busby, 1990; Michelat and Giraudoux, 2000; Mitani, Sanders, Lwanga et al., 2001],
- paleontology [Marcus, 1969; Quinn, 1987],
- parasitology [Pampoulie and Morand, 2002],
- pediatrics [Goldberg et al., 1980; Grossman et al., 2000],
- pharmacology [Plackett and Hewlett, 1963; Oliva, Farina, and Llabres, 2003],
- physics [Penninckx et al., 1996],
- physiology [Zempo et al., 1996],

- psychology [Jennings et al., 1997; Kelly, 1973],
- radiology [Milano, Maggi, and del Turco, 2000; Hossein-Zadeh, Ardekani, and Soltanian-Zadeh, 2003; Raz et al., 2003],
- reliability [Kalbfleisch and Prentice, 1980; Nelson, 1992],
- sociology [Marascuilo and McSweeny, 1977; Pattison et al., 2000],
- surgery [Majeed et al., 1996],
- taxonomy [Alroy, 1994; Gabriel and Sokal, 1969; Fisher, 1936],
- toxicology [Cory-Slechta, Weiss, and Cox, 1989; Farrar and Crump, 1988, 1991],
- vocational guidance [Gliddentracey and Parraga, 1996; Ryan, Tracey, and Rounds, 1996],
- virology [Good, 1979],
- theology [Witztum, Rips, and Rosenberg, 1994].

1.3 Testing a Hypothesis

Shortly after I received my doctorate in statistics, I decided that if I really wanted to help bench scientists apply statistics I ought to become a scientist myself. So I went back to school to learn about physiology and aging in cells raised in petri dishes.

I soon learned there was a great deal more to an experiment than the random assignment of subjects to treatments. In general, 90% of experimental effort was spent mastering various arcane laboratory techniques, another 9% in developing new techniques to span the gap between what had been done and what I really wanted to do, and a mere 1% on the experiment itself. But the moment of truth came finally—it had to if I were to publish and not perish—and I succeeded in cloning human diploid fibroblasts in eight culture dishes: Four of these dishes were filled with a conventional nutrient solution and four held an experimental "life-extending" solution to which vitamin E had been added.

I waited three weeks with fingers crossed that there was no contamination of the cell cultures, but at the end of this test period three dishes of each type had survived. My technician and I transplanted the cells, let them grow for 24 hours in contact with a radioactive label, and then fixed and stained them before covering them with a photographic emulsion.

Ten days passed and we were ready to examine the autoradiographs. Two years had elapsed since I first envisioned this experiment and now the results were in: I had the six numbers I needed.

"I've lost the labels," my technician said as she handed me the results. This was a dire situation. Without the labels, I had no way of knowing which cell cultures had been treated with vitamin E and which had not.

"121, 118, 110, 34, 12, 22." I read and reread these six numbers which represented populations of cells remaining in the dishes, over and over. If the first three counts were from treated colonies and the last three were from untreated, then perhaps I had found the fountain of youth. Otherwise, I had nothing to report.

1.3.1 Five Steps to a Test

How had I reached the conclusion that vitamin E extends cell lifespan? In succeeding chapters, you will learn a wide variety of decision-making techniques ranging from the simple to the complex. In each case, we will follow the same five-step procedure of this problem.

1. Analyze the problem—identify the hypothesis, the alternative hypotheses of interest, and the potential risks associated with a decision.
2. Choose a test statistic.
3. Compute the test statistic.
4. Determine the frequency distribution of the test statistic under the hypothesis.
5. Make a decision using this distribution as a guide.

1.3.2 Analyze the Experiment

For the answer to how I reached this conclusion, let's take a second, more searching, look at the problem of the missing labels. First, we identify the hypothesis and the alternative(s) of interest.

I wanted to assess the life-extending properties of a new experimental treatment with vitamin E. To do this, I divided my cell cultures into two groups: one grown in a standard medium and one grown in a medium containing vitamin E. At the conclusion of the experiment and after the elimination of several contaminated cultures, both groups consisted of three independently treated dishes.

My *null hypothesis* was that the growth potential of a culture would not be affected by the presence of vitamin E in the media: All the cultures would have equal growth potential. The *alternative* of interest was that cells grown in the presence of vitamin E would be capable of many more cell divisions.

Under the null hypothesis, the labels "treated" and "untreated" provide no information about the outcomes: the observations would be expected to have more or less the same values in each of the two experimental groups. If they were to differ, it would be as a result of some uncontrollable random fluctuation alone. Thus, if this null, or no-difference, hypothesis were true, I was free to exchange the labels.

1.3.3 Choose a Test Statistic

The next step was to choose a test statistic that discriminates between the hypothesis and the alternative. The statistic I chose was the sum of the counts in the group treated with vitamin E. If the alternative is true and vitamin E prolongs life span, most of the time this sum ought to be larger than the sum of the counts in the untreated group. If the null hypothesis is true, that is, if it doesn't make any difference which treatment the cells receive, then the sums of the two groups of observations should be approximately the same. One sum might be smaller or larger than the other by chance, but most of the time the two shouldn't be all that different. We formalize this rationale in Chapter 3 via the Fundamental Lemma of Neyman and Pearson.

1.3.4 Compute the Test Statistic

The third step was to compute the test statistic for the observations as originally labeled, thus, $S = 349 = 121 + 118 + 110$.

1.3.5 Determine the Frequency Distribution of the Test Statistic

To obtain a distribution for the test statistic under the null hypothesis, I began to rearrange (permute) the labels on the observations, randomly reassigning one of the six labels, three reading "treated" and three "untreated," to each dish. For example: treated, 121 118 34, and untreated, 110 12 22. In this particular rearrangement, the sum of the observations in the first (treated) group was 273. I repeated this step until all $\binom{6}{3} = 20$ distinct rearrangements had been examined.

	First Group			Second Group			Sum
1	121	118	110	34	22	12	349
2	121	118	34	110	22	12	273
3	121	110	34	118	22	12	265
4	118	110	34	121	22	12	262
5	121	118	22	110	34	12	261
6	121	110	22	118	34	12	253
7	121	118	12	110	34	22	251
8	118	110	22	121	34	12	250
9	121	110	12	118	34	22	243
10	118	110	12	121	34	22	240
11	121	34	22	118	110	12	177
12	118	34	22	121	110	12	174
13	121	34	12	118	110	22	167
14	110	34	22	121	118	12	166
15	118	34	12	121	110	22	164
16	110	34	12	121	118	22	156
17	121	22	12	118	110	34	155
18	118	22	12	121	110	34	152
19	110	22	12	121	118	34	144
20	34	22	12	121	118	110	68

The sum of the observations in what I supposed to be the original vitamin E-treated group, 349, is equaled only once and never exceeded in the 20 distinct random relabelings. If chance alone were operating, then such an extreme value would be a rare, only a 1-in-20 event. If I then were to reject the null hypothesis and embrace the alternative, that the treatment is effective and responsible for the observed difference, I only risk making an error and rejecting a true hypothesis 1 in every 20 times.

1.3.6 Make a Decision

In this instance, I did make just such an error. I found the labels, of course—the coded assignments were in a different notebook, but I was never able to replicate the observed life-promoting properties of vitamin E in repetitions of this experiment. Good statistical methods can reduce and contain the probability of making a bad decision, but they cannot eliminate the possibility.

1.3.7 Variations on a Theme

Will we always make decisions in this fashion? No. Sometimes we will use a different statistic. Or we may be able to obtain the distribution of the test statistic by theoretical means (see Chapter 4) or, for large samples, estimate the distribution via a Monte Carlo scheme (see Chapter 14). We will encounter a situation in the very next chapter in which we will select among several alternatives to the null rather than merely accept or reject a hypothesis. And when we reject a hypothesis of no difference, we will want to obtain some idea of what the difference actually is (see Chapter 3). But invariably, we will use these same five steps to obtain the answers.

> "Actually, the statistician does not carry out this very tedious process but his conclusions have no justification beyond the fact they could have been arrived at by this very elementary method."
>
> R.A. Fisher, 1936.

> "Tests of significance in the randomized experiment have frequently been presented by way of normal law theory, whereas their validity stems from randomization theory."
>
> O. Kempthorne, 1955.

1.4 A Brief History of Statistics in Decision-Making

Virtually every observation Y can be written in the form $f[X] + Z$, where $f[X]$ has a fixed value that depends upon the function f and a vector of concurrent observations X (including sex, race, age, and so forth) and Z is a random variable, one which can take any of a range of possible values with varying probabilities. Throughout the 19th century it was felt that if one could just observe enough concurrent variables, and observe each of these variables with sufficient precision, then Y could be determined exactly, eliminating the stochastic component Z from the equation. The discovery of the quantum basis for physics changed this view, and resulted in a general acceptance of data variation as the norm and a shift in the focus of statistics from seeking to eliminate random variation to characterizing it.

In 1908, W.S. Gossett characterized the distribution of the t-statistic (see Chapters 3 and 4), a function of the mean and standard deviation of a set of independent, identically normally distributed random variables. By 1927, he was expressing doubts about the adequacy of his formula in chemical determinations, noting that observations that are close in time and space are often positively correlated, not independent. A further problem is that most observations come from not a single distribution but from a mixture of distributions (see, for example, Micceri, 1989).

In 1933, J. Neyman and E.S. Pearson provided a method for determining the most efficient test of a hypothesis (see Chapters 3 and 15). In 1935, R.A. Fisher demonstrated the exact analysis of contingency tables (see Chapter 8). In 1937–38, Pitman developed exact permutation methods consistent with the Neyman–Pearson approach for the comparison of k-samples and for bivariate correlation (see Chapter 3).

Unfortunately, by this time, theory had outrun practice. These new permutation methods were beyond the capabilities of the mechanical computing devices of the 1930s–1940s. As a stopgap, E.L. Lehmann and C. Stein [1949] developed the most powerful tests of composite hypotheses based on the parametric approach. Subsequently, Wald and Wolfowitz [1944], Hoeffding [1951, 1952], Kempthorne [1952], and Bickel and VanZwet [1978] showed that for very large samples the parametric and permutation approaches are equivalent.

In a series of journal articles beginning in 1947 and culminating in a textbook in 1950, A. Wald generalized the accept/reject alternatives of testing hypotheses to complete classes of decision procedures. Regrettably, little further progress was made in this area, although, in the next chapter, we show how permutation tests can be applied to a general loss matrix in a practical situation. Wald [1947] also introduced the concept of a sequential test that we discuss in Chapter 6 on experimental design.

When ranks are substituted for the original observations, the permutation distribution of a test statistic can be calculated once for a given set of sample sizes (even if it takes all night as it did with the late 1950-model computers) and the results applied to all k-sample problems with the same number of observations in each sample. Beginning with Wilcoxon [1945] many such tests were developed and catalogued. We note the efforts of Siegel [1956], Hodges and Lehmann [1963], Cox and Kempthorne [1963], Sen [1965, 1967], Bell and Doksum [1965, 1967], Bell and Donoghue [1969], Shane and Puri [1969], Bickel [1969], Puri and Sen [1971], and Lehmann [1975]. Today, of course, it takes only seconds on a desktop computer to obtain a p-value via permutation means using the original, nontransformed observations.

By the middle of the 1970s, most statisticians had mainframe terminals on their desks and conditions were ripe for the development of a wide variety of resampling procedures, including the bootstrap and density estimation, as well as permutation tests (see McCarthy [1969], Hartigan [1969], Efron [1979], Diaconis and Efron [1983], and Izenman [1991]).

The late 1960s and 1970s saw the introduction of cluster analysis (Mantel [1967], Cliff and Ord [1981], Mielke, Berry, and Johnson [1976], Mielke [1978], Mielke, Berry, Brockwell, and Williams [1981]; see also Chapter 10). The late 1970s and early 1980s saw breakthroughs in the analysis of single-case designs (Kazdin [1976, 1980], Edgington [1980a,b, 1996]) and directional data (Hubert et al. [1984]).

The 1990s saw advances in the testing of simultaneous hypotheses (Westfall and Young [1993], Troendle [1995], Blair, Troendle, and Beck [1996]) and sequential analysis (Lefebvre [1982], Lin, Wei, and DeMets [1991]).

The most recent advances in the theory of testing hypotheses have come from F. Pesarin and his colleagues at the University of Padova. Chapter 7 is devoted to their use of symmetric permutations to provide exact solutions for multifactor designs. Their method of nonparametric combination of several independent tests is discussed in Chapter 9.

1.5 Exercises

Take the time to think about the answers to these questions even if you don't answer them explicitly. You may wish to return to them after you've read subsequent chapters.

1. In the simple example analyzed in this chapter, what would the result have been if the experimenter had used as the test statistic the difference between the sums of the first and second samples? the difference between their means? the sum of the squares of the observations in the first sample? the sum of their ranks?

2. How was the analysis of my experiment affected by the loss of two of the cultures due to contamination? Suppose these cultures had escaped contamination and given rise to the observations 90 and 95; what would be the results of a permutation analysis applied to the new, enlarged data set consisting of the following cell counts:

    Treated 121 118 110 90

    Untreated 95 34 22 12

3. Read one or two of the articles that were cited on pages 2 and 3. What were the hypotheses? What was the population to which these hypotheses were applied? Were the observations drawn from this population? Were they representative of this population? What were the sources of variation? Were the observations independent? What led the authors to use the specific statistical procedure they chose?

2
Optimal Procedures

2.1 Defining Optimal

As we saw in the preceding chapter, the professional statistician is responsible for choosing both the test statistic and the testing procedure. An amateur might hope to look up the answers in a book, or, as is all too commonly done, use the same statistical procedure as was used the time before, regardless of whether it continues to be applicable. But the professional is responsible for choosing the best procedure, the optimal statistic. The statistic we selected in the preceding chapter for testing the effectiveness of vitamin E seemed an obvious, intuitive choice. But is it the best choice? And can we prove it is? Intuition can so often be deceptive.

In this chapter, we examine the criteria that define an optimal testing procedure and explore the interrelationships among them.

2.1.1 Trustworthy

The most obvious desirable property of a statistical procedure is that it be trustworthy. If we are advised to make a particular decision, then we should be correct in doing so. Alas, our observations are stochastic in nature, so there may be more than one explanation for any given set of observations. The result is we never can rely 100% on the decisions we make. At best, they can be like politicians, trustworthy up to a point. We ask only that they confine themselves to small bribes and rake-offs, that they not bankrupt or betray the country.

In the example of the missing labels in the preceding chapter, we introduced a statistical test based on the random assignment of labels to treatments. Knowing in advance that the experiment could have any of $\binom{6}{3} = 20$ possible outcomes, we will reject the null hypothesis only if the obtained value of the test statistic is the maximum possible that could arise from only one permutation of the results. The test we derive is valid under very broad assumptions. The data could have been drawn from a normal distribution or they could have come from some quite different distribution. To be valid at a given percent level, all that is required of our permutation test is that (under the hypothesis) the population from which the data in the treatment group are drawn be the same as that from which the untreated sample is taken.

This freedom from reliance on numerous assumptions is a big plus. The fewer the assumptions, the fewer the limitations, and the broader the potential applications of a test. But before statisticians introduce a test into their practice, they need to know a few more things about it, namely:

- Is it *exact*? That is, can we make an exact determination of the probability that we might make an error in rejecting a true hypothesis?
- How *powerful* a test is it? That is, how likely is it to pick up actual differences between treated and untreated populations? Is this test as powerful or more powerful than the test we are using currently?
- Is the test *admissible*? That is, is there no other test that is superior to it under all circumstances?
- How *robust* is the new test? That is, how sensitive is it to violations in the underlying assumptions and the conditions of the experiment?

2.1.2 Two Types of Error

It's fairly easy to reason from cause to effect—that is, if you have a powerful enough computer. Get the right formula (Boyle's Law, say), plug in enough values to enough decimal places, and out pops the answer. The difficulty with reasoning in the opposite direction, from effect to cause, is that more than one set of causes can be responsible for precisely the same set of effects. We can never be completely sure which set of causes is responsible. Consider the relationship between sex (cause) and height (effect). Boys are taller than girls. True? So that makes this new 6'2" person in our lives ... a starter on the women's volleyball team.

In real life, in real populations, there are vast differences from person to person. Some women are tall and some women are short. In Lake Wobegone, Minnesota, all the men are good looking and all the children are brighter than average. But in most other places in the world there is a wide range of talent and abilities. As a further example of this variation, consider that half an aspirin will usually take care of one person's headache while other people take two or three aspirin at a time and get only minimal relief.

Figure 2.1 below depicts the results of an experiment in which two groups were each given a "pain-killer." The first group got buffered aspirin; the second group received a new experimental drug. Each of the participants then provided a subjective rating of the effects of the drug. The ratings ranged from "got worse," to "much improved," depicted below on a scale of 0 to 4. Take a close look at Figure 2.1. Does the new drug represent an improvement over aspirin?

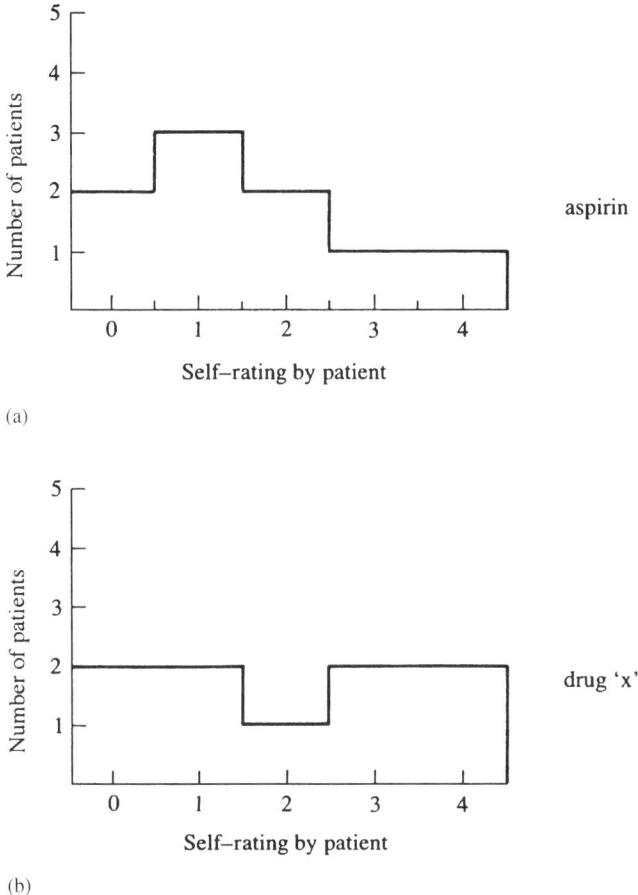

Fig. 2.1. Response to treatment: Self-rating patient in (a) asprin-treated group, (b) drug-'x'-treated group.

Those who took the new experimental drug do seem to have done better on average than those who took aspirin. Or are the differences we observe in Figure 2.1 simply the result of chance? If it's just a *chance* effect—rather than one caused by the new drug—and we opt in favor of the new drug, we've made an error. We also make an error if we decide there is no difference, when, in fact, the new drug really is better. These decisions and the effects of making them are summarized in Table 2.1a below.

We distinguish between the two types of error because they have quite different implications. For example, Fears, Tarone, and Chu [1977] use permutation methods to assess several standard screens for carcinogenicity. Their Type I error, a false positive, consists of labeling a relatively innocuous compound as carcinogenic. Such an action means economic loss for the

Table 2.1a. Decision making under uncertainty.

The Facts	Our Decision	
No difference	No difference	New drug is better Type I error
New drug is better	Type II error	

Table 2.1b. Decision making under uncertainty.

The Facts	Fears et al's Decision	
	Not a carcinogen	Compound a carcinogen
Not a carcinogen (Alternative)		Type I error: manufacturer misses opportunity for profit; public denied access to effective treatment
Carcinogen (Hypothesis)	Type II error: patients die; families suffer; manufacturer sued	

manufacturer and the denial of the compound's benefits to the public. Neither consequence is desirable. But a false negative, a Type II error, would mean exposing a large number of people to a potentially lethal compound.

Because variation is inherent in nature, we are bound to make the occasional error when we draw inferences from experiments and surveys, particularly if, for example, chance hands us a completely unrepresentative sample. When I toss a coin in the air six times, I can get three heads and three tails, but I also can get six heads. This latter event is less probable, but it is not impossible. Variation also affects the answer to the question, "Does the best team always win?"

We can't eliminate risk in making decisions, but we can contain risk through the correct choice of statistical procedure. For example, we can require that the probability of making a Type I error not exceed 5% (or 1% or 10%) and restrict our choice to statistical methods that ensure we do not exceed this level. If we have a choice of several statistical procedures, all of which restrict the Type I error appropriately, we can choose the method that leads to the smallest probability of making a Type II error.

2.1.3 Losses and Risk

The preceding discussion is greatly oversimplified. Obviously, our losses will depend not merely on whether we guess right or wrong, but on how far our guesstimate is off the mark. For example, suppose you've developed a new drug to relieve anxiety and are investigating its side effects. You ask, "Does it raise blood pressure?" You do a study and find the answer is "no." But the truth is your drug raises systolic blood pressure an average of one millibar. What is the cost to the average patient? Negligible, one millibar is a mere fraction of the day-to-day variation in blood pressure.

Now, suppose your new drug actually raises blood pressure an average of 10 mb. What is the cost to the average patient? to the entire potential patient population? to your company in law suits? Clearly, the cost of a Type II error will depend on the magnitude of that error and the nature of the losses associated with it.

Historically, much of the work in testing hypotheses has been limited to zero or one loss function while that of estimation has focused on losses proportional to the square of the error. The result may have been statistics that were suboptimal in nature with respect to the true, underlying loss (see Mielke [1986], Mielke and Berry [1997]).

Are we more concerned with the losses associated with a specific decision or those we will sustain over time as a result of adhering to a specific decision procedure? Which concerns our company the most: reducing average losses over time or avoiding even the remote possibility of a single, catastrophic loss? We return to this topic in Section 2.2.

2.1.4 Significance Level and Power

In selecting a statistical method, statisticians work with two closely related concepts, significance level and power. The *significance level* of a test, denoted throughout the text by the Greek letter α (alpha), is the probability of making a Type I error; that is, α is the probability of deciding erroneously on the alternative when, in fact, the hypothesis is true.

To test a hypothesis, we divide the set of possible outcomes into two or more regions. We accept the primary hypothesis and risk a Type I error when our test statistic lies in the *rejection region* R; we reject the primary hypothesis and risk a Type II error when our test statistic lies in the *acceptance region* A; and we may take additional observations when our test statistic lies in the boundary *region of indifference* I. If H denotes the hypothesis, then

$$\alpha = \Pr\{X \in R | H\}.$$

The power of a test, denoted throughout the text by the Greek letter β (beta), is the complement of the probability of making a Type II error; that is, β is the probability of deciding on the alternative when the alternative is the correct choice. If K denotes the alternative, then

$$\beta = \Pr\{X \in R | K\}.$$

18   2 Optimal Procedures

The ideal statistical test would have a significance level α of zero and a power β of 1, or 100%. But unless we are all-knowing, this ideal cannot be realized. In practice, we will fix a significance level $\alpha > 0$, where α is the largest value we feel comfortable with, and choose a statistic that maximizes or comes closest to maximizing the power for an alternative or set of alternatives important to us.

2.1.4.1 Power and the Magnitude of the Effect

The relationship among power, significance level, and the magnitude of the effect for a specific test is summarized in Figure 2.2, provided by Patrick Onghena. For a fixed significance level, the power is an increasing function of the magnitude of the effect. For a fixed effect, increasing the significance level also increases the power.

2.1.4.2 Power and Sample Size

As noted in Section 2.1.3., the greater the discrepancy between the true alternative and our hypothesis, the greater the loss associated with a Type II error.

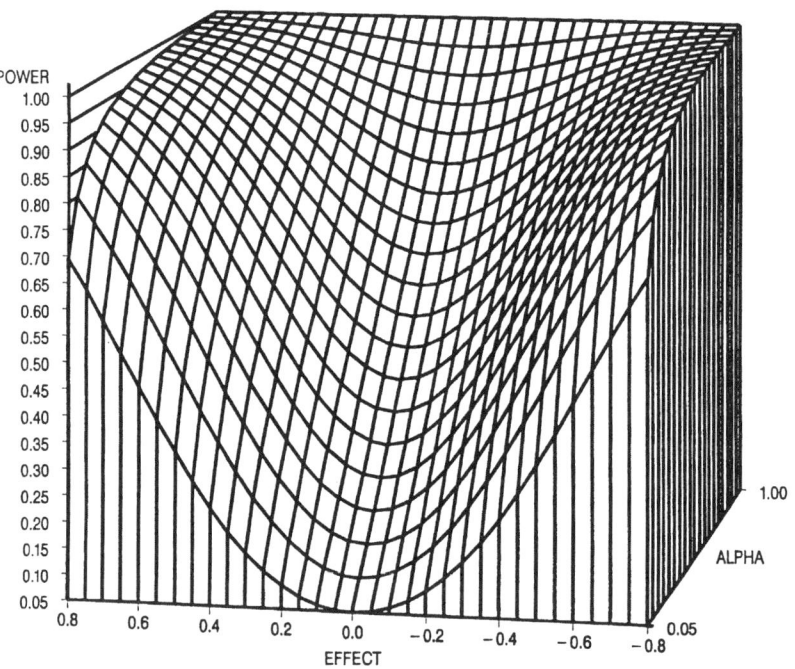

Fig. 2.2. Power of the two-tailed t-test with sample sizes of $n_1 = n_2 = 20$ as a function of the effect size (EFFECT) and the significance level (ALPHA) under the classical parametric assumptions.

Fortunately, in most practical situations, we can devise a test where the larger the discrepancy, the greater the power and the less likely we are to make a Type II error.

The relationship among power, effect magnitude, and number of observations for a specific test is summarized in Figure 2.3, provided by Patrick Onghena.

Figure 2.4a depicts the power as a function of the alternative for two tests based on samples of size 6. In the example illustrated, the test φ_1 is uniformly more powerful than φ_2, hence, using φ_1 in preference to φ_2 will expose us to less risk.

Figure 2.4b depicts the power curve that results from using these same two tests, but for different size samples; the power curve of φ_1 is still based on a sample of size 6, but that of φ_1 now is based on a sample of size 12. The two new power curves almost coincide, revealing the two tests now have equal risks. But we will have to pay for twice as many observations if we use the second test in place of the first.

Moral: A more powerful test reduces the costs of experimentation, and it minimizes the risk.

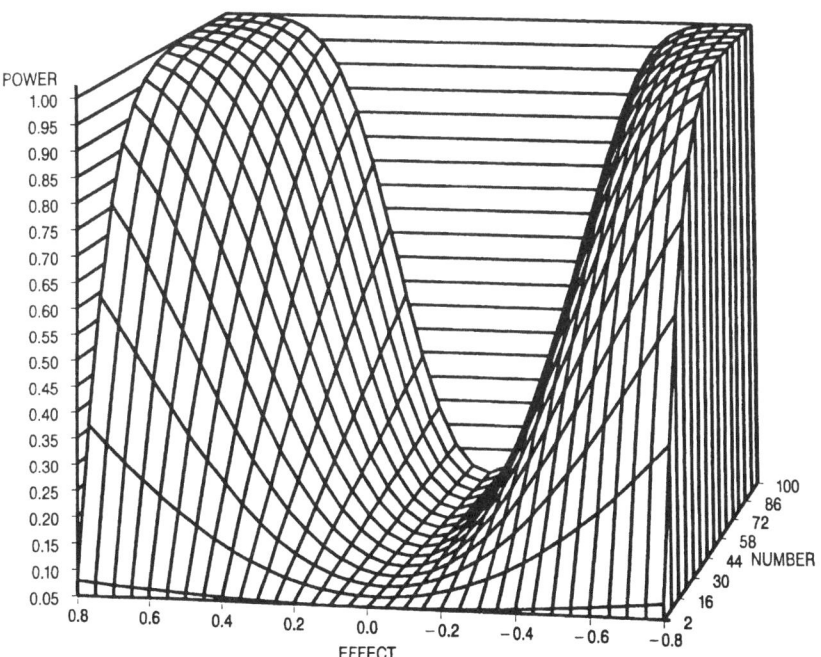

Fig. 2.3. Power of the two-tailed t-test with $p = 0.05$ as a function of the effect size (EFFECT) and the number of observations (NUMBER, $n_1 = n_2$) under the classical parametric assumptions.

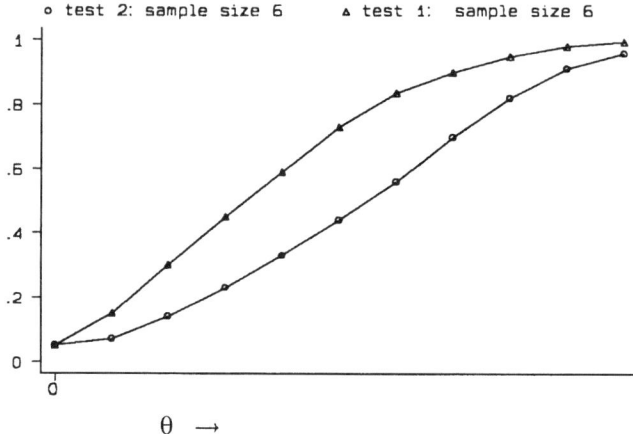

Fig. 2.4a. Power as a function of the alternative. Tests have the same sample size.

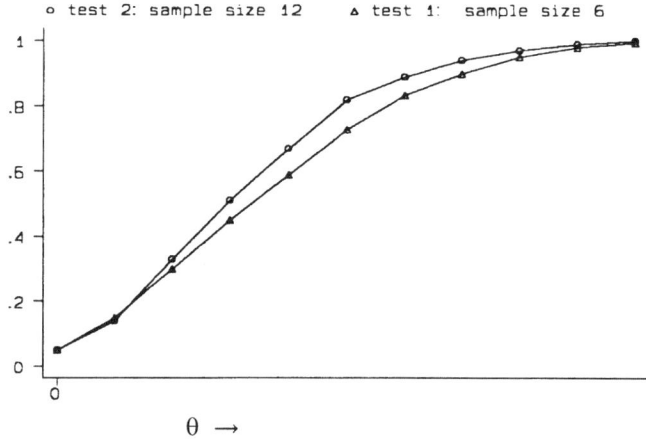

Fig. 2.4b. Power as a function of the alternative. Tests have different sample sizes.

2.1.4.3 Power and the Alternative

If a test at a specific significance level α is more powerful against a specific alternative than all other tests at the same significance level, we term it *most powerful*. But as we see in Figure 2.5, a test that is most powerful for some alternatives may be less powerful for others. When a test at a specific significance level is more powerful against *all* alternatives than all other tests at the same significance level, we term such a test *uniformly most powerful*.

We term a test *admissible*, providing either a) it is uniformly most powerful or b) no other test is more powerful against all alternatives.

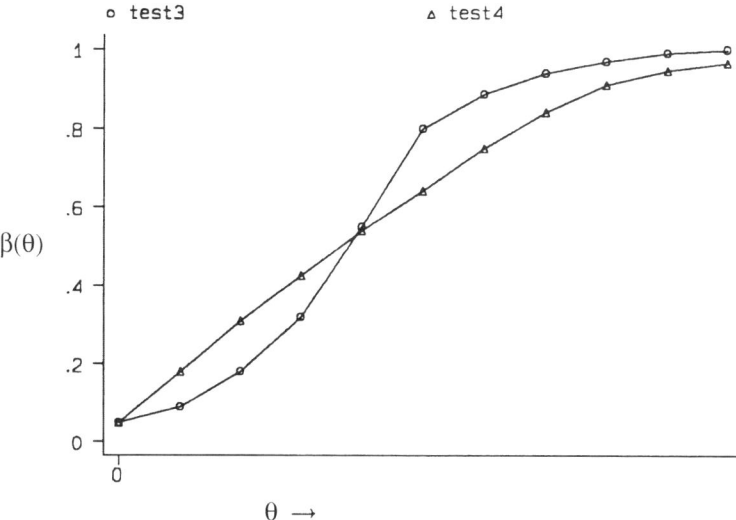

Fig. 2.5. Comparing power curves: For near alternatives, with θ close to zero, test 4 is the more powerful test; for far alternatives, with θ large, test 3 is more powerful. Thus, neither test is uniformly most powerful.

Note: One can only compare the power of tests that have the same significance level. For if the test $\varphi_1[\alpha_1]$ is less powerful than $\varphi_2[\alpha_2]$, where the significance level $\alpha_1 < \alpha_2$, then it may be that the power of $\varphi_1[\alpha_2]$ is greater than the power of $\varphi_2[\alpha_2]$.

The significance level and power may also depend upon how the variables we observe are distributed. For example, does the population distribution follow a bell-shaped normal curve with the most frequent values in the center, as in Figure 2.1a? Or is the distribution something quite different? To protect our interests, we may need to require that the Type I error be less than or equal to some predetermined value for all possible distributions. When applied correctly, permutation tests always have this property. The significance levels of parametric tests and of tests based on the bootstrap are dependent on the underlying distribution.

2.1.5 Exact, Unbiased, Conservative

In practice, we seldom know either the distribution of a variable or the values of any of the distribution's *nuisance* parameters.[1] We usually want to test a *compound hypothesis*, such as H: X has mean value 0. This latter hypothesis includes several *simple hypotheses*, such as H_1: X is normal with mean value

---

[1] A good example of nuisance parameters is a distribution's unknown means when variances are being compared (see Section 3.7.1).

0 and variance 1, H_2: X is normal with mean 0 and variance 1.2, and H_3: X is a gamma distribution with mean zero and four degrees of freedom.[2]

A test is said to be *exact* with respect to a compound hypothesis if the probability of making a Type I error is exactly α for each and every one of the possibilities that make up the hypothesis. A test is said to be *conservative* if the Type I error never exceeds α. Obviously, an exact test is conservative, though the reverse may not be true.

The importance of an exact test cannot be overestimated, particularly a test that is exact regardless of the underlying distribution. If a test that is nominally at level α is actually at level c, we may be in trouble before we start: If $c > \alpha$, the risk of a Type I error is greater than we are willing to bear. If $c < \alpha$, then our test is suboptimal, and we can improve on it by enlarging its rejection region.

A test is said to be *unbiased* and of level α providing its power function β satisfies the following two conditions:

- β is conservative; that is, $\beta \leq \alpha$ for every distribution that satisfies the hypothesis;
- $\beta \geq \alpha$ for every distribution that is an alternative to the hypothesis.

That is, a test is unbiased if you are more likely to reject a false hypothesis than a true one when you use such a test. I find unbiasedness to be a natural and desirable principle, but not everyone shares this view; see, for example, Suissa and Shuster [1984].

Faced with some new experimental situation, our objective always is to derive a uniformly most powerful unbiased test if one exists. But if we can't derive a uniformly most powerful test (and Figure 2.5 depicts just such a situation), then we will look for a test that is most powerful against those alternatives that are of immediate interest.

2.1.6 Impartial

Our methods should be *impartial*. Decisions should not depend on the accidental and quite irrelevant labeling of the samples; nor should decisions depend on the units in which the measurements are made nor when they are made.

To illustrate, suppose we have collected data from two samples and our objective is to test the hypothesis that the difference in location of the two populations from which the samples are drawn is less than or equal to some value. (This is called a *one-tailed* or *one-sided test*.) Suppose further that the first sample includes the values a, b, c, d, and e and the second sample the values f, g, h, i, j, k. If the observations are completely reversed, that is, if the first sample includes the values f, g, h, i, j, k and the second sample the values a, b, c, d, and e, then, if we rejected the hypothesis in the first instance, we ought to reject it in the second.

---

[2] In this example the variance is an example of a nuisance parameter.

The units we use in our observations should not affect our decisions. We should be able to take a set of measurements in feet, convert to inches, make our estimate, convert back to feet, and get absolutely the same result as if we'd worked in feet throughout. Similarly, where we locate the zero point of our scale should not affect the conclusions. Measurements of temperature illustrate both these points.

Finally, if our observations are independent of the time of day, the season, and the day on which they were recorded (facts which ought to be verified before proceeding further), then our decisions should be independent of the order in which the observations were collected.

Such impartial tests are said to be *invariant* with respect to the transformations involved (the conversion of units or the permutation of subscripts).

2.1.7 Most Stringent Tests

Let $\beta_\varphi(\theta)$ denote the power of a test φ against the alternative θ. Let the envelope power function $\beta_\alpha^*(\theta)$ be the supremum of $\beta_\varphi(\theta)$ over all level-α tests of the hypothesis. Then $\beta_\alpha^*(\theta) - \beta_\varphi(\theta)$ is the amount by which a specific test φ falls short of the maximum power attainable. A test that minimizes its maximum shortcoming over all alternatives θ is said to be *most stringent*.

2.2 Basic Assumptions

The parametric and bootstrap tests considered in this text rely on the assumption that successive observations are *independent* of one another. The permutation tests rely on the less inclusive assumption that they are exchangeable. We provide formal definitions of these concepts in this section and in Section 15.5.

2.2.1 Independent Observations

If you and I each flip separate coins, the results are independent of one another. But if the two of us sit together at a table while a poll taker asks us about our preferences, our responses are unlikely to be independent if you or I modify our responses in an effort to please or placate one another.

We say that two observations X_1 and X_2 are *independent* of one another with respect to a collection of events $\mathcal{A}$ if

$$\Pr\{X_1 \in A \text{ and } X_2 \in B\} = \Pr\{X_1 \in A\}\Pr\{X_2 \in B\}$$

where A and B are any two not necessarily distinct sets of outcomes belonging to $\mathcal{A}$.[3]

---

[3] We formalize this definition of independence in Section 15.1.

Suppose I choose to have my height measured by several individuals. These observations may well have a normal distribution with mean μ_{phil}, my height as measured by some "perfect" measuring device. With respect to the set of events leading to such observations, the various measurements on me are independent.

Suppose instead that several individuals including myself are selected from a larger population, one that has a distribution F centered about the value μ. Observations on me may be viewed as including two random components, one that results from selecting me from F and the other the observational error described in the previous paragraph. The result is to generate a much larger set of events with respect to which the observations on my height are no longer independent.[4]

Some additional examples of independent and dependent observations are given in Exercise 5. Some additional properties of independent observations are given in Section 4.1.

2.2.2 Exchangeable Observations

A sufficient condition for a permutation test such as the one outlined in the preceding chapter to be exact and unbiased against shifts in the direction of higher values is the *exchangeability* of the observations in the combined sample.[5] Let $G\{x; y_1, y_2, \ldots, y_{n-1}\}$ be a distribution function in x and symmetric in its remaining arguments—that is, if the remaining arguments were permitted, the value of G would not be affected. Let the conditional distribution function of x_i given $x_1, \ldots, x_{i-1}, x_{i+1}, \ldots, x_n$ be G for all i. Then the $\{x_i\}$ are exchangeable.

Independent, identically distributed observations are exchangeable. So are samples without replacement from a finite population, termed *Polya urn models* [Koch, 1982]. An urn contains **b** black balls, **r** red balls, **y** yellow balls, and so forth. A series of balls is extracted from the urn. After the ith extraction, the color of the ball X_i is noted and k balls of the same color are added to the urn, where k can be any integer, positive, negative, or zero. The set of random events $\{X_i\}$ form an exchangeable sequence.[6]

Also exchangable are dependent normally distributed random variables $\{X_i\}$ for which the variance of X_i is a constant independent of i and the covariance of X_i and X_j is a constant independent of i and j. An additional example of dependent but exchangeable variables is given in Section 3.7.2.

Sometimes a simple transformation will ensure that observations are exchangeable. For example, if we know that X comes from a population with

---

[4] The student in search of greater clarity will find it in the formal exposition of Section 15.1.
[5] If an observation O consists of a deterministic part D and a stochastic part S, $O = D + S$, only the stochastic parts need be exchangeable. We use this more precise definition in Chapter 6.
[6] See, also, Dubins and Freedman [1979].

mean μ and distribution $F(x - \mu)$ and an independent observation Y comes from a population with mean v and distribution $F(x - v)$, then the independent variables $X' = X - \mu$ and $Y' = Y - v$ are exchangeable.

In deciding whether your own observations are exchangeable and a permutation test applicable, the key question is the one we posed in the very first chapter: Under the null hypothesis of no differences among the various experimental or survey groups, can we exchange the labels on the observations without significantly affecting the results?

2.3 Decision Theory

A statistical problem is defined by three elements:

1) the class $F = (F_\theta, \theta \in \Omega)$ to which the probability distribution of the observations belongs; for example, we might specify that this distribution is either unimodal, or symmetric, or normal;
2) the set D of possible decisions $\{d\}$ one can make on observing the sample $X = (X_1, \ldots, X_n)$;
3) the loss $L(d, \theta)$, expressed in dollars, persons' lives or some other quantifiable measure, that results when we make the decision d when θ is true.

The problem is a statistical one when the investigator is not in a position to say that X will take on exactly the value x, but only that X has some probability $P\{A\}$ of taking on values in the set A.

So far in this chapter we've limited ourselves to two-sided decisions in which either we accept a hypothesis H and reject an alternative K or we reject the hypothesis H and accept the alternative K.

One example is H: $\theta \leq \theta_0$ K: $\theta > \theta_0$. In this example we would probably follow up our decision to accept or reject with a confidence interval for the unknown parameter θ. This would take the form of an interval $(\theta_{\min}, \theta_{\max})$ and a statement to the effect that the probability that this interval covers the true parameter value is not less than $1 - \alpha$. This use of an interval can rescue us from the sometimes undesirable all-or-nothing dichotomy of hypothesis vs. alternative.

Our objective is to come up with a decision rule D, such that when we average out over all possible sets of observations, we minimize the associated risk or expected loss,

$$R(\theta, D) = EL(\theta, D(X)).$$

In the first of the preceding examples, we might have

$$L(\theta, d) = 1 \quad \text{if } \theta \in K \text{ and } d = H \text{ (Type II error)},$$
$$L(\theta, d) = 10 \quad \text{if } \theta \in H \text{ and } d = K \text{ (Type I error)},$$
$$L(\theta, d) = 0 \quad \text{otherwise.}$$

Typically, losses L depend on some function of the difference between the true (but unknown) value θ and our best guess θ^* of this value, the *absolute* deviation $L(\theta, \theta^*) = |\theta^* - \theta|$, for example. Other typical forms of the loss function are the *square deviation* $L(\theta^* - \theta)^2$, and the *jump*, that is, no loss occurs if $|\theta^* - \theta| < \delta$, and a big loss occurs otherwise.

Unfortunately, a testing procedure that is optimal for one value of the parameter θ might not be optimal for another. This situation is illustrated in Figure 2.5 with two decision curves that cross over each other. The risk R depends on θ, and we don't know what the true value of θ is! How are we to choose the best decision? This is the topic we now discuss by considering Bayes, mini-max, and generalized decisions.

2.3.1 Bayes' Risk

One seldom walks blind into a testing situation. Except during one's very first preliminary efforts, one usually has some idea of the magnitude and likelihood of the expected effect. This is particularly true of clinical trials that are usually the culmination of years of experimental effort, first on the computer to elicit a set of likely compounds, and then in the laboratory in experiments with inbred mice and, later, dogs or monkeys. The large scale Phase III clinical trial takes place only after several years. And even then after small numbers of humans have been exposed to determine the maximum safe dose and the minimum effective dose.

In the case of a simple alternative, we may start with the idea that the prior probability that the null hypothesis is true is close to 1, while the probability of the alternative is near 0. As we gain more knowledge through experimentation, we can assign posterior odds to the null hypothesis with the aid of Bayes' theorem:

$$\Pr\{(H)E_1, \ldots, E_n, E_{n+1}\}$$
$$= \frac{\Pr\{E_{n+1}|H\}\Pr\{H|E_1, \ldots, E_n\}}{\Pr\{E_{n+1}|H\}\Pr\{H|E_1, \ldots, E_n\} + \Pr\{E_{n+1}|K\}\Pr\{K|E_1, \ldots, E_n\}},$$

where $E_1, \ldots, E_{n+1}$ are the outcomes of various experiments.

We may actually have in mind an *a prior* probability density $\rho(\theta)$ over all possible values of the unknown parameter, and so we use our experiment and Bayes' theorem to deduce a posterior probability density $\rho'(\theta)$.

Here is an example of this approach, taken from a report by D.A. Berry[7]:

> A study reported by Freireich et al.[8] was designed to evaluate the effectiveness of a chemotherapeutic agent 6-mercaptopurine (6-MP) for the treatment of acute leukemia. Patients were randomized to therapy in pairs. Let

---

[7] The full report titled "Using a Bayesian approach in medical device development" may be obtained from Donald A. Berry at the Institute of Statistics & Decision Sciences and Comprehensive Cancer Center, Duke University, Durham NC 27708–025.

[8] *Blood* 1963; **21**: 699–716.

p be the population proportion of pairs in which the 6-MP patient stays in remission longer than the placebo patient. (To distinguish probability p from a probability distribution concerning p, I will call it a *population proportion* or a *propensity*.) The null hypothesis H_0 is $p = 1/2$: no effect of 6-MP. Let H_1 stand for the alternative hypothesis that $p > 1/2$. There were 21 pairs of patients in the study, and 18 of them favored 6-MP.

Suppose that the prior probability of the null hypothesis is 70 percent and that the remaining probability of 30 percent is on the interval (0,1) uniformly.... So under the alternative hypothesis H_1, p has a uniform(0,1) distribution. This is a mixture prior in the sense that it is 70 percent discrete and 30 percent continuous.

The uniform(0,1) distribution is also the beta(1,1) distribution. Updating the beta(a,b) distribution after s successes and f failures is easy, namely, the new distribution is beta($a + s, b + f$). So for $s = 18$ and $f = 3$, the posterior distribution under H_1 is beta(19,4).

If our decision procedure is $\delta(X)$ and our loss function is $L(\theta, \delta(X))$, our *risk* when θ is true is $R(\theta, \delta) = L(\theta, \delta(X))$, and our overall average loss is $r(\rho, \delta) = R(\theta, \delta)\rho(\theta)d\theta$. A decision procedure d that minimizes $r(\rho, d)$ is called a *Bayes' solution* and the resultant r, the *Bayes' risk*.

Suppose Θ, the unobservable parameter, has probability density $\rho(\theta)$, and that the probability density of X when $\Theta = \theta$ is $p_\theta(x)$. Let $p(x) = \rho(\theta') p_{\theta'}(x) d\theta'$. Let $\pi(\theta|x)$ denote the a posteriori probability density of Θ given x, which by Bayes' theorem is $\rho(\theta)p_\theta(x)/p(x)$. Then Bayes' risk can also be written as $L(\theta, \delta(x))\pi(\theta|x)d\theta]p(x)dx$.

In the case of testing a simple alternative against a simple hypothesis, let the cost of each observation be c. This cost could be only a few cents (if, say, we are testing the tensile strength of condoms) or more than \$10,000 in the case of some clinical trials. Let c_1 and c_2 denote the costs associated with Type I and Type II errors, respectively. Then the Bayes' risk of a procedure d is

$$r(\rho, d) = \pi[\alpha c_1 + cE_0 N] + (1 - \pi)[(1 - \beta)c_2 + cE_1 N].$$

2.3.2 Mini-Max

An insurance company uses the expected risk in setting its rates, but those of us who purchase insurance use a quite different criterion. We settle for a fixed loss in the form of the insurance premium in order to avoid a much larger catastrophic loss. Our choice of procedure is the decision rule d, here the decision to pay the premium that *minimizes the maximum risk* for all possible values of the parameter.

Other possible criteria fall somewhere in between these two. For example, we could look for the decision rule that minimizes the Bayes' risk among the class of all decision rules for which $R(\theta, d)$ never exceeds some predetermined upper bound.

2.3.3 Generalized Decisions

The simple dichotomy of hypothesis versus alternative and the associated set of decisions, accept or reject, covers only a few cases. More often, we will have a choice among many decisions.

Recently, a promising treatment was found for a once certain fatal disease. Not all patients were cured completely; for some, there was a temporary remission of the disease, which allowed other cures to be tried, while other patients could only report that they felt better, and, alas, there were still many for whom the inevitable downward progress of the disease continued without interruption. The treatment was expensive and carried its own separate risks for the patient. A university laboratory had come up with a predictive method that could be employed prior to starting the treatment. Still, this method wasn't particularly reliable. The small company for whom I worked as a consultant felt sure its technology would yield a far superior predictive measure. The question for the statistician was how the company could turn this feeling into something more substantial, something that could be used to convince both venture capitalists and regulatory agencies of the new method's predictive value.

A committee was formed consisting of two physicians—specialists in the disease and its treatment, a hospital administrator, and a former senior staff member of a regulatory agency. Each was asked to provide their estimates of the costs or losses, relative or absolute, that would be incurred if a measure predicted one response, while the actual outcome was one of the three alternatives. The result, after converting all the costs to relative values and then averaging them, was a loss matrix that looked like this:

	Cured	Remission	Slight relief	No effect
Cured	0	-1	-3	-6
Remission	-2	0	-2	-4
Slight relief	-5	-2	0	-1.2
No effect	-10	-5	-1	0

We already had records for a number n of patients, including samples of frozen blood that had been drawn prior to treatment. These were tested by each of the proposed prediction methods. For each method, we then had an overall risk given by the formula $\sum_i L[d_i, \delta_i]$, where the sum was taken over the entire sample of patients. Since there were only four outcomes, we might also have written this sum as $\sum_{k=1}^{4} \sum_{j=1}^{4} f_n[k,j] L[d_k, \delta_j]$, where $f_n[k,j]$ is the empirical frequency distribution of outcomes for this sample of patients.

In situations where the objective is to estimate the value of a parameter θ, the further apart the estimate θ^* and the true value θ, the larger our losses are likely to be. Typical forms of the loss function in such a case are the absolute deviation $|\theta^* - \theta|$; the square deviation $(\theta^* - \theta)^2$; and the jump, that is, no loss if $|\theta^* - \theta| < \delta$; and a big loss otherwise. Or the loss function may resemble the square deviation but take the form of a step function increasing in discrete increments.

Where estimation is our goal, our objective may be one of two: either to find a decision procedure $d[X]$ that minimizes the risk function $R(\theta,d) = E_\theta[L(\theta,d[X])]$ or the average loss as in our prediction example, or to find a procedure that minimizes the maximum loss. Note that the risk is a function of the unknown parameter θ, so that an optimal decision procedure based on minimizing the risk may depend upon that parameter unless, as in the example of hypothesis testing, there should exist a uniformly most powerful test.

2.4 Exercises

1. a) Power. Sketch the power curve $\beta(\theta)$ for one or both of the two-sample comparisons described in this chapter. (You already know one of the points for each power curve. What is it?)
   b) Using the same set of axis, sketch the power curve of a test based on a much larger sample.
   c) Suppose that without looking at the data you
      i) always reject;
      ii) always accept;
      iii) use a chance device so as to reject with probability α.

   For each of these three tests, determine the power and the significance level. Are any of these three tests exact? unbiased?
2. Suppose that we are testing a simple hypothesis H against a simple alternative K.
   a) Show that if $\alpha_1 \leq \alpha_2$ then $\beta_1 \leq \beta_2$.
   b) Show that if the test $\varphi_1[\alpha_1]$ is less powerful than $\varphi_2[\alpha_2]$ where the significance level $\alpha_1 < \alpha_2$, it may be that $\varphi_1[\alpha_2] > \varphi_2[\alpha_2]$.
3. a) The advertisement reads, "Safe, effective, faster than aspirin." A picture of a happy smiling woman has the caption, "My headache vanished faster than I thought possible." The next time you are down at the pharmacy, the new drug is there at the same price as your favorite headache remedy. Would you buy it? Why or Why not? Do you think the ad is telling the truth? What makes you think it is?
   b) In the United States, in early 1995, a variety of government agencies and regulations would almost guarantee the ad is truthful—or, if not, that it would not appear in print a second time. Suppose you are part of the government's regulatory team reviewing the evidence supplied by the drug company. Looking into the claim of safety, you are told only "we could not reject the null hypothesis." Is this statement adequate? What else would you want to know?

4. **Unbiasedness.** Suppose a and m denote the arithmetic mean and median of a random variable Y, respectively. Show that
   a) For all real b, c such that $a \leq b \leq c$, $E(Y-b)^2 \leq E(Y-c)^2$.
   b) For all real d, e such that $m \leq d \leq e$, $E|Y-d| \leq E|Y-e|$.

5. Do the following constitute independent observations?
   a) Number of abnormalities in each of several tissue sections taken from the same individual.
   b) Sales figures at Eaton's department store for its lamp and cosmetic departments.
   c) Sales figures at Eaton's department store for the months of May through November.
   d) Sales figures for the month of August at Eaton's department store and at its chief competitor Simpson-Sears.
   e) Opinions of several individuals whose names you obtained by sticking a pin through a phone book, and calling the "pinned" name on each page.
   f) Dow Jones Index and GNP of the United States.
   g) Today's price in Australian dollars of the German mark and the Japanese yen.

6. To check out a new theory regarding black holes, astronomers compare the number of galaxies in two different regions of the sky. Six non-overlapping photographs are taken in each region, and three astronomers go over each photo with each recording his counts. Would the statistical method described in Section 1.3 be appropriate for analyzing this data? If so, how many different rearrangements would there be?

7. a) *Decisions.* Suppose you have two potentially different radioactive isotopes with half-life parameters λ_1 and λ_2, respectively. You gather data on the two isotopes and, taking advantage of a uniformly most powerful unbiased permutation test, you reject the null hypothesis $H: \lambda_1 = \lambda_2$ in favor of the one-sided alternative $\lambda_1 > \lambda_2$. What are you or the person you are advising going to do about it? Will you need an estimate of $\lambda_1 > \lambda_2$? Which estimate will you use? (Hint: See Section 3.2 in the next chapter.)
   b) Review some of the hypotheses you tested in the past. Distinguish your actions after the test was performed from the conclusions you reached. (In other words, did you do more testing? rush to publication? abandon a promising line of research?) What losses were connected with your actions? Should you have used a higher/lower significance level? Should you have used a more powerful test or taken more/fewer observations? Were all the assumptions for your test satisfied?

8. a) Your lab has been gifted with a new instrument offering 10 times the precision of your present model. How might this affect the power of your tests? their significance level? the number of samples you'll need to take?

b) A directive from above has loosened the purse strings so you now can test larger samples. How might this affect the power of your tests? their significance level? the precision of your observations? the precision of your results?

c) A series of lawsuits over silicon implants you thought were harmless has totally changed your company's point of view about the costs of sampling. How might this affect the number of samples you'll take? the power of your tests? their significance level? the precision of your observations? the precision of your results?

9. Give an example (or two) of identically distributed observations that are not independent.

10. Are the residuals exchangeable in a regression analysis? an analysis of variance?

11. Suppose a two-decision problem has the loss matrix $\begin{bmatrix} 0 & a \\ b & 0 \end{bmatrix}$. Show that any mini-max procedure is unbiased.

12. Bayes' solutions. Let Θ be an unobservable parameter with probability density $\rho(\theta)$ and suppose we desire a point estimate of a real-valued function $g(\theta)$.

   a) If $L(\theta, d) = (g(\theta) - d)^2$, the Bayes' solution is $E[g(\Theta)|x]$.
   b) If $L(\theta, d) = |g(\theta) - d|$, the Bayes' solution is median $[g(\Theta)|x]$.

13. Many statistical software packages now automatically compute the results of several tests, both parametric and nonparametric. Show that, unless the choice of test statistic is determined before the analysis is performed, the resultant p-values will not be conservative.

3

Testing Hypotheses

It's understandable that one might elect to specialize in the preparation of Chinese cuisine rather than Greek or vice versa. There's just so much time available. But to decide to eat only the one rather than the other is to act the fool.

Now substitute the words "parametric test" for "Chinese cuisine," "nonparametric" for "Greek," and "use" for "eat" in the above paragraph and read it again. In this chapter, you learn how to approach and resolve a series of testing problems of increasing complexity, specifically, tests for location and scale parameters in one and two samples. You learn how to derive confidence intervals for the unknown parameters.

3.1 Testing a Simple Hypothesis

Our first challenge, that of finding the most powerful test of a simple hypothesis against a simple alternative, is chosen not for its practical applications but because its solution is fundamental to the solution of all other testing problems.

Suppose, first, we are trying to decide between two discrete probability distributions P_0 and P_1 such that $P_i\{X = k\} = P_i[k]$ for $i = 0, 1; k = 0, 1, \ldots$. We would like to designate a set of the possible values of X as our rejection region R such that, if k belongs to R, we will reject the simple hypothesis P_0 in favor of the simple alternative P_1. We specify a significance level α and require that $\sum_{k \in R} P_0[k] \leq \alpha$. Given this restriction, our objective is to choose the values of k to include in R so that the power, $\sum_{k \in R} P_1[k]$, is a maximum.

Let $r(k) = P_1[x]/P_0[x]$. To maximize the power against P_1, we need to include in R all the points with the highest values of r until we attain the desired significance level. That is, there exists a constant c such that, if $r(k) > c$, k is to be included in R. If $r[k] < c$, then we will accept the hypothesis.

What if $r[k] = c$? That depends. The answer according to theoreticians is that if $\sum_{k>c} P_0[k] = \alpha' < \alpha$, then, when $r[k] = c$, we pick a random number from 0 to 1. If this random number is less than or equal to $(\alpha - \alpha')/P_0[c]$ we reject the hypothesis; otherwise we accept it.

In Section 15.2, we show that a similar result holds when $P_0[x]$ and $P_1[x]$ are continuous distributions (or a mixture of continuous and discrete). This latter result is the fundamental lemma of Neyman and Pearson [1928].

3.2 One-Sample Tests for a Location Parameter

One of the simplest of practical testing problems would appear to be that of testing for the value of the location parameter of a distribution $F(\theta)$ using a series of observations $x_1, x_2, \ldots, x_n$ from that distribution.

3.2.1 A Permutation Test

This *semiparametric*[1] testing problem is a simple one *if* we can assume that the underlying distribution is symmetric about the unknown parameter θ, that is, if

$$\Pr\{X \leq \theta - x\} = F(\theta - x) = 1 - F(\theta + x) = \Pr\{X \geq \theta + x\}, \text{for all } x.$$

The normal distribution, with its familiar symmetric bell-shaped curve, the double exponential, Cauchy, and uniform distributions are examples of symmetric distributions.[2] The difference of two independent observations drawn from the same population also has a symmetric distribution, as you will see when we come to consider experiments involving matched pairs in Section 5.2.2.2.

Suppose now we wish to test the hypothesis that $\theta \leq \theta_0$ against the alternative that $\theta > \theta_0$. As in Chapter 1, we proceed in four steps:

First, we choose a test statistic that will discriminate between the hypothesis and the alternative. As one possibility, consider the sum of the deviations of θ about θ_0. Under the hypothesis, positive and negative deviations ought to cancel and this sum should be close to zero. Under the alternative, positive terms should predominate and this sum should be large. But how large should the sum be for us to reject the hypothesis?

We saw in Chapter 1 that we can use the permutation distribution to obtain the answer; but what should we permute? The principle of *sufficiency* can help us here.

Suppose we had lost track of the signs (plus or minus) of the deviations. We could attach new signs at random, selecting a plus or a minus with equal

---

[1] A problem is *parametric* if the form of the underlying distribution is known, and it is *nonparametric* if we have no knowledge concerning the distribution(s) from which the observations are drawn.

[2] These distributions are described in more detail in Chapter 5.

probability. If we are correct in our hypothesis that the variables have a symmetric distribution about θ_0, the resulting values should have precisely the same distribution as the original observations. That is, the absolute values of the deviations are *sufficient* for regenerating the sample. (You'll find more on the topic of sufficiency in Section 3.5.2 and in Appendix 1.1.)

Under the alternative of a location parameter larger than θ_0, randomizing the signs of the deviations should reduce the sum from what it was originally. As we consider one after another in a series of random reassignments, our original sum should be revealed as an extreme value.

Before implementing this permutation procedure, we note that the sum of just the deviations with plus signs attached is related to the sum of *all* the deviations by the formula

$$\sum_{x_i>0} x_i = \frac{\sum x_i + \sum |x_i|}{2},$$

because the +1 values get added twice, once in each sum on the right hand side of the equation, while the values of -1 and $|-1|$ cancel. Thus, we can reduce the number of calculations by summing only the positive deviations. As an illustration, suppose that θ_0 is 0 and that the original observations are -1, 2, 3, 1.1, 5. Our second step is to compute the sum of the positive deviations, which is 11.1.

Among the 2^5 possible reassignments of plus and minus signs are

$$+1, -2, 3, 1, 5$$
$$+1, 2, 3, 1, 5$$

and

$$-1, -2, 3, 1, 5.$$

Our third step is to compute the sum of the positive deviations for each rearrangement. For the three rearrangements shown above, this sum would be 10, 12, and 9, respectively.

Our fourth step is to compare the original value of our test statistic with its permutation distribution. In Section 3.1, we showed that the most powerful test would set aside in the rejection region those outcomes that have the greatest value of the ratio $P_1[x]/P_0[x]$. Under the null hypothesis, all labelings of the observations $(x_{L1}, \ldots x_{Ln})$ from which x may be composed are equally likely, and $P_0[x]$ is a constant equal to one divided by the number of labelings. Our rejection region will consist of those outcomes for which $P_1[x]$ is a maximum.

Only 2 of the 32 rearrangements have sums as large as the sum 11.1 of the original observations. Is $2/32 = 1/16 = .0625$ statistically significant? Perhaps not at the 5% level, but surely a *p*-value of .0625 is suggestive enough that we might want to look at additional data or perform additional experiments before accepting the hypothesis that 0 is the true value of θ.

36   3 Testing Hypotheses

Sidebar

Although software to execute parametric tests is plentiful, programs with which to do permutation tests are few and far between. *Resampling Stats* is one package specifically designed to help with bootstrap and permutation tests. Available from www.statistics.com, its drawback is that it does not come with complete routines, but requires the user do the programming. Here is one example for use in testing for the location parameter of a population.

```
'Perform a one-sided test for matched pairs
DATA (1 2 3 4 10 6) Before
DATA (2 2 4 6 8 8) After
SUBTRACT After Before Change
'The next two instructions zero out the negative changes and double the positive
ABS Change Total
ADD Change Total Work
SUM Work sumorig
DIVIDE sumorig 2 sumorig
Data (0 1) Basis
LET n = 6
LET cnt = 0
REPEAT 400
    RANDOM 6 Basis Sample
    'select random values from absolute changes
    MULTIPLY Sample Total Temp
    SUM Temp sumperm
    IF sumorig <= sumperm
        LET cnt = cnt + 1
    END
END
DIVIDE cnt 400 pvalue
PRINT pvalue
```

3.2.2 A Parametric Test

Suppose we know more about the distribution F. In particular, suppose we know F is a normal distribution with mean θ and variance 1, $N(0,1)$, so that the cumulative distribution function can be written in the form

$$F[x] = \int_{-\infty}^{x} \frac{1}{\sqrt{2\pi}} \exp[-(y-\theta)^2/2] dy \qquad (3.1)[3]$$

A property of this distribution is that the mean of n independent normally distributed random variables with mean θ and variance 1 is normally distributed with mean θ and variance $1/n$ (see Exercise 3.1). Knowing this, an alternate

---

[3] The use of the constant $1/\sqrt{2\pi}$ ensures that $F[\infty] = 1$.

approach to a test of the hypothesis that $\theta \leq \theta_0$ is to reject the hypothesis only if the mean of the observations $\bar{X}$ is greater than the 95th percentile of a normal distribution with mean θ_0 and variance $1/n$.

Tables of the normal distribution with mean 0 and variance 1 are readily available. To take advantage of these tables, we can use the test statistic $(\bar{X} - \theta_0)\sqrt{n}$ (Exercise 3.2), rejecting the hypothesis when it is greater than the 95th percentile of a normal distribution with mean 0 and variance 1.

3.2.3 Properties of the Parametric Test

Most Powerful Test. Using the fundamental lemma of Neyman and Pearson we can show immediately that our parametric test is most powerful for testing the hypothesis that our observations are normally distributed with mean $\theta = \theta_0$ and variance 1 against the alternative that the observations are normally distributed with expectation $\theta = \theta_1 > \theta_0$ and variance 1 (Exercise 3.3). But this applies to any alternative $\theta_1 > \theta_0$; thus, this same test is uniformly most powerful for testing the hypothesis $\theta = \theta_0$ against the set of alternatives $\theta > \theta_0$ when the observations are normally distributed with variance 1. It is also uniformly most powerful for testing the hypothesis $\theta \leq \theta_0$ against the set of alternatives $\theta > \theta_0$ among all tests for which $\beta(\theta_0) = \alpha$.

We can generalize this result in stages. Suppose, first, we know F is a normal distribution with mean θ and variance σ^2, $N(\theta, \sigma^2)$, so that F can be written in the form

$$\int_{-\infty}^{x} \frac{1}{\sqrt{2\pi\sigma^2}} \exp[-(y-\theta)^2/2\sigma^2] dy. \tag{3.2}$$

Then it is obvious we can attain a uniformly most powerful (UMP) test of the hypothesis that our observations are normally distributed (with mean $\theta \leq \theta_0$ and variance σ^2) against the alternatives that the observations are normally distributed (with expectation $\theta > \theta_0$ and variance σ^2) among all tests for which $\beta(\theta_0)$ is some fixed percentage α using the statistic $(\bar{X} - \theta)\sqrt{n}/\sigma$.

In both equations (3.1) and (3.2), we have written $F[x]$ in the form

$$F[x] = \int_{-\infty}^{x} f(y) dy.$$

The function f is called a *probability density*.[4] Note that for any $\theta < \theta'$ the distributions F_θ and $F_{\theta'}$ are distinct and the ratio of probability densities $f_{\theta'}[y]/f_\theta[y]$ is a nondecreasing function of y. Such a family of distributions is said to have *monotone likelihood ratio in y*.

Theorem 3.1. *Let θ be a real-valued parameter and let the random variable X have probability density $f_\theta[y]$ with monotone likelihood ratio in $T[y]$.*

---

[4] See Appendix, Section 1 for a mathematically precise definition.

For testing the hypothesis that $\theta \leq \theta_0$ *against the alternative that* $\theta > \theta_0$, *there exists a UMP test given by*

$$\varphi(y) = \begin{cases} 1 & \text{when } T[y] > C \\ \gamma & \text{when } T[y] = C \\ 0 & \text{when } T[y] < C \end{cases}$$

where C *and* γ *are determined by* $E(\varphi(y)|\theta_0) = \alpha$.

Here, $\varphi(y) = 1$ means to reject the hypothesis, $\varphi(y) = 0$ means to accept, and $\varphi(y) = \gamma$ means that the decision is to be left to an unfair coin that lands reject side up with probability γ. By the fundamental lemma of Neyman and Pearson, we know we can find a C and γ for use in testing $\theta = \theta_0$ against the alternative $\theta = \theta_1 > \theta_0$. Noting that T[y] will be greater than C if and only if there is a C' such that the likelihood ratio $f[y|\theta']/f[y|\theta_0] > C'$ where $\theta' > \theta_0$, we see that a critical function that satisfies the conditions of the theorem is uniformly most powerful for testing $\theta = \theta_0$ against any alternative $\theta > \theta_0$. In fact, $\beta(\theta)$ is strictly increasing for all values of the parameter θ for which $0 < \beta(\theta) < 1$. Consequently, the test satisfies $E(\varphi(y)|\theta) \leq \alpha$ for $\theta \leq \theta_0$ and among all tests which satisfy this condition is UMP for testing $\theta \leq \theta_0$ against the alternative that $\theta > \theta_0$.

3.2.4 Student's t

The existence of such a uniformly most powerful test would be exciting, but for two drawbacks: What if the variance of the observations is not σ^2 as assumed? And what if the data are not normally distributed?

The first of these objections is addressed directly through the use of an alternate statistic, known as *Student's t*. The second objection is "almost" resolved via this same choice.

The statistic is almost the same as the one we used previously, except that we have substituted the sample variance for the population variance $(\bar{X} - \theta)^2$. Thus,

$$t[X] = \frac{(\bar{X} - \theta)\sqrt{n}}{\sqrt{\sum(X_i - \bar{X})^2/(n-1)}}.$$

We look up the cut-off values for this statistic by consulting tables of the Student's t distribution with $n - 1$ degrees of freedom. The resultant test is not uniformly most powerful for all values of σ and θ (see Exercise 3.4), but as we will show in Section 3.6.1, it is uniformly most powerful when we restrict ourselves to the class of unbiased tests.

The attractiveness of the t-test, which we recommend for the comparison of two populations, is two-fold. First, it does not depend upon the value of the unknown variance σ^2. Second, even for moderate sample sizes of 6 or more, it is almost but not quite distributed in accordance with the theoretical distribution of Student's t, even if the original observations are not distributed like this.[5] From the central limit theorem, we know that $(\bar{X} - \theta)\sqrt{n}/\sigma$ has the limiting distribution $N(0,1)$. And from a convergence theorem of Cramer [1946] we know that $\sum(x_i - \bar{x})^2/(n-1)\sigma^2$ tends to 1 in probability.

3.2.5 Properties of the Permutation Test

Our permutation test is exact whether the observations are normally distributed or not. It is applicable even if the different observations come from different distributions, providing, that is, these distributions are all symmetric and all have the same location parameter or median. (If these distributions are symmetric, then, if the mean exists, the mean of the distribution is identical with its median.) If you are willing to specify their values through the use of a parametric model, then the medians needn't be the same! (See Exercise 3.10.)

Asymptotic consistency. What happens if the underlying distributions are almost but not quite symmetric? Romano (1990) shows that the permutation test for a location parameter is asymptotically exact providing the underlying distribution has finite variance. His result applies whether the permutation test is based on the mean, the median, or some statistical functional of the location parameter. If the underlying distribution is almost symmetric, the test will be almost exact even when based on as few as 10 or 12 observations. See Appendix 10.1 for the details of a Monte Carlo procedure to use in deciding when "almost" means "good enough."

3.2.6 Exact Significance Levels: A Digression

Many of us are used to reporting our results in terms of significance levels of 0.01, 0.05, or 0.10, so significance levels of .0625 or .03125 resulting from the application of a permutation tests may seem confusing at first. These "oddball" significance levels often occur in permutation tests with small sample sizes. Five observations means just 32 possibilities, and 1 extreme observation out of 32 corresponds to .03125. Things improve as sample sizes get larger. With seven observations, we can test at a significance level of .049. Is this close enough to 0.05?

Lehmann [1986] describes a method called "randomization on the boundary" for obtaining a significance level of exactly 5% (or exactly 1%, or exactly 10%). But this method isn't very practical. In the worst case, "on the boundary," a chance device is allowed to make your decision for you.

What is the practical solution? We agree with Kempthorne [1975, 1977, 1979]. Forget tradition. There is nothing sacred about a significance level of

---
[5] We'll consider some worst-case exceptions in Chapter 5.

5% or 10%. Report the exact significance level, whether it is .065 or .049. Let your colleagues reach their own conclusions based on the losses they associate with each type of error.

In reporting the results of parametric and other types of tests, do not confuse the *p-value* with the significance level. The *p-value* is a random variable whose value varies from sample to sample. A *significance level* is a fixed value associated with a testing procedure, one *always* determined in advance of considering the actual observations.

3.3 Confidence Intervals

After rejecting a hypothesis, we will normally want to go further and say something about the most likely values the parameter under investigation might take. Confidence intervals can be derived from the rejection regions of our hypothesis tests, whether the latter are based on parametric, semiparametric, or nonparametric methods. Confidence intervals include all values of a parameter for which we would accept the hypothesis that the parameter takes the value in question.

If $A(\theta')$ is a $1 - \alpha$ level acceptance region for testing the hypothesis $\theta = \theta'$, and $S(\boldsymbol{X})$ is a $1 - \alpha$ level confidence interval for θ based on the vector of observations $\boldsymbol{X}$, then for the confidence intervals defined here, $S(\boldsymbol{X})$ consists of all the parameter values θ^* for which $\boldsymbol{X}$ belongs to $A(\theta^*)$, while $A(\theta)$ consists of all the values of the statistic x for which θ belongs to $S(x)$.

$$\Pr(S(x) \supset \{\theta\}|\theta) = \mathrm{P}\{X \in A(\theta)|\theta\} \geq 1 - \alpha.$$

Note that the relationship extends in both directions so that the rejection regions of our hypothesis tests can be derived from confidence intervals. Suppose our hypothesis is that the odds ratio for a 2×2 contingency table is 1, a problem considered at greater length in Chapter 8. Then we would accept this hypothesis if and only if our confidence interval for the odds ratio includes the value 1. Confidence intervals based on the bootstrap (our third example below) are the standard basis for tests of hypotheses using the bootstrap.

As our confidence level increases, from 90% to 95%, for example, the width of the resulting confidence interval increases. Thus, a 95% confidence interval is wider than a 90% confidence interval.

It is easy to show that if $A(\theta)$ is the acceptance region of a uniformly most powerful unbiased test, the correct value of the parameter is more likely to be covered by the confidence intervals we've constructed than would any incorrect value.

Theorem 3.2. *Let $x = \{X_1, X_2, \ldots, X_n\}$ be an exchangeable sample from a distribution F_θ that depends upon a parameter $\theta \in \Omega$. For each $\theta' \in \Omega$, let $A(\theta')$ be the acceptance region of the level-α test for $H(\theta')$: $\theta = \theta'$, and for*

3.3 Confidence Intervals

each x let $S(x)$ denote the set of parameter values $\{\theta: x \in A(\theta), \theta \in \Omega\}$. Then $S(x)$ is a family of confidence sets for θ at confidence level $1 - \alpha$.

Theorem 3.3. *If, for all θ', $A(\theta')$ is UMP unbiased for testing $H(\theta')$ at level a against the alternatives $K(\theta')$, then for each θ' in Ω, $S(X)$ minimizes the probability $\Pr(S(x) \supset \{\theta'\}|\theta)$ for all $\theta \in K(\theta')$ among all unbiased level-$1 - \alpha$ families of confidence sets for θ.*

Proof 3.2. By definition, $\theta \in S(x)$ if and only if $x \in A(\theta)$, hence $\Pr(S(x) \supset \{\theta\}|\theta) = P_\theta\{X \in A(\theta)\} \geq 1 - a$. □

Proof 3.3. If $S^*(x)$ is any other family of unbiased confidence sets at level $1 - a$ and if $A^*(\theta) = \{x: \theta \in S^*(x)\}$, then

$$P_\theta\{X \in A^*(\theta')\} = \Pr(S^*(x) \supset \{\theta'\}|\theta) \geq 1 - \alpha \text{ for all } \theta \in H(\theta'),$$

and

$$P_\theta\{X \in A^*(\theta')\} = \Pr(S^*(x) \supset \{\theta'\}|\theta) \leq 1 - \alpha \text{ for all } \theta \in K(\theta'),$$

so that $A^*(q')$ is the acceptance region of a level-α unbiased test of $H(\theta')$. Since A is UMPU, $P_\theta\{X \in A^*(\theta')\} \geq P_\theta\{X \in A(\theta')\}$ for all $\theta \in K(\theta')$, hence $\Pr(S^*(x) \supset \{\theta'\}|\theta) \geq \Pr(S(x) \supset \{\theta'\}|\theta)$ for all $\theta \in K(\theta')$, as was to be proved. □

3.3.1 Confidence Intervals Based on Permutation Tests

In the first step of a permutation test for a non-zero value θ_0 of the location parameter, we subtract this value from each of the observations. We could test a whole series of hypotheses involving different values for θ_0 in this fashion until we found a θ_1 such that as long as $\theta_0 \geq \theta_1$, we accept the null hypothesis, but for any $\theta_0 < \theta_1$ we reject it. Then a $100(1 - \alpha)\%$ confidence interval for θ is given by the interval $\{\theta > \theta_1\}$.

Note that the preceding interval based on a one-sided test is also one-sided. In this interval, θ_1 is referred to as the *lower confidence bound*.

Suppose the original observations are $-1, 2, 3, 1.1, 5$, and we want to find a confidence interval that will cover the true value of the parameter $31/32$ of the time. In the first part of this chapter, we saw that $1/16$ of the rearrangements of the signs resulted in samples that were as extreme as these observations. Thus, we would accept the hypothesis that $\theta \leq 0$ at the $1/16$ level and any smaller level, including the $1/32$. Similarly, we would accept the hypothesis that $\theta \leq -0.5$ at the $1/32$ level, or even that $\theta \leq -1+\varepsilon$ where ε is an arbitrarily small but still positive number. But we would reject the hypothesis that $\theta \leq -1 - \varepsilon$, as after subtracting $-1 - \varepsilon$ the transformed observations are $\varepsilon, 2, 3, 1.1, 5$.

Our one-sided confidence interval is $(-1, \infty)$ and we have confidence that $31/32$ of the time the method we've used yields an interval that includes the true value of the location parameter θ.

Our one-sided test of a hypothesis gives rise to a one-sided confidence interval. But knowing that θ is larger than −1 may not be enough. We may want to pin θ down to a more precise two-sided interval, say, that θ lies between −1 and +1.

To accomplish this, we need to begin with a two-sided test. Our hypothesis for this test is that $\theta = \theta_0$ against the two-sided alternatives that θ is smaller or larger than θ_0. We use the same test statistic—the sum of the positive observations, which we used in the previous one-sided test. Again, we look at the distribution of our test statistic over all possible assignments of plus and minus signs to the observations. But this time we reject the hypothesis if the value of the test statistic for the original observations is either one of the largest or one of the smallest of the possible values.

In our example we don't have enough observations to find a two-sided confidence interval at the 31/32 level, so we'll try to find one at the 15/16 level. The lower boundary of the new confidence interval is still −1. But what is the new upper boundary? If we subtract 5 from every observation, we would have the values $-6, -3, -2, -4.9, -0$; their sum is -15.9. Only the current assignment of signs to the transformed values, that is, only 1 out of the 32 possible assignments, yields this small a sum for the positive values. The symmetry of the permutation test requires that we set aside another 1/32 of the arrangements at the high end. Thus we would reject the hypothesis that $\theta = 5$ at the $1/32 + 1/32$, or 1/16 level. Consequently, the interval $(-1, 5)$ has a 15/16 chance of covering the unknown parameter value.

These results are readily extended to a confidence interval for a vector of parameters θ that underlies a one-sample, two-sample, or k-sample experimental design with single- or vector-valued variables. In each case, the 100(1 − α)%-confidence interval consists of all values of the parameter vector θ for which we would accept the hypothesis at level α. Remember, one-sided tests produce one-sided confidence intervals and two-sided tests produce two-sided intervals.

For further information on deriving confidence intervals using the permutation approach see Lehmann [1986, pp. 246–263], Gabriel and Hsu [1983], John and Robinson [1983], Maritz [1981, p. 7, p. 25], and Tritchler [1984].

3.3.2 Confidence Intervals Based on Parametric Tests

The derivation of confidence intervals based on parametric tests is relatively simple. Let us suppose we are testing a hypothesis concerning the location parameter of a normal distribution whose variance is known to be 4. We make a series of nine independent observations whose mean is 11. The 95th percentile of a $N(0, 1)$ distribution is 1.64. The 5th percentile is −1.64.

$(\bar{X} - \theta)\sqrt{9}/2$ is distributed as $N(0, 1)$. This means that the interval $\bar{X} \pm \sqrt{4/9} * 1.64$ will cover the unknown population mean θ about 90% of the time.

3.3.3 Confidence Intervals Based on the Bootstrap

The bootstrap can help us obtain an interval estimate for any aspect of a distribution—a median, a variance, a percentile, or a correlation coefficient—*if* the observations are independent and all come from distributions with the same value of the parameter to be estimated. This interval provides us with an estimate of the precision of the corresponding point estimate.

We resample with replacement repeatedly from the original sample, 1000 times or so, computing the sample statistic for each bootstrap sample.

For example, here are the heights of a group of 22 adolescents, measured in centimeters and ordered from shortest to tallest:

137.0 138.5 140.0 141.0 142.0 143.5 145.0 147.0 148.5 150.0 153.0
154.0 155.0 156.5 157.0 158.0 158.5 159.0 160.5 161.0 162.0 167.5

The median height lies somewhere between 153 and 154 centimeters. If we want to extend this result to the population, we need an estimate of the precision of this estimate.

Our first bootstrap sample, arranged in increasing order of magnitude for ease in reading, might look like this:

138.5 138.5 140.0 141.0 141.0 143.5 145.0 147.0 148.5 150.0 153.0
154.0 155.0 156.5 157.0 158.5 159.0 159.0 159.0 160.5 161.0 162.0

Several of the values have been repeated, which is not surprising as we are sampling with replacement, treating the original sample as a stand-in for the much larger population from which the original sample was drawn. The minimum of this bootstrap sample is 138.5, higher than that of the original sample; the maximum at 162.0 is less than the original, while the median remains unchanged at 153.5.

137.0 138.5 138.5 141.0 141.0 142.0 143.5 145.0 145.0 147.0 148.5
148.5 150.0 150.0 153.0 155.0 158.0 158.5 160.5 160.5 161.0 167.5

In this second bootstrap sample, again we find repeated values; this time the minimum, maximum, and median are 137.0, 167.5 and 148.5, respectively.

The medians of 50 bootstrapped samples drawn from our sample ranged between 142.25 and 158.25 with *their* median being 152.75 (see Figure 3.1). These numbers provide an insight into what might have been had we sampled repeatedly from the original population.

Fig. 3.1. Scatterplot of 50 bootstrap medians derived from a sample of heights.

44   3 Testing Hypotheses

We can improve on the interval estimate [142.25, 158.25] if we are willing to accept a small probability that the interval will fail to include the true value of the population median. We will take several hundred bootstrap samples instead of a mere 50, and use the 5th and 95th percentiles of the resulting bootstrap distribution to establish the boundaries of a 90% confidence interval.

This method might be used equally well to obtain an interval estimate for any other population attribute: the mean or variance, the 5th percentile or the 25th, and the inter-quartile range. When several observations are made simultaneously on each subject, the bootstrap can be used to estimate covariances and correlations among the variables. The bootstrap is particularly valuable when trying to obtain an interval estimate for a ratio or for the mean and variance of a nonsymmetric distribution.

Unfortunately, such intervals have two deficiencies:

1. They are biased, that is, they are more likely to contain certain false values of the parameter being estimated than the true one [Efron, 1988];
2. They are wider and less efficient than they could be [Efron, 1988].

Two methods have been proposed to correct these deficiencies; let us consider each in turn.

The first is the bootstrap-t in which the bootstrap estimate is Studentized. For the one-sample case, we want an interval estimate based on the distribution of $(\hat{\theta}_b - \hat{\theta})/SE(\hat{\theta})_b$, where $\hat{\theta}$ and $\hat{\theta}_b$ are the estimates of the unknown parameter based on the original and bootstrap sample, respectively, and $SE(\hat{\theta})_b$ denotes an estimate, derived from the bootstrap sample, of the standard error of the original estimate. An estimate $\hat{\sigma}$ of the population variance is required to transform the resultant interval into one about θ (see Carpenter and Bithell, 2000).

For the two-sample case we want a confidence interval based on the distribution of

$$\frac{(\hat{\theta}_{nb} - \hat{\theta}_{mb}) - (\hat{\theta}_n - \hat{\theta}_m)}{\sqrt{\frac{(n-1)s_{nb}^2 + (m-1)s_{mb}^2}{n+m-2}(1/n + 1/m)}},$$

where n, m, and s_{nb}, s_{mb} denote the sample sizes and standard deviations, respectively, of the bootstrap samples. Applying the Hall–Wilson corrections, we obtain narrower interval estimates that are as or more likely to contain the true value of the unknown parameter.[6]

The *bias-corrected and accelerated* BC_a *interval* due to Efron and Tibshirani [1986] also represents a substantial improvement, though for samples under size 30 the interval is still suspect. The idea behind these intervals comes from the observation that percentile bootstrap intervals are most accurate when the estimate is symmetrically distributed about the true value of

---

[6] Unlike the original bootstrap interval, this new interval may contain "impossible" values; see Exercise 3.15.

the parameter and the tails of the estimate's distribution drop off rapidly to zero. In other words, when the estimate has an almost normal distribution.

Suppose θ is the parameter we are trying to estimate, $\hat{\theta}$ is the estimate, and we are able to come up with a monotone increasing transformation t such that $t(\theta')$ is normally distributed about $t(\theta)$. We could use this normal distribution to obtain an unbiased confidence interval, and then apply a back-transformation to obtain an almost unbiased confidence interval.[7]

Even with these modifications, we do not recommend the use of the nonparametric bootstrap with samples of fewer than 100 observations. Simulation studies suggest that with small sample sizes, the coverage is far from exact and the endpoints of the intervals vary widely from one set of bootstrap samples to the next. For example, Tu and Zhang [1992] report that with samples of size 50 taken from a normal distribution, the actual coverage of an interval estimate rated at 90% using the BC_a bootstrap is 88%. When the samples are taken from a mixture of two normal distributions (a not uncommon situation with real-life data sets) the actual coverage is 86%. With samples of only 20 in number, the actual coverage is 80%.

A more serious problem that arises when one tries to apply the bootstrap is that the endpoints of the resulting interval estimates may vary widely from one set of bootstrap samples to the next. For example, when Tu and Zhang drew samples of size 50 from a mixture of normal distributions, the average of the left limit of 1000 bootstrap samples taken from each of 1000 simulated data sets was 0.72 with a standard deviation of 0.16, the average and standard deviation of the right limit were 1.37 and 0.30, respectively.

3.3.4 Parametric Bootstrap

When we know the form of the population distribution, the use of the *parametric bootstrap* to obtain interval estimates may prove advantageous either because the parametric bootstrap provides more accurate answers than textbook formulas or because no textbook formulas exist.

Suppose we know the observations come from a normal distribution and want an interval estimate for the standard deviation. We would draw repeated bootstrap samples from a normal distribution the mean of which is the sample mean and the variance of which is the sample variance. (As a practical matter, we would program our computer to draw an element from an $N(0,1)$ population, multiply by the sample standard deviation, then add the sample mean to obtain an element of our bootstrap sample.) By computing the standard deviation of each bootstrap sample, an interval estimate for the standard deviation of the population may be derived. Because the resultant

---

[7] Stata^{TM} provides for bias-corrected intervals via its .bstrap command. R and S-Plus both include BC_a functions. A SAS macro is available at www.asu.edu/it/fyi/research/helpdocs/statistics/SAS/tips/jackboot.html.

3.3.5 Better Confidence Intervals

In deriving a confidence interval, we look first for a *pivotal quantity* or *pivot* $Q(X_1, \ldots, X_n, \theta)$ whose distribution is independent of the parameters of the original distribution. One example is a pivotal quantity is $\bar{X} - \theta$, where $\bar{X}$ is the sample mean, and the observations X_i are drawn independently from identically distributions $F(x - \theta)$. A second example is $Q = \bar{X}/\sigma$ where the X_i are independent and identically distributed as $F(x/\sigma)$. If the $\{X_i\}$ are independent from the identical exponential distribution $1 - \exp[-\lambda t]$ (see Exercise 2 in Chapter 2), then $T = 2\sum t_i/\lambda$ is a pivotal quantity whose distribution does not depend on λ. We can use this distribution to find an L and a U such that

$$\Pr(L < T < U) = 1 - \alpha.$$

But then

$$\Pr\left(\frac{1}{2b\sum t_i} < \lambda < \frac{1}{2a\sum t_i}\right) = 1 - \alpha.$$

We use a pivotal quantity in Chapter 9.4 to derive a confidence interval for a regression coefficient. For a discussion of the strengths and weaknesses of pivotal quantities, see Berger and Wolpert [1984].

3.4 Comparison Among the Test Procedures

We are now able to make an initial comparison of the five types of statistical hypothesis test—permutation, rank, bootstrap, parametric bootstrap, and parametric.

Recall from Chapter 1 that with all tests we need to complete the same five basic steps. The only differences are in how we arrive at the distribution of the test statistic and in any subsequent limitations upon the choice of a statistic.

We obtain the permutation distribution of a test statistic T by repeatedly rearranging the observations. With two or more samples, we combine all the observations into a single large sample before we rearrange them. There are no limitations upon the test statistic. If the observations are exchangeable then the resultant test is exact and unbiased.

As noted in the brief history provided in the opening chapter, though permutation tests were among the very first statistical tests to be developed, they were beyond the computing capacities of the 1930s. One alternative, which substantially reduces the amount of computation required, is the *rank test*. To form a *rank test* (e.g., Mann–Whitney or Friedman's test), we replace the original observations by their ranks, and then obtain the permutation distribution of sample S by repeatedly rearranging the ranks and recomputing the test statistic. This approach is appropriate when outliers are suspected, and it is also useful for combining observations that may have been taken on different scales (see Chapters 5, 7, and 11). But throwing away this information has the same effect as if we were to reduce the sample size by roughly 3% for very large samples and much more for smaller ones.

The bootstrap, like the permutation test, requires a minimum number of assumptions and derives its critical values from the data at hand. To obtain a nonparametric bootstrap, we obtain the bootstrap distribution of T or $\hat{\theta}$ by repeatedly resampling from the observations. We need not combine the samples, but may resample separately from each sample. We resample with replacement.

The bootstrap is neither exact nor conservative. Generally, but not always, a nonparametric bootstrap is less powerful than a permutation test. If the observations are independent and come from distributions with identical values of the parameter of interest, then the bootstrap is asymptotically exact [Liu, 1988]. And it may be possible to bootstrap when no other statistical method is applicable (see Section 7.6 and Chernick, 1999).

To obtain a parametric bootstrap, we obtain the bootstrap distribution of T by repeatedly resampling from the observations and then using the bootstrap sample to estimate the parameters of the distribution. This method is more powerful than the nonparametric bootstrap and yields narrower confidence intervals providing we are correct in our choice of distribution. This method is limited to statistics T whose distribution is known.

To obtain a parametric test, we compare the observed value of $T(X)$ with the percentage points of a known distribution F and accept or reject the null hypothesis according to whether $T(X)$ is smaller or larger than this value.

If T is distributed as F, then the parametric test is exact and, often, the most powerful test is available. In order for T to have the distribution F, in most cases the observations $\{X_i\}$ need to be independent and, with small samples, identically distributed with a specific distribution G. If the observations are not identically distributed or have some distribution other than G, the T may have some distribution other than F, so that the test based on T may not be conservative, and its claimed power may be suspect.

When a choice of statistical methods exists, the best method is the one that yields the shortest confidence interval for a given significance level. Robinson [1989] finds approximately the same asymptotic coverage probabilities for three sets of confidence intervals for the slope of a simple linear regression based, respectively, on 1) the standardized bootstrap, 2) parametric theory, and 3) a permutation procedure. With large samples, the permutation test is usually as powerful as the most powerful parametric test [Bickel and Van Zwet, 1978].

3.5 One-Sample Tests for a Scale Parameter

3.5.1 Semiparametric Tests

Suppose now that $F[x] = G[x/\sigma]$, where σ is generally referred to as a *scale parameter*. Recall that the standard deviation σ of a normal population serves as its scale parameter. To obtain a bootstrap test of the hypothesis $H: \sigma = \sigma_0$ we proceed as we did in the preceding sections, taking a series of bootstrap samples from the original sample, computing an estimate of the standard deviation σ each time, and using the percentiles of the resulting bootstrap distribution as the cut-off point(s) of a one- or two-sided test.

While we can improve the accuracy of the bootstrap test by using either the Hall–Wilson corrections or the BC_a interval, its Type I error cannot be guaranteed.

To obtain a permutation test of the scale parameter, one ought begin by taking the logarithms of the observations, $Y = \log[X]$, so that the distribution of Y may be represented as $H[y - \mu]$ where $\mu = \log \sigma$.[8] If H is symmetric, then we can immediately derive a test based on Y. But how likely is such a property? An exact permutation test may not exist in this instance.

3.5.2 Parametric Tests: Sufficiency

Fortunately, a uniformly most powerful parametric test exists, that is, a UMP exists *if* we can assume the observations are normally distributed. And, unfortunately, this test, unlike the *t*-test, is not robust to departures from normality, so that it may not be of value if our observations are drawn from some other distribution.

Our test statistic is $\sum_{i=1}^{n}(x_i - \bar{x})^2/\sigma_0^2$ and when the observations are independent and normally distributed $N(\mu, \sigma^2)$, this statistic has the chi-square distribution with $n - 1$ degrees of freedom.[9]

This test is UMP in part because $\bar{x}$ and $\sum_{i=1}^{n}(x_i - \bar{x})^2$ are sufficient statistics for the normal family of distributions. A statistic T is said to be *sufficient* for a family of distributions based on the parameter θ if the conditional distribution of X given t is independent of θ. A necessary and sufficient condition for T to be sufficient is that we can factor the probability density f in the form $f_\theta[x] = g_\theta[T(x)]h(x)$ where the first factor g *may* depend upon the parameter, but depends on x *only* through $T(x)$, and the second factor h is independent of the parameter.

---

[8] If the data are all positive, then taking logarithms will result in an improved bootstrap interval.

[9] Although, we had n degrees of freedom in making the original n independent observations, a little algebra reveals that the first $n - 1$ deviations about the mean uniquely determine the nth deviation. (Exercise 3.7).

3.5 One-Sample Tests for a Scale Parameter

In the discrete case, for example, suppose that

$$P(x|\theta) = g_\theta[T(x)]h(x).$$

$\Pr\{T = t|\theta\} = \sum P(x'|\theta)$ where the sum extends over all values x' for which $T(x') = t$ and the conditional probability that $X = x$ given that $T = t$ is $P(x|\theta)/\Pr(T = t|\theta)$ or $h(x)/\sum h(x')$, which is independent of θ. On the other hand, if this conditional distribution does not depend on θ and is equal to some function of x and t, say $k(x,t)$, then $P(x|\theta)$ can be written as the product of $\Pr\{T = t|\theta\}$ and $k(x,t)$.

$\bar{x}$ and $\sum_{i=1}^{n}(x_i - \bar{x})^2$ are sufficient statistics for the normal family, whose distribution is given in Equation (3.2). Noting that $(x_i - \mu)^2 = (x_i - \bar{x} + \bar{x} - \mu)^2$ and $\sum(x_i - \bar{x}) = 0$, we can rewrite the normal probability density as

$$(2\pi\sigma^2)^{-n/2} \exp\left[-\frac{1}{2\sigma^2}\sum(x_i - \bar{x})^2 - \frac{1}{2\sigma^2}\sum(\mu - \bar{x})^2\right]. \quad (3.3)$$

Unfortunately, the hypothesis that our observations come from a normally distributed population with standard deviation $\sigma \leq \sigma_0$ is not a simple one (if it were, we could apply the Neyman–Pearson lemma), but a *composite hypothesis* consisting of a large number of possible distributions $N(\mu, \sigma^2)$ with $\sigma \leq \sigma_0$ and $-\infty < \mu < \alpha$. Without further restriction, the maximum power that can be obtained for this composite hypothesis against any simple alternative is limited to the maximum power that can be attained by testing the least favorable hypothesis in this family against that same alternative.[10]

In the present case, it seems reasonable to assume that the least favorable distribution Λ of the parameters in the (μ, σ)-plane is concentrated on the boundary between the hypothesis and the alternative $\sigma = \sigma_0$. The joint density of the sufficient statistics $Y = \bar{x}$ and $U = \sum_{i=1}^{n}(x_i - \bar{x})^2$ under H_Λ is

$$C_0 u^{(n-3)/2} \exp[-u/2\sigma_0^2] \int \exp[-n(y-\mu)^2/2\sigma_0^2] d\Lambda(\mu)$$

Now suppose our interest is in a simple alternative $K: \mu = \mu_1, \sigma = \sigma_1$. The least favorable distribution of the parameters would be $N(\mu_1, (\sigma_1^2 - \sigma_0^2)/n)$, as then the distribution of Y under H_Λ is $N(\mu_1, \sigma_1^2/n)$, the same as under the alternative. The likelihood ratio reduces to a constant times U and does not depend upon the specific alternative. The test based upon the distribution of U is UMP.

Curiously, an analogous test of the hypothesis $\sigma \geq \sigma_0$ would depend on the value of μ and is not UMP (Exercise 3.10).

---

[10] The proof of this seemingly intuitive result occupies most of Chapter 3 of Lehmann [1986].

3.5.3 Unbiased Tests

We can find a most powerful test of the hypothesis $\sigma \geq \sigma_0$ if we restrict ourselves to the set of tests that are *unbiased*, that is, whose power function satisfies

$$\beta_\varphi(\theta) \leq \alpha \text{ if } \theta \in H,$$
$$\beta_\varphi(\theta) \geq \alpha \text{ if } \theta \in K.$$

Such a test is said to be *uniformly most powerful unbiased* or UMPU. Any such test is *admissible* in the sense that there cannot exist another test φ' that is at least as powerful against all alternatives and more powerful against some; whenever a UMP test exists it is UMPU (Exercise 3.12). All the permutation tests we have looked at so far have been unbiased.

When $\beta_\varphi(\theta)$ is a continuous function of θ (that is, no jumps), unbiasedness implies that $\beta_\varphi(\theta) = \alpha$ for all θ on the boundary ω between the hypothesis and the alternative. Tests that satisfy this latter condition are called *similar*. If we can find a level-α test that is uniformly most powerful among all similar tests, it must be unbiased, since it is uniformly at least as powerful as $\varphi(x) \equiv \alpha$.

Let T be a sufficient statistic for $\mathcal{P}^X = \{P_\theta, \theta \in \omega\}$ and let $\mathcal{P}^T = \{P_\theta^T, \theta \in \omega\}$. A test φ is said to have *Neyman structure* if $E[\varphi(x)|t] = \alpha$ almost everywhere with respect to $\mathcal{P}^T$.[11] A test that has Neyman structure is similar to $\mathcal{P}^X$, as then $E[\varphi(x)|\theta] = E\{E[\varphi(x)|t]|\theta\} = \alpha$ for all $\theta \in \omega$.

Referring to equation 3.3, we see that for any value of the sample mean $\bar{X}$, the conditional distribution $\sum_i (X_i - \bar{X})^2$ given $\bar{X}$ is the same as the unconditional probability.[12] As a result, our test of the hypothesis $\sigma \geq \sigma_0$ based on the sum of squares about the mean has Neyman structure, is similar, and, as the normal distribution has monotone likelihood ratio in this statistic, is UMPU.

3.5.4 Comparison Among the Test Procedures

A permutation test of the scale parameter might exist if the distribution of the log-transformed observations were symmetric. A UMPU test based on the chi-square distribution might exist if the data were normally distributed. If the data are not normally distributed, then for small samples, the significance level of such a test will be in error. Even for large samples, the true significance level will depend upon the fourth moment of X, (the variance of X^2), and may take any value from 0% to 50% [Lehmann, 1986, p 206].

The best approach to this problem will be via the bootstrap, though for small samples (less than 25 observations) even with the Hall–Wilson

---

[11] This statement is true almost everywhere (a.e.) except on a set N with $P(N) = 0$ for all $P \in \mathcal{P}^T$.
[12] A formal definition of conditional probability is given in Section 16.2.

and BC$_a$ corrections, the significance level may not be exact. A preliminary transformation of the observations (see Chapter 11) that will make the resultant distribution more symmetric will improve the bootstrap's accuracy [Efron and Tibshirani, 1993, Chapter 14].

3.6 Comparing the Location Parameters of Two Populations

3.6.1 A UMPU Parametric Test: Student's t

Suppose now that we have taken samples from two populations, $x_1, x_2, \ldots, x_n$ from a $N(\mu, \sigma^2)$ distribution and $y_1, y_2, \ldots, y_m$ from a $N(\nu, \sigma^2)$ distribution. To find a UMPU statistic to test the hypothesis that $\mu = \nu$, we know we can confine our attention to the sufficient statistics $\bar{X}, \bar{Y}, \sum(X_i - \bar{X})^2$, and $\sum(Y_i - \bar{Y})^2$. If $\mu = \nu$, then the distribution of the statistic $t(X, Y)$, given by

$$t(X,Y) = \frac{(\bar{Y} - \bar{X})/\sqrt{1/m + 1/n}}{\sqrt{[\sum(X_i - \bar{X})^2 + \sum(Y_i - \bar{Y})^2]/(m+n-2)}},$$

does not depend on the common population mean μ or the common population variance σ^2. The arguments raised in preceding sections lead us to the UMPU test of the compound hypothesis $\mu \geq \nu$ against the alternatives $\mu < \nu$, rejecting the hypothesis in favor of the alternative if $t(X, Y) > C$ where C is the $(1 - \alpha)$th percentile of the Student's t distribution with $m - 1$, $n - 1$ degrees of freedom. As this distribution is symmetric, we may also make use of its percentiles to create a two-tailed UMPU test of the hypothesis that $\mu = \nu$ against a two-sided alternative.[13]

3.6.2 A UMPU Semiparametric Procedure

We tested the equality of the location parameters of two samples via permutation means in Chapter 1. Recall that our test statistic is the sum of the observations in one of the samples. In this section we show that a permutation test[14] based on this statistic is exact and unbiased against stochastically increasing alternatives of the form $K: F_2[x] = F_1[x - \delta]$, $\delta > 0$. In fact, we show that this permutation test is a uniformly most powerful unbiased test of the null hypothesis $H: F_2 = F_1$ against normally distributed shift alternatives.

---

[13] We defer the proof of this statement to the Appendix.
[14] The terminology "permutation test" is a misnomer, albeit a well-established one. For all practical testing purposes, the permutations may be divided into equivalence classes termed "rearrangements." Within each equivalence class, the only differences among the permutations lie in the arrangement of the observations within samples.

Against normal alternatives and for large samples, its power is equal to that of the standard t-test [Bickel and van Zwet, 1978].

A family of cumulative distribution functions is said to be *stochastically increasing* if the distributions are distinct and if $\theta < \theta'$ implies $F[x|\theta] \geq F[x|\theta']$ for all x. One example is the location parameter family for which $F_\theta[x] = F[x-\theta]$. If X and X' have distributions F_θ and $F_{\theta'}$, then $P\{X > x\} \leq P\{X' > x\}$, that is, X' tends to have larger values than X. Formally, we say that X' is *stochastically larger* than X.

Lemma 3.1. $F_1[x] \leq F_0[x]$ for all x only if there exist two nondecreasing functions f_0 and f_1 and a random variable V such that $f_0[v] \leq f_1[v]$ for all v, and the distributions of f_0 and f_1 are F_0 and F_1, respectively.

Proof. Set $f_i[y] = \inf\{x: F_i(x-0) \leq y \leq F_i(x)\}$, $i = 0, 1$. These functions are nondecreasing and, for $f_i = f$, $F_i = F$, satisfy $f[F(x)] \leq x$ and $F[f(y)] \geq y$ for all x and y. Thus, $y \leq F(x_0)$ implies $f[y] \leq f[F(x_0)] \leq x_0$, and $f(y) \leq x_0$ implies $F[f(y)] \leq F(x_0)$ implies $y \leq F(x_0)$.

Let V be uniformly distributed on (0,1). Then $P\{f_i(V) \leq x\} = P\{V \leq F_i(x)\} = F_i(x)$, which completes the proof. □

We can apply this result immediately.

Lemma 3.2. Let $X_1, \ldots, X_m; Y_1, \ldots, Y_n$ be samples from continuous distributions F, G, and let $\varphi[X_1, \ldots, X_m; Y_1, \ldots, Y_n]$ be a test such that,

a) whenever $F = G$, its expectation is α;
b) $y_i \leq y'_i$ for $i = 1, \ldots, n$ implies $\varphi[x_1, \ldots, x_m; y_1, \ldots, y_n] \leq \varphi[x_1, \ldots, x_m; y'_1, \ldots, y'_n]$.

Then the expectation of φ is greater than or equal to α for all pairs of distributions for which Y is stochastically larger than X.

Proof. From our first lemma, we know there exist functions f and g and independent random variables $V_1, \ldots, V_{m+n}$ such that the distributions of $f(V_i)$ and $g(V_i)$ are F and G, respectively, and $f(z) \leq g(z)$ for all z.

$$E\varphi[f(V_1), \ldots, f(V_m); f(V_1), \ldots, f(V_n)] = \alpha$$

and

$$E\varphi[f(V_1), \ldots, f(V_m); g(V_1), \ldots, g(V_n)] = \beta.$$

From condition b) of the lemma, we see that $\beta > \alpha$, as was to be proved. □

3.6 Comparing the Location Parameters of Two Populations

We are now in a position to state the principal result of this section:

Theorem 3.4 (Unbiased). Let $X_1, \ldots, X_m$; $Y_1, \ldots, Y_n$ be samples from continuous distributions F, G. Let $\beta(F, G)$ be the expectation of the critical function φ defined as $\varphi[X_1, \ldots, X_m; Y_1, \ldots, Y_n] = 1$ only if $\sum Y_j$ is greater than the equivalent sum in α of the $\binom{n+m}{n}$ possible rearrangements. Then $\beta(F, F) = \alpha$ and $\beta(F, G) \geq \alpha$ for all pairs of distributions for which Y is stochastically larger than X; $\beta(F, G) \leq \alpha$ if X is stochastically larger than Y.

Proof. Exactness, that is $\beta(F, F) = \alpha$, follows from Theorem 3.1 and the definition of φ. We can apply our lemmas and establish that the two-sample permutation test is unbiased if we can show that $y_i \leq y'_i$ for $i = 1, \ldots, n$ implies

$$\varphi[x_1, \ldots, x_m; y_1, \ldots, y_n] \leq \varphi[x_1, \ldots, x_m; y'_1, \ldots, y'_n].$$

□

Rename the observations so that $z_i = x_i$ for $i = 1, \ldots, m$ and $z_{i+m} = y_i$ for $i = 1, \ldots, n$. $\varphi = 1$ if the sum of the observations in the second sample as originally labeled $\left(\sum_{i=1}^{n} y_i = \sum_{i=m+1}^{m+n} z_i\right)$ exceeds sufficiently many of the sums after relabeling, that is, if sufficiently many of the differences

$$\sum_{i=m+1}^{m+n} z_i - \sum_{i=m+1}^{m+n} z_{\pi(i)} = \sum_{i=1}^{p} z_{s_i} - \sum_{i=1}^{p} z_{r_i}$$

are positive, where the r_i label those observations in the second sample after permutation that were not present in the original sample, and the s_i label those observations from the second sample that were rearranged to the first. If such a difference is positive and $y_i \leq y'_i$, that is $z_i \leq z'_i$ for $i = m+1, \ldots, m+n$, then the difference $\sum z'_{s_i} - \sum z_{r_i}$ is also positive, so that $\varphi(z') \geq \varphi(z)$. But then we may apply the lemmas to obtain the desired result. The proof is similar for the case in which X is stochastically larger than Y.

Suppose, as in the preceding section, that we have taken samples from two populations, $z_1, z_2, \ldots, z_n$ from a $N(\mu, \sigma^2)$ distribution, and $z_{n+1}, z_{n+2}, \ldots, z_{n+m}$ from a $N(\nu, \sigma^2)$ distribution. Their joint distribution may be written as

$$(2\pi\sigma^2)^{-(n+m)/2} \exp\left[-\frac{1}{2\sigma^2}\left(\sum_{j=1}^{m}(z_j - \mu)^2 + \sum_{j=m+1}^{n+m}(z_j - \mu - \delta)^2\right)\right]$$

or

$$(2\pi\sigma^2)^{-(n+m)/2} \exp\left[-\frac{1}{2\sigma^2}\left(\sum_{j=1}^{n+m}(z_j - \mu)^2 - 2\delta\sum_{j=m+1}^{n+m}(z_j - \mu)^2 + n\delta^2\right)\right].$$

Before making use of this expression to select terms for inclusion in the rejection region, we may eliminate all factors that remain constant under permutation of the subscripts. These include $(2\pi\sigma^2)^{-(n+m)/2}$ and $\sum_{j=1}^{n+m}(z_j-\mu)^2$, and $n\delta(\delta+n)$. The resulting test rejects when $\exp[\delta\sum_{j=m+1}^{n+m}z_j]>C$, or, equivalently, if $\delta>0$ when the sum of the observations in the second sample is large. Our permutation test is the same whatever the unknown values of μ and σ and thus is uniformly most powerful against normally distributed alternatives among all unbiased tests of the hypothesis that the two samples come from the same population.

3.6.3 An Example

Suppose we have two samples: The first, the control sample, takes values 0, 1, 2, 3, and 19. The second, the treatment sample, takes values 3.1, 3.5, 4, 5, and 6. Does the treatment have an effect?

The answer would be immediate if it were not for the value 19 in the first sample. The presence of this extreme value changes the mean of the first sample from 1.5 for values 0, 1, 2, and 3 to 5. To dilute the effect of this extreme value on the results, we convert all the data to ranks, giving the smallest observation a rank of 1, the next smallest the rank of 2, and so forth. The first sample includes the ranks 1, 2, 3, 4, and 10, and the second sample includes the ranks 5, 6, 7, 8, and 9. Is the second sample drawn from a different population than the first?

Let's count. The sum of the ranks in the first sample is 15. All the rearrangements with first samples of the form 1, 2, 3, 4, k, where k is chosen from $\{5, 6, 7, 8, 9,$ or $10\}$, have sums that are as small or smaller than that of our original sample. That's six rearrangements. The four rearrangements whose first sample contains 1, 2, 3, 5, and a fifth number chosen from the set $\{6, 7, 8, 9\}$ also have smaller sums. That's $6+4=10$ rearrangements so far.

Continuing in this fashion—I leave the complete enumeration as an exercise—we find that 24 of the $\binom{10}{5}=252$ possible rearrangements have sums that are as small or smaller than that of our original sample. Two samples with sums as different as this will be drawn from the same population just under 10% of the time by chance.

This example also reveals that a rank test is simply a permutation test applied to the ranks rather than the original observations.

3.6.4 Comparison of the Tests: The Behrens–Fisher Problem

The permutation test offers the advantage over the parametric t-test that it is exact even for very small samples whether or not the observations come from a normal distribution. The parametric t-test relies on the existence of

---

R Program to Compute p-Value via a Monte Carlo

```
A <- c (1, 2, 3, 4, 10)
sumorig <- sum(A)
B <- c(5, 6, 7, 8, 9)
cnt<- 0
A <- c(A,B)
for (i in 1:400){
    D<- sample (A,n)
    sumperm <- sum(D)
    if (sumperm <= sumorig)cnt<-cnt+1
    }
cnt/400      #pvalue
```

---

a mythical infinite population from which all the observations are drawn. The permutation test is applicable even to finite populations such as all the machines in a given shop or all the supercomputers in the world.

Against specific normal alternatives as we saw in the preceding section, this permutation test provides a most powerful unbiased test of the distribution-free hypothesis $H: F_Y = F_X$. For large samples, its power is almost the same as Student's t-test [Albers, Bickel, and van Zwet, 1976].

Both tests rely on the assumption that the variances of the two samples are the same. But what if F_Y is not equal to F_X under the hypothesis? What if only the location parameters of F_Y and F_X are equal, while the other parameters of these two distributions are not the same? Are our tests still exact or almost exact? Are they still efficient for testing against normal alternatives? When the sample sizes are equal, the t-test will be almost exact even if the variances are quite different; otherwise, the actual size of the test can differ greatly from the declared level even for large samples [Ramsey, 1980; Posten, Yeh, and Owen, 1982]. Many, many parametric alternatives have been proposed; see, for example, Tiku and Singh [1981], O'Brien [1988], Manly and Francis [1999], and Weerahandi [1995, Chapter 7].

Romano [1990] shows that the permutation test based on the sum of the observations in the first sample is asymptotically exact for testing whether the expectations of F_Y and F_X are equal, even if the two distributions are not identical, providing both distributions have finite variances and the samples are of equal size.

But suppose the first sample comes from a population all of whose members are very close to zero, while the second comes from a population all of whose members are very large in absolute value. The sum of the observations in the first sample is close to zero in the original sample. When we rerandomize, mixing elements of the second sample with those of the first, the test statistic will either be a very large positive number or a very large negative one. In this

Table 3.1. Significance level. Effect of unequal variance on the permutation test, bootstrap, and Student's for various sample sizes and distributions. Rejections in 1000 Monte Carlo simulations.

σ_1/σ_2	1.0						4.0						10.0					
	p-test		boot		t-test		p-test		boot		t-test		p-test		boot		t-test	
Ideal	100	50	100	50	100	50	100	50	100	50	100	50	100	50	100	50	100	50
Normal																		
6, 6	100	57	108	57	98	45	98	50	84	36	115	58	111	56	83	44	119	67
8, 8	100	47	103	54	102	52	112	67	94	45	117	68	106	64	94	47	106	65
8, 12	99	52	111	47	101	50	69	29	98	48	72	34	51	23	83	35	64	27
12, 8	115	59	103	47	95	37	138	75	80	36	75	37	168	111	95	42	113	52
12, 12	101	53	107	55	100	50	105	62	110	57	87	48	101	43	107	55	83	50
Double exponential																		
6, 6	102	57	130	69	92	44	104	57	119	68	109	61	138	87	148	82	112	62
8, 8	108	45	119	65	116	59	114	62	136	83	132	66	126	70	151	88	139	81
8, 12	89	51	102	48	101	50	61	27	102	44	72	25	65	27	109	53	102	50
12, 8	113	44	129	72	93	28	146	95	121	72	106	52	134	78	99	47	77	34
12, 12	100	44	132	78	80	38	85	46	117	76	85	40	132	77	179	122	87	37

worst case example, for any choice of significance level less than 50% and any finite sample size, the null hypothesis cannot be rejected.

The preceding example is a worst case. In a series of Monte Carlo simulations with small sizes, I found that in many common applications including the Behrens–Fisher problem, the permutation test remains close to exact even for very small equal-sized samples (see Table 3.1). This result is in line with the findings of Box and Anderson [1955] and Brown [1982]. But see, also, Boik [1987].

A second resampling method, the nonparametric bootstrap, provides asymptotically exact solutions, whether or not $F_Y \equiv F_X$ and whether or not the sample sizes are equal; but see Gine and Zinn [1989]. In a bootstrap we resample separately from each of the two original samples. The underlying populations need not be the same even under the null hypothesis.

The primitive, uncorrected bootstrap is far from exact except for very large samples. But if we modify the bootstrap, using pivotals, Studentization, bias and higher-order correction as in Hall [1992], we can derive an almost exact bootstrap, even for samples with as few as eight observations.

This result is not unexpected: Liu [1988] shows the bootstrap test of the hypothesis of equal means retains the second-order convergence properties it has in the case $F_Y \equiv F_X$.[15]

3.7 Comparing the Dispersions of Two Populations

Precision is essential in a manufacturing process. Items that are too far out of tolerance must be discarded and an entire production line brought to a halt if too many items exceed (or fall below) designated specifications. With some testing equipment, such as that used in hospitals, precision can be more important than accuracy. For *accuracy* (closeness to the correct value) can always be achieved through the use of standards with known values, while a lack of *precision* may render an entire sequence of tests invalid.

3.7.1 The Parametric Approach

There is no shortage of parametric methods to test the hypothesis that two samples come from populations with the same inherent variability. Sukhatme [1958] lists four alternative approaches and adds a fifth of his own; Miller [1968] lists ten alternatives and compares four of these with a new test of his own; Conover, Johnson, and Johnson [1981] list and compare 56 tests; and Balakrishnan and Ma [1990] list and compare nine tests with one of their own.

None of these tests can be relied on. Many promise an error rate or significance level of 5% but in reality make errors as frequently as 8% to 20% of the time. Others have severe restrictions. The F-ratio test [Fisher, 1925]

---

[15] For more on convergence properties, see Appendix, Section A.7.5.

is exact only if the observations come from a normal distribution, and as noted in Section 3.5.4, unlike the t-test, it is very sensitive to deviations from normality.

Each of these tests requires that two or more of the following four conditions be satisfied:

1. The observations are normally distributed.
2. The location parameters of the two distributions are the same or differ by a known quantity.
3. The two samples are equal in size.
4. The samples are large enough that asymptotic approximations to the distribution of the test statistic are valid.

As an example, the first published solution to this classic testing problem is the z-test proposed by Welch [1937] based on the ratio of the two sample variances. If the observations are normally distributed, this ratio has the F-distribution, and the test whose critical values are determined by the F-distribution is uniformly most powerful among all unbiased tests [Lehmann, 1986, Section 5.3]. But with even small deviations from normality, significance levels based on the F-distribution are grossly in error [Lehmann, 1986, p. 207]; the magnitude of the error will depend on the fourth moment of the distribution from which the samples are drawn.

Box and Anderson [1955] propose a correction to the F-distribution for "almost" normal data, based on an asymptotic approximation to the permutation distribution of the F-ratio. Not surprisingly, their approximation is close to correct only for normally distributed data or for very large samples. The Box–Anderson statistic results in an error rate of 21%, twice the desired value of 10%, when two samples of size 15 are drawn from a gamma distribution with four degrees of freedom.

The test proposed by Miller [1968] yields conservative Type I errors, less than or equal to the declared error, unless the sample sizes are unequal. A 10% test with samples of size 12 and 8 taken from normal populations yielded Type I errors 14% of the time.

3.7.2 The Permutation Approach

At first glance, the permutation test for comparing the variances of two populations would appear to be an immediate extension of the test we use for comparing location parameters, in which we use the squares of the observations rather than the observations themselves. But these squares are actually the sum of two components, one of which depends upon the unknown variance, the other upon the unknown location parameter. That is, $EX^2 = E(X - \mu + \mu)^2 = E(X - \mu)^2 + 2\mu E(X - \mu) + \mu^2 = \sigma^2 + 0 + \mu^2$.

A permutation test based upon the squares of the observations is appropriate only if the location parameters of the two populations are known, or if they are known to be equal [Bailer, 1989].

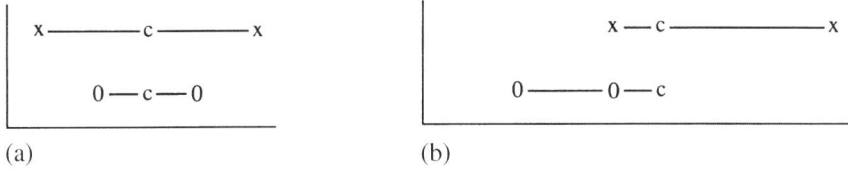

Fig. 3.2. Comparison of two samples: a) original data, b) after first sample is shifted to the right. c common center, x—x first sample, 0—0 second sample.

We cannot eliminate the effects of the location parameters by working with the deviations about each sample mean, as these deviations are interdependent [Maritz, 1981]. The problem is illustrated in Figure 3.2. In the sketch on the left, the observations in the first sample are both further from the common center than either of the observations in the second sample. Moreover, of the four possible rearrangements of four observations between two samples, this arrangement is the most extreme. In the sketch on the right, the observations in the first sample have undergone a shift to the right; this shift has altered the relative ordering of the absolute deviations about the common center, and at least one other rearrangement is more extreme.

Still, we needn't give up; we can obtain an exact permutation test with just a few preliminary calculations. First, we compute the median for each sample; next, we replace each of the remaining observations by the square of its deviation about its sample median; last, we discard certain redundant values.

Suppose the first sample contains the observations $x_1, \ldots, x_n$ whose median is mdn$\{x_i\}$; we begin by forming the deviates $x'_j = |x_j - \text{mdn}\{x_i\}|$ for $j = 1, \ldots, n$. Similarly, we form the set of deviates $\{y'_j\}$ using the observations in the second sample and their median.

If there are an odd number of observations in the sample, then one of these deviates must be zero. We can't get any information out of a zero, so we throw it away.[16] If there is an even number of observations in the sample, then two of these deviates must be equal. We can't get any information out of the second one that we didn't already get from the first, so we throw it away.

Our test statistic S is the sum of the remaining deviations in the first sample, that is, $S = \sum_{j=1}^{n-1} x'_j$. We obtain its permutation distribution and the cut-off point for the test by considering all possible rearrangements of the deviations that remain in both the first and second samples.

To illustrate the application of this statistic, suppose the first sample consists of the measurements 121, 123, 126, 128.5, and 129, and the second sample of the measurements 153, 154, 155, 156, and 158. Thus $x'_1 = 5$, $x'_2 = 3$, $x'_3 = 2.5$, $x'_4 = 3$, and $S_0 = 13.5$. While $y'_1 = 2$, $y'_2 = 1$, $y'_3 = 1$, $y'_4 = 3$.

---

[16] In the event of ties, should there be more than one zero, we still throw only one away.

There are $\binom{8}{4}$ arrangements in all, of which only three yield values of the test statistic as, or more, extreme than our original value. This proportion is $3/70 = 0.043$, so we conclude that the difference between the dispersions of the two manufacturing processes is statistically significant at the 5% level.

There is a weak dependency among these deviates and thus they are only asymptotically exchangeable. (See Section 15.7.4.) The associated test is alternately conservative and liberal [Baker, 1995]. For this test to be even approximately exact, it requires that

a) the ratio of the sample sizes n, m be close to 1;
b) the population variances exist and are equal (as they would be under the null hypothesis);
c) the only other difference between the two populations from which the samples are drawn is that they might have different means.

We can derive a permutation test for comparing variances that is free of even these restrictions if, instead of working with the original observations, we replace them with the differences between successive order statistics and then permute the labels.[17] The test statistic proposed by Aly [1990] is

$$\delta = \sum_{i=1}^{m-1} i(m-i)(X_{(i+1)} - X_{(i)}),$$

where $X_{(1)} < X_{(2)} < \cdots < X_{(m)}$ are the order statistics of the first sample. That is, $X_{(1)}$ is the smallest of the observations in the first sample (the minimum), $X_{(2)}$ is the second smallest, and so forth, up to $X_{(m)}$, the maximum.

To illustrate the application of Aly's statistic, suppose the first sample consists of the measurements 121, 123, 126, 128.5, and 129, and the second sample of the measurements 153, 154, 155, 156, and 158. $X_{(1)} = 121$, $X_{(2)} = 123$, and so forth.

Set $z_{1i} = X_{(i+1)} - X_{(i)}$ for $i = 1, \ldots, 4$. In this instance, $z_{11} = 123 - 121 = 2$, $z_{12} = 3$, $z_{13} = 2.5$, $z_{14} = 0.5$.

The original value of Aly's test statistic is $8 + 18 + 15 + 2 = 43$. To compute the test statistic for other arrangements, we also need to know the differences $z_{2i} = Y_{(i+1)} - Y_{(i)}$ for the second sample; $z_{21} = 1$, $z_{22} = 1$, $z_{23} = 1$, $z_{24} = 2$.

Only certain exchanges are possible. Rearrangements are formed by first choosing either z_{11} or z_{21}, next either z_{12} or z_{22}, and so forth until we have a set of four differences.

One possible rearrangement is $\{2, 1, 1, 2\}$, which yields a value of $\delta = 28$. There are $2^4 = 16$ rearrangements in all, of which only one—$\{2, 3, 2.5, 2\}$—yields a more extreme value of the test statistic than our original observations. With 2 out of 16 rearrangements yielding values of the statistic as, or more, extreme than the original, we should accept the null hypothesis. Better still,

---

[17] As shown in Section A.3.3, the order statistics are sufficient for the set of all absolutely continuous distributions.

given the limited number of possible rearrangements, we should gather more data before we make a decision.[18]

If our second sample is larger than the first, we have to resample in two stages. First, we select a subset of m values $\{Y_i^*, I = 1, \ldots, m\}$ without replacement from the n observations in the second sample, and compute the order statistics $Y_{(1)}^* < Y_{(2)}^* < \cdots < Y_{(m)}^*$ and their differences $\{z_{2i}\}$. Last, we examine all possible values of Aly's measure of dispersion for permutations of the combined sample $\{\{z_{1i}\}, \{z_{2i}\}\}$, as we did when the two samples were equal in size, and compare Aly's measure for the original observations with this distribution.

A similar procedure, first suggested by Baker [1995], should be used with the first test in the case of unequal samples.

3.7.3 The Bootstrap Approach

In order to use permutation methods to compare the variances of two populations, we have to sacrifice two of the observations. The resultant test is exact and distribution-free, but it is not most powerful. A potentially more powerful test is provided by the bootstrap confidence interval for the variance ratio. To derive this test, we resample repeatedly without replacement, drawing independently from the two original samples, until we have two new samples the same size as the originals. Each time we resample, we compute the variances of the two new independent subsamples and calculate their ratio. The resultant bootstrap confidence interval is asymptotically exact [Efron, 1981] and can be made close to exact with samples of as few as eight observations (see Table 3.2a). As Table 3.2b shows, this bootstrap is more powerful than the permutation test we described in the previous section. One caveat also revealed in the table: This bootstrap is still only "almost" exact.

Table 3.2a. Significance level for variance comparisons for BC_a method, Efron and Tibshirani [1986]. For various underlying distributions by sample size. 500 simulations.

	6,6	8,8	8,12	12,8	12,12	15,15
Ideal	50	50	50	50	50	50
Normal (0,1)	44	52	53	56	45	49
Double (0,1)	53	51	63	70	55	54
Gamma (4,1)	48	55	60	65	52	52
Exponential	54	58	56	70	46	63

---

[18] How much data? See Chapter 6.

Table 3.2b. Power as a function of the ratio of the variances. For various distributions and two samples each of size 8. Rejections in 500 Monte Carlo simulations.

	Permutation Test					Bootstrap*				
$\phi = \sigma_2/\sigma_1$	1.	1.5	2.	3.	4.	1.	1.5	2.	3.	4.
Ideal	50				500	50				500
Normal	52	185	312	438	483	52	190	329	444	482
Double	55	153	215	355	439	53	151*	250*	379*	433
Gamma	44	158	255	411	462	49	165	288	426	464
Exponential	51	132	224	323	389	54	150*	233*	344*	408

*Bootstrap interval shortened so actual significance level is 10%.

3.8 Bivariate Correlation

Often we observe pairs of variables—weight and cholesterol level, SAT score and subsequent GPA—and naturally question whether the members of a pair are related. If we have two sets of fencing of varying lengths and are asked to build a fixed number of animal pens from them so as to maximize the area within, we know from the formula for the area of a rectangle, $A = W \times L$, that we can maximize the area if we match the largest lengths in the first group with the largest lengths from the second group and so on down the line until the smallest lengths are matched with the smallest.[19]

Suppose now we've observed n pairs of observations (X_i, Y_i), $i = 1, \ldots, n$. If the labels are meaningless, the variables uncorrelated, and every combination of values equally likely, then every value of the statistic $S = \sum X_i Y_i$ obtained by permuting the subscripts for one variable, but not the other, is equally likely, also. On the other hand, if the two variables are positively correlated then, as we saw from our geometric argument, large values of the statistic are more probable than small ones. In particular, a permutation test based on the sum of cross products S is uniformly most powerful against normally distributed alternatives among all unbiased tests of the hypothesis that pairs of observations are uncorrelated.

A bivariate normal distribution has the density

$$h(z) = (2\pi\sigma\tau\sqrt{1-\rho^2})^{-n} \exp[-J/2(1-\rho^2)], \qquad (3.4)$$

where $J = \dfrac{1}{2\sigma^2} \sum (x_j - \mu)^2 + \dfrac{2\rho}{\sigma\tau} \sum (x_j - \mu)(y_j - \nu) + \dfrac{1}{\tau^2} \sum (y_j - \nu)^2$.

To find a most powerful test of the null hypothesis that is unbiased against alternatives with probability density $h(z)$ and $\rho > 0$, we need to maximize the expression

$$\sum_{z \in R(s)} \phi(z) \frac{h(z)}{\sum_{z' \in R(s)} h(z')}.$$

---

[19] Check this out on a piece of graph paper if it's not immediately obvious.

Many of the sums that occur in this expression are invariant under permutations of the subscripts j. These include the four sums $\sum x_j$, $\sum y_j$, $\sum x_j^2$, $\sum y_j^2$. Eliminating all these invariant terms leaves us with the test statistic S, establishing the desired result.

3.9 Which Test?

"Every statistical procedure relies on certain assumptions for correctness. Errors in testing hypotheses come about either because the assumptions underlying the chosen test are not satisfied, or because the chosen test is less powerful than other competing procedures." [Good and Hardin, 2003].

All the testing methods described in this chapter require that the observations be independent. All require that at least one of the following successively weaker assumptions be satisfied under the null hypothesis:
1. The observations be identically distributed.
2. The observations be exchangeable, that is, their joint distribution be the same for any relabeling.
3. The observations be drawn from populations in which, under the null hypothesis, a specific parameter is the same across all the populations.

Parametric procedures require that all three of these assumptions be satisfied and, moreover, that the explicit form of the population distribution be known. Permutation methods require that the second and third assumptions be satisfied. The bootstrap requires only that the third assumption be satisfied.

For a comparison of permutation and parametric tests, see Bradbury [1987]. In most cases the optimum permutation and parametric tests are asymptotically equivalent. For discussions on this, see Albers, Bickel, and Van Zwet [1976] and Bickel and Van Zwet [1978]. For a comparison of permutation tests with the bootstrap, see Romano [1989], ter Braak [1992], and Good [2001].

3.10 Exercises

1. Prove all of the following:
    a) The sum of two independent normally distributed random variables each with mean 0 and variance 1 is a normally distributed random variable with mean 0 and variance 2.
    b) The sum of two independent normally distributed random variables with means μ_1, μ_2 and variances σ_1^2, σ_2^2 is a normally distributed random variable with mean $\mu_1 + \mu_2$ and variance $\sigma_1^2 + \sigma_2^2$.
    c) The sum of n independent normally distributed random variables with mean μ and variance σ^2 is a normally distributed random variable with mean μ and variance $n\sigma^2$.

d) If S is the sum of n independent normally distributed random variables with mean μ and variance σ^2, then $(S - \mu)/\sqrt{n\sigma^2}$ is a normally distributed random variable with mean 0 and variance 1.
2. Under what conditions are the sample mean and sample variance uncorrelated?
3. Suppose that under the conditions of Theorem 3.1, φ is UMP for testing $\theta \leq \theta_0$ against the alternative that $\theta > \theta_0$. Show that for any $\theta < \theta_0$, φ minimizes $\beta(\theta)$ among all tests that satisfy $\beta(\theta_0) = \alpha$.
4. Show that our parametric test is most powerful for testing the hypothesis that our observations are normally distributed with mean $\theta = \theta_0$ and variance σ^2 against the alternative that the observations are normally distributed with expectation $\theta = \theta_1 > \theta_0$ and variance σ^2.
5. Prove that Student's t-test cannot be uniformly most powerful for testing the hypothesis that our observations are normally distributed with mean $\theta = \theta_0$ against the alternative that they are normally distributed with expectation $\theta = \theta_1 > \theta_0$.
6. Find a 90% confidence interval for the mean of a normal distribution with variance 9 when a sample of 16 independent observations has a mean of 8. [Hint: Confidence intervals get smaller with more observations; they are larger when the variance is larger.]
7. Find a 90% confidence interval for the mean of a normal distribution with unknown variance when a sample of 16 independent observations has a mean of 7 and a sample variance of 7.2. The 95th percentile of a t-distribution with 15 degrees of freedom is 1.753. (Hint: The t-distribution, like the normal, is symmetric.)
8. If asked to provide a single point estimate instead of an entire interval, would the center of the confidence interval be your best choice?
9. Show that if you know the values of the first $n - 1$ deviations about the mean of a sample of size n, you can immediately calculate the nth.
10. Let $\{X_1, \ldots X_n\}$ denote a set of exchangeable random variables whose individual distributions, though not necessarily the same, belong to the family of distributions $\mathcal{P}$ that are symmetric about 10. Show via the factorization criterion that the absolute values of the observations are sufficient for $\mathcal{P}$.
11. Find the least favorable distribution for testing $H: N(\mu, \sigma^2); \sigma \geq \sigma_0$ against the alternative $K: N(\mu, \sigma^2); \sigma < \sigma_0$.
12. Prove that i) any UMPU test is admissible; ii) whenever a UMP test exists it is UMPU.
13. Show that the following statistics lead to equivalent permutation tests for the equality of two location parameters: a) the sum of the observations in the smallest sample; b) the difference between the sample means; c) the t-statistic. (Hint: The sum of all the observations, the sum of the squares of all the observations, and the sample sizes are invariant under permutations.)

14. Show that if T' is a monotonic function of T, then a test based on the permutation distribution of T' will accept or reject only if a permutation test based on T also accepts or rejects.
15. In the example of Section 3.3.2, list all rearrangements in which the sum of the ranks in the first sample is less than or equal to the original sum.
16. Suppose we'd planned to look at two samples, each of size four, but when the data are finally in hand, one of the observations is missing. One alternative is to treat the data as if we'd planned on two unequal samples from the beginning, thus leading to 7 choose 3 possibilities. Another alternative is use the mean of the first sample as our test statistic, include the missing observation in our rearrangements, and examine all 8 choose 4 possibilities. Which alternative is preferable?
17. Show that the permutation test introduced in Section 3.7.2 for comparing variances based on deviations about the sample median is asymptotically exact.
18. Although the Hall–Wilson corrections are widely accepted, sometimes they can produce idiotic results. Use the following real-world data to obtain a Hall–Wilson corrected bootstrap interval estimate for the population mean:
0 0 0 0 7.53 0 0 0 15.77 0 0 0 0 7.53 6.16 0 0 0 0 18 0 5.71 5.71 0 7.78 0 7.03 0 10.22 0 12 19.07 15.50 0 0 0 0 0 3.81 6.10 3 10.78 0 10.44 0 0 0 0 0 0 4 0 0 0 103.05 0 0 0 0 12 0 0 0.
19. Provide a formal step-by-step proof that a permutation test based on the sum of the cross-products is uniformly most powerful against normally distributed alternatives among all unbiased tests of the hypothesis that the pairs of observations are uncorrelated. (Hint: Find the sufficient statistics for the bivariate normal distribution. Does the bivariate normal density $h(x)$ have monotone likelihood ratio in the cross-product statistic? Does our permutation test have Neyman structure in this statistic?)
20. Is a permutation test based on the sum of the cross-products of X and Y most powerful among unbiased tests for testing the hypothesis that X and Y are independent against alternatives of the form $F[Y|X=b] \leq F[Y|X=a]$ if $a<b$? Is it uniformly most powerful among all tests of this hypothesis against these alternatives?
21. Does a UMP test always minimize the risk? Or would it depend on the risk function? In particular, consider the two losses associated with a testing problem concerning a parameter θ, where $L_0(\theta) = 0$ for $\theta \leq \theta_0$ and $L_0(\theta)$ is strictly increasing otherwise, and $L_1(\theta)$ is increasing for $\theta \leq \theta_0$ and 0 otherwise. The risk function of any test φ is $R(\theta, \varphi) = \varphi\{(x)L_1(\theta) + (1-\varphi(x))L_0(\theta)\}\, dF_\theta(x)$. If the random variable X has a cumulative distribution function $F_\theta[x]$ with monotone likelihood ratio in $T[x]$, show that the associated family of one-sided tests minimizes $R(\theta, \varphi)$.

4
Distributions

In many problems such as the analysis of data from radioactive decay, the distribution of observations can be determined on theoretical grounds, and the optimal decision procedure is one that takes advantage of knowledge obtained this way. In this chapter, we consider optimal tests for data drawn from the binomial, Poisson, exponential, uniform, and exponential family of distributions.

4.1 Properties of Independent Observations

All the tests that we have considered so far require that the individual observations be independent of one another. We will continue to make such an assumption in succeeding chapters, even in cases when the "observation" consists of a vector of interdependent variables. This is because independent observations have many desirable properties. Recall that if X and Y are independent, then $\Pr\{X \in A \text{ and } Y \in B\} = \Pr\{X \in A\}\Pr\{Y \in B\}$. In consequence, if $E|X| < \infty$ and $E|Y| < \infty$, we easily can show that $E(aX + b) = aEX + b$ where a and b are constants, and $E(X+Y) = E(X) + E(Y)$ (Exercise 4.1). If Var $X < \infty$, and Var $Y < \infty$, then Var $(aX + b) = a^2$ Var X, Var $(X+Y) = $ Var $X +$ Var Y, and Var $(X - Y) = $ Var $X +$ Var Y. We take advantage of all these properties in what follows.

In particular, because Var $(aX+b) = a^2$ Var X, the mean of 100 independent identically distributed observations has $1/100$ the variance of a single observation; the standard deviation of the mean of 100 observations, termed the *standard error*, has $1/10$ the standard deviation of a single observation. We'll use this latter property of the mean in Sections 13.1.3 and 14.10 when we try to determine how large a sample should be.

4.2 Binomial Distribution

A binomial frequency distribution, written $B(n,p)$, results from a series of n independent trials each with the same probability p of success, and the same probability $q = 1 - p$ of failure. This distribution arises in any context in which there are only two possible outcomes, such as "the patient recovers or the patient dies." The distribution that arises when we can prefer Coke to Pepsi, Pepsi to Coke, have no preference, or prefer some other beverage entirely, is termed a *multinomial* and will be considered in Chapter 8.

If X is $B(n,p)$, then the expected value of X denoted by EX is np and the variance of X about its expected value, $EX - E(X)^2$, denoted by Var X, is npq.

Suppose we've flipped a coin in the air seven times, and six times it has come down heads. Do we have reason to suspect the coin is not fair, that $p > 1/2$?

To answer this question, we need to look at the frequency distribution of the binomial observation. If X is $B(n,p)$, then $\Pr\{X = k\} = \binom{n}{k} p^k (1-p)^{n-k}$ for $k = 0, 1, \ldots, n$, and is zero otherwise. Note that we needn't keep track of the individual observations; the number of successes k is *sufficient* for testing hypotheses concerning p.

If $n = 7$ and $k = 6$, this probability is $7p^6(1-p)$. If $p = 1/2$, this probability is $7/128 = 0.0547$, just slightly more than 5%. But before we reject the hypothesis that ours is a fair coin, consider that if six heads out of seven tries seems extreme to us, seven heads out of seven would seem even more extreme. Adding the probability of this more extreme event to what we have already, we see the probability of throwing six or more heads in seven tries is $8/128 = 0.0625$. While not significant at the 5% level, six or more heads out of seven tries does seem suspicious. If you were a mad scientist and observed that six times out of seven your assistant Igor began to bay like a wolf when there was a full moon, wouldn't you get suspicious?

Warning: In the example of the unfair coin, we formulated our hypothesis *after* we observed the data. While we might have reason to be suspicious and enough reason to justify further testing, we would be in error if we reported a significance level of 0.0625. When we are observing more or less at random, without a specific hypothesis in mind, human beings being what they are, the probability that sooner or later we will see something interesting is one. (Next time you play pool, try naming both the ball and the pocket.)

4.3 Poisson: Events Rare in Time and Space

The decay of a radioactive element, an appointment to the United States Supreme Court and a cavalry officer trampled by his horse have in common that they are relatively rare but inevitable events. They are inevitable, that is, if there are enough atoms, enough seconds or years in the observation period, and enough horses and momentarily careless men. Their frequency of occurrence has a Poisson distribution.

The number of events in a given interval has the Poisson distribution *if* it is the cumulative result of a large number of independent opportunities, each of which has only a small chance of occurring. The interval can be in space as well as time. For example, if we seed a small number of cells into a Petri dish that is divided into a large number of squares, the distribution of cells per square follows the Poisson.

If the number of events X has a Poisson distribution such that we may expect an average of λ events per unit interval, then $\Pr\{X = k\} = \lambda^k e^{-\lambda}/k!$ for $k = 0, 1, 2, \ldots$. For the purpose of testing hypotheses concerning λ, we needn't keep track of the times or locations at which the various events occurred; the number of events k is sufficient.

As the Poisson distribution arises when events in nonoverlapping intervals of space or time are independent of one another it is not surprising that the sum of a Poisson with expected value λ_1 and a second independent Poisson with expected value λ_2 is also a Poisson with expected value $\lambda_1 + \lambda_2$ (Exercise 4.7).

4.3.1 Applying the Poisson

John Ross of the Wistar Institute held there were two approaches to biology: the analog and the digital. The analog was served by the scintillation counter: one ground up millions of cells then measured whatever radioactivity was left behind in the stew after centrifugation; the digital was to be found in cloning experiments where any necessary measurements would be done on a cell-by-cell basis.

John was a cloner and, later, as his student, so was I. We'd start out with 10 million or more cells in a 10-milliliter flask and try to dilute them down to one cell per milliliter. We were usually successful in cutting down the numbers to 10,000 or so. Then came the hard part. We'd dilute the cells down a second time by a factor of 1:100 and hope we'd end up with 100 cells in the flask. Sometimes we did. Ninety percent of the time we'd end up with between 90 and 110 cells, just as the binomial distribution predicted. But just because a mixture is cut in half (or a dozen, or 100 parts) doesn't mean equal numbers will be present in each part. It does mean that the probability of getting a particular cell is the same for all the parts. With large numbers of cells, things seem to even out. With small numbers, chance seems to predominate.

Things got worse when I went to seed the cells into culture dishes. These dishes, made of plastic, had a rectangular grid cut into their bottoms, so they were divided into approximately 100 equal size squares. Dropping 100 cells into the dish meant an average of 1 cell per square. Unfortunately, for cloning purposes this average didn't mean much. Sometimes 40% or more of the squares would contain two or more cells. It didn't take long to figure out why. Planted at random, the cells obey the Poisson distribution in space.

An average of one cell per square means

$$\Pr\{\text{No cells in a square}\} = 1 * e^{-1}/1 = 0.32$$
$$\Pr\{\text{Exactly one cell in a square}\} = 1 * e^{-1}/1 = 0.32$$
$$\Pr\{\text{Two or more cells in a square}\} = 1 - 0.32 - 0.32 = 0.36.$$

Two cells was one too many. A clone or colony must begin with a single cell. I had to dilute the mixture a third time to ensure the percentage of squares that included two or more cells was vanishingly small. Alas, the vast majority of squares were now empty; I was forced to spend hundreds of additional hours peering through the microscope looking for the few squares that did include a clone.

4.3.2 A Poisson Distribution of Poisson Distributions

The stars in the sky do not have a Poisson distribution, although it certainly looks this way. Stars occur in clusters, and these clusters, in turn, are distributed in superclusters. The centers of these superclusters do follow a Poisson distribution in space, and the stars in a cluster follow a Poisson distribution around the center almost as if someone had seeded the sky with stars and then watched the stars seed themselves in turn. See Neyman and Scott [1965] for a discussion of this phenomenon.

4.3.3 Comparing Two Poissons

Suppose in designing a new nuclear submarine (or that unbuilt wonder, a nuclear spacecraft) you become concerned about the amount of radioactive exposure that will be received by the crew. You conduct a test of two possible shielding materials. During 10 minutes of exposure to a power plant using each material in turn as a shield, you record 14 counts with material A, and only 4 with experimental material B. Can you conclude that B is safer than A?

Or, suppose you wish to assess the preventive value of a new vaccine against a relatively rare disease. You inoculate 100,000 individuals in a double blind study so that one-half receive the new vaccine and one-half an innocuous saline solution. You follow these individuals for a year and record the numbers in each group who come down with the disease. How can you determine from these numbers whether the new vaccine is of value?

Our hypothesis is that $\lambda_1 = \lambda_2$ or $\lambda_1 < \lambda_2$ against the alternative $\lambda_1 > \lambda_2$. We are not interested in the absolute values of these parameters, though we could estimate these separately, but in their relative values. The equivalent hypothesis is that $\theta = \lambda_1/(\lambda_1+\lambda_2) \leq 1/2$ against the alternatives $\theta = \lambda_1/(\lambda_1+\lambda_2) > 1/2$.

Let X_1 and X_2 denote the counts associated with each distribution. It is easy to see that the conditional distribution of X_2 given $X_1+X_2 = t$ is $B(t,\theta)$. The corresponding test has Neyman structure and thus, by the arguments of the preceding chapter, is UMP unbiased for testing $\theta \leq 1/2$ against the alternatives $\theta > 1/2$ (Exercise 4.9).

If the shielding materials in our first example are equal in their capabilities, then each of the 18 recorded counts is as likely to be obtained through the first material as through the second. Under the least favorable null hypothesis you would be observing a binomial distribution with 18 trials each with probability 1/2. The numeric answer is left as an exercise (see Exercise 4.3).

Unfortunately, the power of this test is not merely a function of the relative values of the two population parameters θ, but of their absolute values λ_1 and λ_2. This became evident in an actual vaccine study, when even 100,000 subjects proved too few. The disease proved much rarer than was expected, that is, λ_1 was small, so that t was small and the power of the test inadequate.

What can one do in such a case? A solution lies in the *sequential analysis* with multiple outcomes described in Section 6.6.

4.4 Time Between Events

If the number of events in a given interval has the Poisson distribution with λ the expected number of events, then the time t between events has the exponential distribution $F[t|\lambda] = 1 - \exp[-\lambda t]$ for $t \geq 0$.

Now, imagine a system, one on a spacecraft, for example, where various critical components have been duplicated, so that k consecutive failures are necessary before the system as a whole fails. If each component has a lifetime that follows an exponential distribution with the same value of the parameter $\lambda = 1/2$, then twice the lifetime of the system as a whole obeys the *chi-square* distribution with $2k$ degrees of freedom, that is,

$$p(y) = \frac{y^{k-1} e^{-y/2}}{2^k \Gamma(k)} \quad \text{for } y > 0.$$

In many instances the exponential distribution does not lend itself to parametric analysis, see Exercise 4.12, for example, and the permutation method is recommended. Another alternative, applicable in some instances, is to apply a variance-equalizing transformation before analyzing the data (see Section 11.3).

4.5 The Uniform Distribution

One might ask why a statistician, the most applied of all mathematicians, would be interested in the uniform distribution $U(a, b)$, which assigns equal weight to all values in the closed interval $[a, b]$, so that the density $f[x] = 1/(b-a)$ if $a \leq x \leq b$ and is zero, otherwise. It's because if $F[G[x]] = \Pr\{u \leq G[x]\}$ where G is *any* distribution, F is $U[0, 1]$.

By using the rand() function found in Excel or in the stdlib of any C or C++ compiler one can generate $U[0,1]$ random variables u. The equivalent command in R is unif(0, 1). For any distribution G for which one knows the inverse distribution function G^{-1}, one can obtain a random variable x with this distribution from the formula $G^{-1}[u]$.

A major aid to the generation of random permutations is the C++ function

```
int Choose (int lo, int hi)
{
    int z = rand()%(hi - lo + 1) + lo;
    return (z);
}
```

which yields random integers between lo and hi using the distribution $U[0,1]$.

The mean of 12 independent random variables, each distributed as $U[-1/2, +1/2]$ is approximately $N(0, 1/12)$.

The impractical uniform distribution is very practical, indeed.

4.6 The Exponential Family of Distributions

The members of the exponential family all have distributions whose probability densities (with respect to some measure μ) take the form

$$p_\theta[x] = C_\theta \exp\left[\sum \theta_j T_j(x)\right] h(x).$$

The $\{T_j\}$ are sufficient for the parameters $\{\theta_j\}$, and the probability densities of the exponential family have monotone likelihood ratio in any of the individual T_j.

Probability densities belonging to the one-parameter exponential family

$$p_\theta[x] = C_\theta \exp[\theta T(x)] h(x)$$

have monotone likelihood ratio in $T(x)$. As shown in the previous chapter there exists a UMP test for testing $H_1: \theta \leq \theta^*$ against the alternatives $K: \theta > \theta^*$, that is, a constant C, such that the test $\varphi(x) = 1, \gamma, 0$ according to whether $T(x) >, =, < C$, respectively, where C and γ are determined by the equality $E(\varphi(x) \mid \theta^*) = \alpha$.

As we shall prove in Section 15.3: *There exists a UMP procedure for testing the hypothesis $H_2: \theta \leq \theta_1$ or $\theta \geq \theta_2$ against the alternatives $K: \theta_1 < \theta < \theta_2$ in the one-parameter exponential family given by*

$$\Phi(x) = \begin{cases} 1 & \text{when } C_1 < T(x) < C_2, \\ \gamma_i & \text{when } T(x) = C_i, \\ 0 & \text{when } T(x) < C_1 \text{ or } T(x) > C_2. \end{cases}$$

Where the $\{\gamma_i, C_i; i = 1, 2\}$ are determined by

$$E\Phi(x)|\theta_1 = E\Phi(x)|\theta_2 = \alpha.$$

Rewriting the expression for the density of the multiparameter exponential family in the form

$$p_\theta[x] = C_\theta \exp[\theta U(x) + \sum \psi_j T_j(x)]h(x), \qquad (4.1)$$

in the next section we shall derive UMP unbiased level-α tests of the four hypotheses H_1, H_2, H_3: $\theta_1 \leq \theta \leq \theta_2$, and H_4: $\theta \leq \theta_0$.

Examples of one-parameter exponential families include the binomial distribution, as can be seen by rewriting it in the form,

$$(1-p)^n \exp[k \log(p/(1-p))] \binom{n}{x},$$

the Poisson, the normal (see Exercise 3.14), Student's t, and the gamma distribution

$$\frac{1}{\Gamma(g)b^g} x^{g-1} e^{-x/b}, \text{ with } 0 < x \text{ and } 0 < b, g.$$

Examples of a gamma distribution include the chi-square and exponential distributions. Note that the generalized exponential distribution,

$$F[t|\lambda, a] = 1 - \exp[-\lambda(t-a)] \text{ for } t > a,$$

where the boundary value a is also a parameter, is *not* a member of the exponential family.

4.6.1 Proofs of the Properties

In Section 15.4, we shall prove that for the multiparameter exponential family whose form is given in (4.1), the conditional distribution of U given t constitutes an exponential family with density

$$p_\theta[u|t] = C_\theta(t) \exp[\theta u] dv_t.$$

Consequently, we can use the results of the previous chapter, first to derive a UMP test φ_1 of H_1 versus K_1: $\theta > \theta_0$ depending on t, such that

$$\varphi_1(u, t) = \begin{cases} 1 & \text{when } u > C(t), \\ \gamma(t) & \text{when } u = C(t), \\ 0 & \text{when } u = C(t), \end{cases}$$

where the functions γ and C are determined by $E\varphi_1(U, T)|t = \alpha$. As this test has Neyman structure, it is UMP unbiased.

A similar argument will be used to show that the test of H_2: $\theta \leq \theta_1$ or $\theta \geq \theta_2$ versus K_2: $\theta_1 < \theta < \theta_2$ based on u given t has Neyman structure and is UMP unbiased.

To test H_3 and H_4, we introduce the critical function

$$\varphi(u,t) = \begin{cases} 1 & \text{when } u < C_1(t) \text{ or } u > C_2(t), \\ \gamma_i(t) & \text{when } u = C_i(t), i = 1 \text{ or } 2, \\ 0 & \text{when } C_1(t) < u < C_2(t), \end{cases}$$

with the $\{\gamma_i, C_i; i = 1, 2\}$ for H_3 determined by

$$E[\varphi(u,t)|t,\theta_1] = E[\varphi(u,t)|t,\theta_2] = \alpha,$$

and for H_4 by the equations

$$E[\varphi(u,t)|t,\theta_0] = \alpha,$$
$$E[U\varphi(u,t)|t,\theta_0] = \alpha E[U|t,\theta_0].$$

Using the results in the Appendix, it is easy to show that these tests, too, are UMP unbiased among all tests satisfying the corresponding boundary equations.

4.6.2 Normal Distribution

Many of our measurements appear to be the sum of the effects of a large number of hard-to-pinpoint factors, each of which makes only a small contribution to the total. For example, the price you are willing to pay for an automobile depends upon the sticker price, how much you paid for your last car, and the prices of comparable cars, but it also depends to a varying degree on the salesperson's personality, your past experience with this brand and model of car, ads you've seen on TV, and how badly you need a new car.

In many though not all cases, the resulting frequency distribution of measurements is that of the normal or Gaussian (expression 3.1). Note that this distribution, a member of the exponential family, is

a) symmetrical about its single mode;
b) its mean, median, and mode are the same;
c) most of its values, approximately 68%, lie within one standard deviation of the median; about 95% lie within two standard deviations,
d) yet arbitrarily large values are possible, although with vanishingly small probability.

If we make repeated independent measurements (a classmate's height, the length of a flower petal), we will seldom get the same value twice (if we measure to sufficient decimal places) but rather a series of measurements that are normally distributed about a single central value.

An interesting and useful property of independent normally distributed observations is that their sum also has a normal distribution. In particular, if X is $N(\mu, \sigma^2)$ and Y is $N(v, \tau^2)$ then $X + Y$ is $N(\mu + v, \sigma^2 + \tau^2)$. More important, the mean of a sample of n independent identically distributed observations that are $N(\mu, \sigma^2)$ will be distributed as $N(\mu, \sigma^2/n)$.

Under fairly general conditions, for example, when a set of independent observations all come from distributions whose means and variances exist and are bounded, the limiting distribution of their mean will be normal.

The square of a normally distributed random variable $N(0, 1)$ yields a variable we first encountered in Section 4.4, the chi-square random variable with one degree of freedom (Exercise 4.2.2).

4.7 Which Distribution?

Counting the number of successes in N independent trials? Use binomial.

Counting the number of rare events that occur in a given time interval or a given region when events in nonoverlapping intervals are independent? Use Poisson.

Recording the length of the interval that elapses between rare events? Use exponential or chi-square.

You observe the sum of a large number of factors, each of which makes only a small but independent contribution to the total? Use normal.

Relationships among these and other common univariate distributions are described in Leemis [1986]. Mixtures of these distributions are discussed in McLachlan [2000]. The *CRC Handbook of Tables for Probability and Statistics* can help you in deriving the appropriate cut-off values.

4.8 Exercises

(Warning: The following questions are more than academic exercises. Some require a rigorous mathematical proof, others have answers that are readily calculated, while still others cannot be answered precisely.)

1. a) Show that if Var $X < \infty$ then $E(aX + b) = aE(X) + b$ and Var $(aX + b) = a^2$ Var (X).
   b) Show that if X and Y are independent, Var $(X) < \infty$, and Var $(Y) < \infty$, then $E(X + Y) = E(X) + E(Y)$ and Var $(X \pm Y) =$ Var $(X) +$ Var (Y).
2. 70% of the registered voters in Orange County are Republican. What is the probability the next two people in Orange County you ask about their political preferences will say they are Republican? The next two out of three people? What is the probability that at most one of the next three people will admit to being a Democrat?

3. Given a binomial distribution of 18 trials each with probability 1/2; of success, what is the probability of observing a) 14 successes? b) 14 or more successes?
4. Two thousand bottles of aspirin have just rolled off the assembly line, 15 have ill-fitting caps, 50 have holes in their protective covers, and 5 are defective for both reasons. What is the probability that a bottle selected at random will be free from defects? What is the probability that a group of 10 bottles selected at random will all be free of defects?
5. A poll taken in January 1998 of 1000 people in the United States revealed 64% felt President Clinton should remain in office despite certain indiscretions. What would you estimate the variance of this estimate to be? Suppose 51% were the true value of the proportion supporting Clinton. What would the true value of the variance of your original estimate be?
6. How many students in your class have the same birthday as the instructor? In what proportion of classes, all of size 20, would you expect to find a student who has the same birthday as the instructor? Two students who have the same birthday?
7. Show that the sum of a Poisson with expected value λ_1 and a second independent Poisson with expected value λ_2 is also a Poisson with expected value $\lambda_1 + \lambda_2$.
8. What is the largest number of cells you can drop into a Petri dish divided into 100 squares and be sure the probability a square contains two or more cells is less than 1%? What percentage of the squares would you then expect would contain exactly one cell?
9. Show that if X is Poisson with parameter λ and Y is Poisson with parameter μ, that a) the conditional distribution of X given $X+Y$ is binomial, and b) the resulting test is UMP unbiased for testing $\theta = \lambda/(\lambda + \mu) \leq 1$ against $\theta > 1$.
10. Bus or subway? The subway is faster, but there are more buses on the route. One day, you and a companion count the number of arrivals at the various terminals. Fourteen buses arrive in the same period as only four subway trains pull in. Use the binomial distribution to determine whether the bus is a significantly better choice than the subway. (The Poisson distribution would not be appropriate. Why?)
11. Inspectors threaten to close a hospital because of a series of near fatal incidents that occurred there during surgery last June. Investigating, you find that the incidents arose from a multitude of causes and involved several different surgeons. What alternate explanation might account for the hospital's problems?
12. I was given this problem as a graduate assistant at UC Berkley's Donner Laboratory. Unfortunately, it took me almost 40 years to come up with the answer.

    Margaret White had been conducting a series of experiments in which animals were given radioactive isotopes. Samples of their blood and various other tissues were examined for residual radioactivity. The problem was Berkeley's cyclotron located just up the hill from the laboratory. How could one distinguish sample from background when the background changed by several orders of magnitude whenever the cyclotron was in use?

One way to do this experiment is to put two vials into the scintillation counter simultaneously, one with the sample and one without, and record the number of counts observed during the next minute (or two minutes, or five). The drawback of this method is that one does not know in advance what period to set the timer for.

The alternative employed by Ms. White was to record the time required for each of the vials to register a fixed number of counts. Because of the random nature of the counts, she used six vials each time, three with sub-aliquots of the sample and three without; these were labeled $\{t_{b,1}, t_{b,2}, t_{b,3}, t_{b+s,1}, t_{b+s,2}, t_{b+s,3}\}$. How would you test the hypothesis that there was no additional radioactivity in the sample?

13. For each of the following indicate whether the observation is binomial, Poisson, exponential, normal or almost normal, comes from some other distribution, or is predetermined (that is, not random at all).
    a) Number of books in your local public library.
    b) Guesses as to the number of books in your local public library.
    c) Heights of Swedish adults.
    d) Weights of Norwegian women.
    e) Alligators in an acre of Florida swampland.
    f) Vodka drinkers in a sample of 800 Russians.
    g) Messages on your answer machine.
    h) The distance from the point of impact to the target's center when you shoot at a target.

14. a) Compute the mean and variance of the uniform distribution.
    b) Find the sufficient statistics for a and b if $U_1, \ldots, U_n$ are independent identically distributed as $U[a, b]$.
    c) Suppose you mounted a pointed stick one foot in length on a center pivot so that when spun, it traced a circle one foot in diameter. Now suppose you place a second board, also one foot in length, below this circle. Each time you spin the stick, you drop a projection from the point of the stick to the board below after the stick comes to rest and make a mark. Would the distribution of these marks be uniform?

15. Let $X_1, \ldots, X_n$ be a set of independent identically distributed observations from the exponential distribution with density $\exp[-(x-a)/b]/b$ for $x \geq a$ and zero otherwise; $0 < b$. Find the UMP test of the hypothesis $a = a_0$.

16. Show that the normal distribution $N(\mu, \sigma^2)$ is a member of the exponential family with parameters $(\theta = -1/2\sigma^2, \psi = \mu/\sigma^2)$. Is the bivariate normal distribution (Equation 3.4) a member of the exponential family?

17. The probability density of a log-normal distribution is $1/x\sqrt{2\pi}\sigma \exp[-(\log x - \mu)/2\sigma^2]$ if $x > 0$, and 0 otherwise. Is the log-normal distribution a member of the exponential family?

18. One method used to simulate a mixed normal distribution that we'll encounter in Chapter 6 is to generate a uniformly distributed variable $U[0,1]$ and then to select from the appropriate normal distribution, say, $N(0,1)$, $N(2,1)$, and $N(2,2)$, based on the value of the uniform variable. Is the resultant mixture a member of the exponential family?

19. Show that if T is a sufficient statistic and $T = f(S)$ where f is continuous (measurable) and S is another statistic, S is sufficient.

20. Suppose the observations $(X_1, \ldots, X_K)$ are distributed in accordance with the multivariate normal probability density

$$\frac{\sqrt{|D|}}{(2\pi)^{K/2}} \exp[-\frac{1}{2} \sum \sum d_{ij}(x_i - \mu_i)(x_j - \mu_j)],$$

where the matrix $D = (d_{ij})$ is positive definite; $|D|$ denotes its determinant; $EX_j = \mu_j$; $E(X_j - \mu_j)(X_j - \mu_j) = \sigma_{ij}$; and $(\sigma_{ij}) = D^{-1}$. If $\sigma_{ij} = \sigma^2$ when $i = j$ and $\sigma_{ij} = \sigma_{12}$ when $i \neq j$, are the observations independent? Exchangeable?

21. Many statistics packages have a feature that allows you to generate simulated data. Stata^{TM} can be used to generate binomial, normal, chi-square, gamma variables, or mixtures of any of these (see Section 6.2.2). R (S, S-Plus) has built-in functions for generating beta, binomial, gamma, normal, and log-normal variables. Of course, any program, which like both Stata^{TM} and R can generate uniform random variables, can be used with just a bit of additional programming to generate any distribution you desire. If you enjoy programming, conduct the following experiment:

   a) Generate two samples from any distribution(s).
   b) Use the t-test, a permutation test, and a bootstrap to test for differences in the location parameters of the two distributions.
   c) Repeat steps (a) and (b) 400 times. Was the significance level of each of the tests in step (b) exact?
   d) Add one to each of the observations in the second sample before you perform the tests. Which test is the most powerful?

22. Let X and Y be independent variables, each distributed as $N(0,1)$. Let W be a Poisson variable with parameter $1/2$. Show that T, the time to the first Poisson event, has the same distribution as $Z = X^2 + Y^2$.

5
Multiple Tests

In this chapter we consider methods to control the overall error rate when multiple tests are performed and, if the tests are independent, methods to combine them.

5.1 Controlling the Overall Error Rate

One of the difficulties with clinical trials and other large-scale studies is that frequently so many variables are under investigation that one or more of them is practically guaranteed to be statistically significant by chance alone. If we perform 20 tests at the 5% level, we expect at least one significant result in twenty on the average. If the variables are related (and in most large-scale medical and sociological studies the variables have complex interdependencies), the number of falsely significant results could be many times greater.

David Freeman [1983] conducted a simulation study in which he generated 100 values each of 51 independent normally distributed variables. He designated one of the variables as the "dependent" variable and using a multiple regression technique found that 15 of the remaining variables made significant contributions as predictors at the 25% level. In a second multiple regression confined to these 15 variables, he found that 14 of the 15 made significant contributions as predictors at the 25% level and 6 of the 15 made contributions that were significant at 5% level. Clearly, "significance" in a multiple regression context is deceptive.

One way, and not a very good one, to ensure that the probability of making at least one Type I error is less than some predesignated value α is to make the k different comparisons each at level α/k. This method, attributed to Carl Bonferroni, is conservative, so that it can result in increased Type II error and, in consequence, has been widely criticized (see, for example, Perneger, 1998).

A better method, first described by Holm [1979], first orders the p-values from smallest to largest (or the corresponding standardized test statistics from largest to smallest). One begins with the most significant result and decides whether to accept or reject. Obviously, once a hypothesis is accepted then all hypotheses with larger p values are accepted as well. If a hypothesis is rejected, then a new critical value is determined, and the next p-value inspected. Permutation procedures utilizing this step-down approach were developed independently by Westfall and Young [1993], Blair and Karniski [1994], and Troendle [1995]; a test based on the latter's work is described in the next subsection.

The chief weakness of the step-down procedure is its dependence on the rejection criteria used to test the smallest p-value, normally $p_{(1)} \leq \alpha/k$. An alternative developed by Hochberg [1988] begins with the largest p-value at the first step. If a hypothesis is rejected then all hypotheses with smaller p-values are rejected as well. If a hypothesis is accepted, then a new critical value is determined, and the next p-value inspected. Blair, Troendle, and Beck [1996] report that this step-up method, an example of which is provided in Section 5.1.2, is slightly more powerful than the step-down.

5.1.1 Standardized Statistics

As an alternative to the analytic step-up method of Dunnett and Tarnhane [1992], which requires a specific distribution and correlation structure, we may apply the following permutation method due to Troendle [1996]. Suppose we have measured k variables on each subject, and are now confronted with k test statistics $s_1, s_2, \ldots, s_k$. To make these statistics comparable, we need to standardize them and render them dimensionless, dividing each by its respective L_1 or L_2 norm. For example, if one variable, measured in centimeters, takes values like 144, 150, and 156, and the other, measured in meters, takes values like 1.44, 1.50, and 1.56, we might set $t_1 = s_1/4$ and $t_2 = s_2/0.04$.

Next, we order the standardized statistics by magnitude so that $t_{(1)} \leq \cdots \leq t_{(k)}$. Denote the corresponding hypotheses as $H_{(1)}, \ldots, H_{(k)}$. The probability that at least one of these statistics will be significant by chance alone at the α level is $1 - (1 - \alpha)^k$, which is approximately $k\alpha$. But once we have rejected one hypothesis (assuming it was false), there will be only $k - 1$ true hypotheses to guard against rejecting.

Begin with $i = 1$ and

1. repeatedly resample the data (with or without replacement), estimating the cut-off value $\varphi(\alpha, k - i + 1)$ such that $\alpha = \Pr\{T(k - i + 1) \leq \varphi(\alpha, k - i + 1)\}$, where $T(k - i + 1)$ is the largest of the $k - i + 1$ test statistics $t_{(1)} \leq \cdots \leq t_{(k-i+1)}$ for a given resample;
2. if $t_{(k-i+1)} \leq \varphi(\alpha, k - i + 1)$, then accept all the remaining hypotheses $H_{(1)}, \ldots, H_{(k-i+1)}$ and STOP.
Otherwise, reject $H_{(k-i+1)}$, increment i, and RETURN to step 1.

5.1.2 Paired Sample Tests

Suppose we observe before and after values of k attributes for each of N subjects and wish to test the k null hypotheses that the before and after means are the same. We perform a set of k tests, which may be parametric, permutation, or bootstraps, and order the significance levels from largest to smallest, $p_{(1)} \geq \cdots \geq p_{(k)}$.

At the jth step of our multiple comparison procedure, following Blair, Troendle, and Beck [1996], we compute the critical level

$$\gamma_j = \frac{1}{2^N} \sum_{i=1}^{2^N} I[\min_{j \leq m \leq k} p_m(\pi_{mi}) \leq p_{(j)}],$$

where I is an indicator function, taking the value 1 if its argument is true, and the value 0 otherwise, and π_{mi} denotes the ith of the 2^N possible permutations of the before/after data for the mth variable. Thus, the sum counts the number of permutations for which the inequality is true.

In the step-up procedure, if $\gamma_j < \alpha$, we accept the remaining hypotheses and stop. Otherwise, we accept the hypothesis $H_{(j)}$ increment j and continue.

For example, suppose we have collected the following observations:
subject 1: before = $\binom{5}{6}$, after = $\binom{3}{7}$, showing a decline in both variables,
subject 2: before = $\binom{4}{4}$, after = $\binom{3}{4}$, showing a decline in only the first variable.

A t-test of variable 1 yields a p-value of 10%; a permutation test of variable 2 yields a p-value of 50%.

In rearranging the labels (before, after), we treat each vector of observations as an indivisible entity. There are only three possible rearrangements of the data labels in addition to the original, for example, the rearrangement that leaves the before and after values of subject 1 unchanged, while swapping the before and after readings of subject 2. All rearrangements yield the same p-value for variable 2, but two of the four rearrangements yield p-values of 85% for variable 1.

The step-up test starts with the largest of the original p-values, 50%, for variable 2. $\gamma_1 = 1$ and we accept the hypothesis concerning variable 2. $\gamma_2 = 0.5$ and with an experiment-wide error of 0.5, we also accept the null hypothesis for variable 1.

5.2 Combination of Independent Tests

Suppose now that k experiments have been performed to detect an effect. The experiments may have been performed under different conditions, and thus the magnitude of the effect, if any, may have varied from experiment to experiment. Three methods are in common use to combine the results of the various experiments. We shall consider each of the methods in turn in what follows. The combination of data from experiments in which different or interdependent observations have been made is considered in Chapter 7.

5.2.1 Omnibus Statistics

Suppose the ith experiment yielded a test statistic t_i with an associated p-value under the null hypothesis of p_i. The combination method of Fisher would reject the null hypothesis if $\Pi_i p_i$ were small or, equivalently, if $F = -2\log[\Pi_i p_i] = -2\sum_i \log[p_i]$ were large. If the $\{t_i\}$ have continuous distribution functions and the null hypothesis is true, then F has a chi-square distribution with $2k$ degrees of freedom [Fisher, 1925]. If the $\{t_i\}$ have discrete distributions, the chi-square distribution is not applicable, though it may be possible to find the distribution by exact enumeration [Wallis, 1942].

Among the alternate combining methods that have been proposed are the Liptak combining function $L = \sum_i \Phi^{-1}(1 - p_i)$, where Φ is the $N(0,1)$ cumulative distribution function, and the Tippett combining function $T = \max_{1 \leq i \leq I}(1 - p_i)$.

If the I test statistics are independent and continuous, then under the null hypothesis L is normally distributed with mean 0 and variance I [Liptak, 1958] and T has the same distribution as would the largest of I uniform random values chosen from $U[0, 1]$.

Birnbaum [1954] analyzed the case $I = 2$ and found that Tippet's solution is preferable when one but not both of the single-variable alternatives is true, and Liptak's solution is preferable when both alternatives are true.

Pesarin [1992] notes that to be suitable for test combination, the combining function must:

- be nonincreasing in each argument;
- attain its supremum value even when only one argument is zero.

Symmetry would be a prerequisite for impartiality.

As their name suggests, omnibus statistics provide omnibus tests against all possible alternatives. Almost always, one can find a test that is more powerful against a specific simple alternative. We give just such an example in the next section.

5.2.2 Binomial Random Variables

Suppose we've observed k independent binomial random variables, where the ith b_i is distributed as $B(p_i, n_i)$. The hypothesis to be tested is that $p_i = 0.5$ for all i, against the alternative that $p_i \geq 0.5$ for all i, with strict inequality being true for at least one value of i.

Set $\theta_i = p_i/(1-p_i)$. The distribution of the random variables is a member of an exponential family with respect to θ_i. Applying the fundamental lemma of Neyman and Pearson we see that the most powerful combination procedure for testing H against the simple alternative $\boldsymbol{p} = \boldsymbol{p}^*$ rejects for large values of the test statistic $\sum_i b_i \log[\theta_i^*]$.

5.2.3 Bayes' Factor

The previous two testing methodologies have supposed that we have all the experimental results available to us at one time. If instead the results are acquired sequentially as each experiment is performed, Bayes' method recommends itself.

We may start with the idea that the *prior* odds that the null hypothesis is true are close to one, while that of the alternative is near zero. As we gain more knowledge by observing first E_1 and, later, E_2, we can assign *posterior* odds to the null hypothesis with the aid of Bayes' theorem:

$$\Pr\{H|E_1, E_2\} = \frac{\Pr\{E_2|H\}\Pr\{H|E_1\}}{\Pr\{E_2|H\}\Pr\{H|E_1\} + \Pr\{E_2|\sim H\}\Pr\{\sim H|E_1\}}.$$

Steven Goodman (2001) demonstrates the correspondence between p-values, t-scores, and the shift in odds that results. He shows that even the strongest evidence against the null hypothesis does not lower its odds as much as the p-value might lead us to believe. More importantly, he makes it clear that we cannot estimate the credibility of the null hypothesis without considering evidence outside the study. (Also, see the article by Hodges, 1987.)

Utilizing Table B.1 of Goodman's article, we find that p-value of 0.01 represents a weight of evidence for the null hypothesis of somewhere between 1/25th and 1/8th. That is, the relative odds of the null hypothesis versus any alternative are at most 8–25 times lower than they were before the study. In order to claim the existence of an effect at the 95% significance level. Table B.1 tells us we need to know that the prior probability of the null is no greater than 60%. Absent any evidence that the prior probability of the null hypothesis is this large, even observing a p-value of 0.01 in our latest study would be insufficient to justify a conclusion that a non-zero effect exists. On the other hand, given 'substantial prior evidence that a non-zero effect exists, even a p-value as large as 10% out current might be enough to make a convincing case.

One caveat: Bayesian methods cannot be used in support of after-the-fact-hypotheses for, by definition, an after-the-fact-hypothesis has zero a priori probability and, thus, by Bayes' rule, zero a posteriori probability.

5.3 Exercises

1. Given k independent binomial random variables, where the ith b_i is distributed as $B(p_i, n_i)$, derive a most powerful test of the hypothesis $p_i = 0.5$ for all i against the alternative $p_i = p^*$ for all i where $p^* > 0.5$.
2. If F is the cumulative distribution function associated with any continuous univariate random variable unbounded on the right, would $T = F^{-1}(1 - p_i)$ be useful as a combining function?
3. Would this be true if the random variable were discrete?

6
Experimental Designs

In this chapter and the next, you learn to analyze the results of complex experimental designs that may involve multiple control variables, covariates, and restricted randomization.

The material is advanced and the discussion presupposes you have already completed Chapters 1–3.

6.1 Invariance

In Section 2.1.6, we discussed the importance of impartiality in a test. We shall apply the principle of impartiality repeatedly in subsequent sections.

Let X be distributed according to a probability distribution $P_\theta, \theta \in \Omega$ and let G be a group of transformations of the sample space, such as transformations of scale or zero point. Denote by gX the random variable that takes on the value gx when $X = x$ and suppose the distribution of gX is P_{θ^*}. Suppose that $\theta * \in \Omega$; that is, the transformation g on the sample space induces a transformation g^* on the parameter space Ω with $P_{g^*\theta}(gA) = P_\theta(A)$ for all events A belonging to the sample space. If the distributions P_θ corresponding to different values of θ are distinct, then a group of transformations G on the sample space induces a group of transformations G^* on the parameter space (Exercise 6.1).

We shall write that the problem of testing $H: \theta \in \Omega_H$ against $H: \theta \in \Omega_K$ remains invariant with respect to a group of transformations G, providing the induced group of transformations G^* leaves Ω_H and Ω_K distinct. In such a case, any UMP invariant test will also be most stringent (Exercise 6.2).

A function M will be said to be *maximal invariant* with respect to G if it is invariant with respect to G and if $M(x) = M(y)$ implies $y = gx$ for some $g \in G$.

Theorem 6.1. *Let X be distributed according to a probability distribution P_θ and let G be a group of transformations of the sample space. If $M(x)$ is* invariant under G and if $v(\theta)$ is maximal invariant under the induced group G^*, then the distribution of $M(x)$ depends only on $v(\theta)$.

Proof. Let $v(\theta_1) = v(\theta_2)$. Then $\theta_2 = g^*\theta_1$ so

$$P\{M(x) \in A|\theta_2\} = P\{M(x) \in A|g^*\theta_1\} = P\{M(gx) \in A|\theta_1\}$$
$$= P\{M(x) \in A|\theta_1\}.$$

□

We make use of this theorem in what follows to reduce the number of potential statistics to those that involve only a single maximal invariant with respect to one or more groups of transformations.

6.1.1 Some Examples

If $X_1, \ldots, X_n$ is a sample from $N(\mu, \sigma^2)$, the hypothesis $\sigma \geq \sigma_0$ remains invariant under transformations of the zero point $X'_i = X_i + c$. $U = \sum_{i=1}^n (x_i - \bar{x})^2$ is a maximal invariant (Exercise 6.3) and in accordance with the arguments presented in Section 3.5.2, the test that rejects when $U \leq C$ is UMP among tests that remain invariant under transformations of the zero-point. The power of this test is a constant on each of the sets $\{(\mu, \sigma): -\infty < \mu < \infty, \sigma = \sigma'\}$. As it is the most powerful test on each such set, it also is most stringent.

Suppose now that we conduct a series of side-by-side comparisons under varying conditions, that is, the members of each pair differ only in the treatment that is applied, but the various pairs may be handled quite differently. Let the probability that the first or control member of each pair prove superior be $1 - p_i$. We often wish to test the hypothesis that the treatment has no effect, that is, $p_i = \frac{1}{2}$, against the alternative that the treatment is superior, that is, $p_i > \frac{1}{2}$ for all i.

This problem remains invariant under all permutations of the n observations, and a maximal invariant is the total number of instances S in which the treatment is superior. The distribution of S is $P\{S = k\} = \prod_i q_i \sum_{\pi_j} \left(\prod_i (p_{i_j}/q_{i_j})\right)$ where the summation extends over all $\binom{n}{k}$ choices of subscripts $i_1, \ldots, i_n$. The most powerful invariant test against the specific alternative $(p'_1, \ldots, p'_n)$ rejects the hypothesis when

$$f(k) = \sum_{\pi_j} \left(\prod_i (p'_{i_j}/q'_{i_j})\right) / \binom{n}{k} > C.$$

f is an increasing function of k (Exercise 6.4) so that regardless of the alternative, the test rejects when $k > C'$ and is UMP among all tests invariant under permutations of the subscripts.

Of course, a much more powerful test would be in a matched pairs comparison which utilizes the actual results of each experiment as in Section 6.4.2.2.

6.2 k-Sample Comparisons—Least-Squares Loss Function

There is no single best test for all k-sample problems. The optimal choice will depend upon both the alternative hypothesis and the loss function. In this section we derive a class of parametric tests for the k-sample problem that are uniformly most powerful among tests that are impartial and that minimize a least-squares loss function under fairly broad conditions on the underlying distribution.

6.2.1 Linear Hypotheses

The k-sample comparison is the simplest example of a *univariate linear hypothesis*. Let $X_1, \ldots, X_n$ be independent and distributed with means $\mu_1, \ldots, \mu_n$ and common variance σ^2. The vector of means $\boldsymbol{\mu}$ is known to lie in a given s-dimensional subspace Ω. This would be the case if, for example, the $\{X_i\}$ were derived from k independent samples each taken from one of k distinct populations, that is, $s = k$. The hypothesis to be tested is that $\boldsymbol{\mu}$ lies in an $s - r$-dimensional subspace of Ω. In the k-sample case, the null hypothesis $\mu_1 = \cdots = \mu_k$ imposes $r = k - 1$ restrictions (see Exercise 6.5).

In the general case, the hypothesis can be given a particularly simple form by making an orthogonal transformation[1] to variables $Y_1, \ldots, Y_n$,

$$Y = CX,$$

such that the first s row vectors of C span Ω, and the first r row vectors of C span the subspace under the hypothesis.

Let $\eta_i = E(Y_1)$ so that $\boldsymbol{\eta} = C\boldsymbol{\mu}$. Then $\eta_i = 0$ for $i = s+1, \ldots, n$, and if the hypothesis is true, then $\eta_i = 0$ for $i = s - r, \ldots, s$.

The testing problem expressed in terms of the Y_i remains invariant with respect to the group G_1 of additive transformations (zero-point settings) such that $Y_i' = Y_i + c_i$ for $i = 1, \ldots, s - r$ and $Y_i' = Y_i$, otherwise. Consequently, for testing purposes, we can neglect the set $\{Y_i, i = 1, \ldots, s - r\}$ completely.

The group G_2 of all orthogonal transformations of the set $\{Y_i, i = s - r + 1, \ldots, s\}$ also leaves the problem invariant as does the group G_3 of all orthogonal transformations of the set $\{Y_i, i = s+1, \ldots, n\}$. Orthogonal transformations preserve distances. Thus, the sums of squares

$$U = \sum_{i=s-r}^{s} Y_i^2 \quad \text{and} \quad V = \sum_{i=s+1}^{n} Y_i^2$$

are *maximal invariants* with respect to G_2 and G_3, in the sense that: (1) U is invariant with respect to G_2, and (2) $U(\mathbf{Z}) = U(\mathbf{Z}')$ implies $\mathbf{Z}' = g\mathbf{Z}$ for

---

[1] An $n \times n$ matrix C is orthogonal if all pairs of row vectors are orthogonal. Two n-vectors $\boldsymbol{x}, \boldsymbol{y}$ are orthogonal if $\boldsymbol{x}'\boldsymbol{y} = 0$.

some $g \in G_2$ (Exercise 6.6). Consequently, for testing purposes, we can focus on the statistics U and V.

The testing problem also remains invariant with respect to the group G_4 of changes in scale $Y_i' = cY_i$ for $i = 1, \ldots, n$. This group induces the changes $U' = c^2 U$ and $V' = c^2 V$ with respect to which $W = U/V$ is maximal invariant.

The four groups also induce changes in the parameter space. If the observations $\{X_i\}$ in the jth sample are from a normal distribution $N(\mu_j, \sigma^2)$, then a maximal invariant with respect to the totality of transformations is $\varphi^2 = \sum_{i=1}^{k} \mu_i^2 / \sigma^2$ (Exercise 6.7). It follows from Theorem 6.1 that the distribution of W depends only on φ^2, so the problem reduces to that of testing the simple hypothesis $\varphi = 0$. If we can show that the probability density of W has monotone likelihood ratio in φ as outlined by Lehmann [1986, pp. 368, 427, and 428] then it would follow from the Neyman–Pearson fundamental lemma that the uniformly most powerful test of $\varphi = 0$ versus $\varphi > 0$ among all those tests that are invariant with respect to the four groups of transformations G_1, G_2, G_3, and G_4 rejects when W is too large or, equivalently, when

$$F = \frac{\sum_{i=s-r}^{s} Y_i^2 / r}{\sum_{i=s+1}^{n} Y_i^2 / (n-s)} > C.$$

If the observations are drawn from normal distributions $N(\mu, \sigma^2)$, then under the null hypothesis, F has the F-distribution with r and $n-s$ degrees of freedom.

Rewriting the test statistic in terms of the original observations $\{X_i\}$ we can show that this UMP invariant test minimizes squared losses. Since the transformation $\boldsymbol{Y} = \boldsymbol{CX}$ is orthogonal, it leaves distances unchanged, so that $\sum_{i=1}^{n}(Y_i - EY_i)^2 = \sum_{i=1}^{n}(X_i - \mu_i)^2$. Choose the $\{\hat{\mu}_i\}$ so as to minimize $\sum_{i=1}^{n}(X_i - \mu_i)^2$ under Ω. Then $\sum_{i=s+1}^{n} Y_i^2 = \sum_{i=1}^{n}(X_i - \hat{\mu}_i)^2$ for $\sum_{i=s+1}^{n} Y_i^2$ is the minimum value of $\sum_{i=1}^{n}(Y_i - EY_i)^2 = \sum_{i=s}^{s}(Y_i - \eta_i)^2 + \sum_{i=s+1}^{n} Y_i^2$ under unrestricted variation of the k population means. Let the $\{\hat{\hat{\mu}}_i\}$ be chosen so as to minimize $\sum_{i=1}^{n}(X_i - \mu_i)^2$ under the hypothesis. Then it is easy to see that $\sum_{i=s-r+1}^{n} Y_i^2 = \sum_{i=1}^{n}(X_i - \hat{\hat{\mu}}_i)^2$. We may rewrite the test statistic in the form

$$\frac{\left(\sum_{i=1}^{n}(X_i - \hat{\hat{\mu}}_i)^2 - \sum_{i=1}^{n}(X_i - \hat{\mu}_i)^2\right)/r}{\sum_{i=1}^{n}(X_i - \hat{\mu}_i)^2/(n-s)}.$$

In the case of k samples when we are testing the hypothesis that $\boldsymbol{\mu} = \mu \mathbf{1}$, the test statistic becomes the F-ratio

$$\frac{\sum_{i=1}^{k} n_i (\bar{X}_{i.} - \bar{X}_{..})/(k-1)}{\sum_{i=1}^{k}\sum_{j=1}^{n_i}(X_{ij} - \bar{X}_{i.})^2/(N-k)}. \tag{6.1}$$

This statistic is invariant under changes in zero-point and scale, and among all such statistics minimizes losses that are a function of $\sum(\mu - \mu_i)^2$ if the error terms are normally distributed. If the error terms are normally distributed, this statistic has the F-distribution with $I-1$ and $N-I$ degrees of freedom, tables for which are incorporated in most standard statistical software.

6.2.2 Large and Small Sample Properties of the F-ratio Test

In our derivation of the F-ratio test, we relied in several places upon the observations coming from a normal distribution. Even if we cannot assume normality, providing the observations $\{X_i\}$ are independent and drawn from distributions $F[x-\mu_i]$ where F has finite variance σ^2, then for large samples the size and power of this parametric test is the same as it would be if the observations had been independent and normally distributed. The within-cells sum of squares in the denominator of the test statistic divided by its degrees of freedom tends in probability to σ^2 as the sample size grows larger, and the between-samples sum of squares in the numerator divided by its degrees of freedom and by σ^2 tends to a chi-square distribution.

In a recent series of simulations with observations drawn from mixtures of normal distributions and sample sizes as small as 3 observations per cell, we found that the significance levels provided by the assumption of normality for the F-test were accurate to within the level of precision provided by the 10,000 replications used in the simulation.

We used a three-step procedure to generate the simulated observations:

1. Generate an $N(0,1)$ variable z.
2. Generate a $U(0,1)$ variable v.
3. Replace z with $z = \sigma^* z + \mu$ if $v < p$.

Since our objective was to assess the behavior of the analysis of variance with observations likely to be observed in practice, we generated samples with $1 \leq \sigma \leq 2, 0 \leq \mu \leq 4$, and $0.1 \leq p \leq 0.5$.

A total of 18 to 32 simulated observations was generated for each replication and assigned to $1 \times 3, 1 \times 6, 1 \times 8$ and 2×3 designs. If the null hypothesis were true and all the assumptions of the analysis of variance (ANOVA) were satisfied, the expected number of rejections at the 5% level would be 500 and the mean standard error of this number would be 22.

In one extreme case, a mixture of two normal distributions, $N(0,1)$ and $N(2,4)$ in the ratio of 0.7 to 0.3, the population effect was found to be significant in 464 cases out of 10,000 at the 5% level in a one-way analysis with 4 observations per cell and 6 cells. With 3 observations per cell and 6 cells, the population effect was found to be significant in 527 cases out of 10,000 at the 5% level.

Other slight deviations to normality, such as one that results from distorting a normal distribution by setting $z = 2^* z$ if $z > 0$, or censoring it by setting $z = -0.5$ if $z < -0.5$, also left significance levels unaffected.

Not all data are drawn from continuous distributions. Typical survey questions limit responses. The question, "How do you feel about such and such?"

may limit choices to, "Strongly like, like, indifferent, dislike, strongly dislike." These responses are often translated as $X = 2, 1, 0, -1$, or -2. We generated sets of normally distributed observations, then rounded them to the nearest integer so that the values ranged from -3 to $+4$. The significance level was not affected by the rounding, although, as one might expect, there was a slight drop in power. For example, with 24 observations divided among 6 cells, in 10,000 simulations, the results were significant at the 5% level in 537 instances. When 1 was added to each of the observations in one of the cells, the addition was declared significant 19% of the time with the original observations, and 17.6% of the time with those that had been discretized.

6.2.3 Discrete Data and Time-to-Event Data

The means of large samples of discrete observations from the binomial or the Poisson or of time-to-event data have close-to-normal distributions. But the variances of such means will depend upon their expectations. The results of Section 6.2. are applicable only when the observations are identically distributed in all groups. An initial variance-equalizing transformation is required if we are to use tests based on the F-distribution.

That is, we want to find a function f such that the transformed observations $\{f(X_i)\}$ are identically distributed, even if the original observations $\{X_i\}$ are not. Fortunately, we have the following theoretical result:

Theorem 6.2. *If the distribution of $\sqrt{n}(T_n - \theta)$ converges with increasing sample size n to a normal distribution with mean 0 and variance τ^2, and the derivative of the function f exists and is nonzero at θ, the distribution of $\sqrt{n}(f[T_n] - f[\theta])$ converges with increasing sample size n to a normal distribution with mean 0 and variance $\tau^2 (df[\theta]/d\theta)^2$.*

Proof. Applying Taylor's theorem,

$$f[T_n] = f[\theta] + (T_n - \theta)f'[\theta] + o(T_n - \theta)$$

or

$$\sqrt{n}(f[T_n] - f[\theta]) = \sqrt{n}(T_n - \theta)f'[\theta] + o[\sqrt{n}(T_n - \theta)].$$

□

The first term on the right converges to $N(0, \tau^2 (df[\theta]/d\theta)^2)$. We may neglect the second term provided that, as the sample size n increases, the probability that T_n differs from its expectation θ so that $\sqrt{n}(T_n - \theta)$ is larger than some arbitrary fixed value goes to zero. But this is precisely what is implied by the convergence in distribution of $\sqrt{n}(T_n - \theta)$ to a normal distribution. For a formal proof of this latter result, see Lehmann [1999, Section 2.3].

Let T_n denote the mean of n exponentially distributed observations. Then $T_n/\sqrt{n}$ converges to a normal distribution $N(\lambda, 2\lambda^2)$. This result holds whether each observation is distinct or if there are k observations, each consisting of m separate exponentially distributed events where $km = n$. To equalize variances for different values of λ, we set $f'[\lambda] = 1/(\lambda\sqrt{2})$, or $f[\lambda] = \log[\lambda]/\sqrt{2}$. Similarly, to approximate a Poisson by a normal distributed variable, we set $f[\lambda] = \sqrt{\lambda}$. The binomial we leave as Exercise 6.8. □

6.3 k-Sample Comparisons—Other Loss Functions

As noted in Chapter 2, the loss function is the primary consideration in the selection of an appropriate decision-making procedure. More precisely, our objective to minimize the risk associated with a decision procedure, $r(d, \lambda) = L(d(X), \theta)dP_\theta d\lambda[\theta] = EL(\theta)d\lambda[\theta]$, where L is our loss function and λ stands for the subjective probability we assign to the various possible underlying states of nature θ.

Unfortunately, the uniformly most powerful invariant test derived in the previous section is limited to a single loss function and a single broad alternative. By contrast, for any loss function and any set of alternatives, we can almost always find an optimal unbiased and invariant permutation decision procedure.

6.3.1 *F-ratio*

Suppose we wanted to assess the effect on crop yield of hours of sunlight, observing the yield X_{ij} for I different levels of sunlight $i = 1, \ldots, I$ with n_i observations at each level. Our model is that $X_{ij} = \mu_i + e_{ij}$, where the $\{e_{ij}\}$ are exchangeable random elements.

Let $\bar{X}_{i.}$ and $\bar{X}_{..}$ denote the mean of the ith group and the grand mean, respectively.[2] Note that the sum of squares of the deviations about the grand mean can be decomposed into two sums, the first of which represents the *within-group* sum of squares, and the second, the *between-group* sum of squares:

$$\sum_{i=1}^{I}\sum_{j=1}^{n_i}(X_{ij} - \bar{X}_{..})^2 = \sum_{i=1}^{I}\sum_{j=1}^{n_i}(X_{ij} - \bar{X}_{i.})^2 + \sum_{i=1}^{I}n_i(\bar{X}_{i.} - \bar{X}_{..})^2.$$

The F-ratio of the between-group variance to the within-group variance that we derived in the preceding section is the classic parametric statistic for testing the hypothesis that the means of k normal distributions are the same [Welch, 1937].

---

[2] Here and in similar expressions, the use of a dot as in $\bar{X}_{i.}$ denotes summation over the missing subscript(s).

Before deriving the data's permutation distribution, we should eliminate all terms that are invariant under permutations. Rewriting expression (6.1), eliminating constants, gives

$$\frac{\sum_{i=1}^{I} n_i(\bar{X}_{i.} - \bar{X}_{..})^2}{(\sum_{i=1}^{I} \sum_{j=1}^{n_i}(X_{ij} - \bar{X}_{..})^2 - \sum_{i=1}^{I} n_i(\bar{X}_{i.} - \bar{X}_{..})^2}$$

and noting that the sum of squares of all observations about the grand mean is invariant under rearrangements of the observations among samples, we see that we may focus attention on the between-sample sum of squares $\sum_{i=1}^{I} n_i(\bar{X}_{i.} - \bar{X}_{..})^2$.

We may further reduce the time required for its computation on noting that this sum may be written as

$$\sum_{i=1}^{I} n_i \bar{X}_{i.}^2 - 2\bar{X}_{..} \sum_{i=1}^{I} n_i \bar{X}_{i.} + N\bar{X}_{..}^2 = \sum_{i=1}^{I} \left(\sum_{j=1}^{n_i} X_{ij} \right)^2 - N\bar{X}_{..}^2.$$

As the grand mean, too, is invariant under permutations, our test statistic is $F_2 = \sum_i (\sum_j X_{ij})^2 / n_i$. If the loss function is $\sum(\mu - \mu_i)^2$, tests based on F_2 will be invariant under all transformations that leave the expected losses (risk) unchanged whether or not the data are normally distributed [Jagers, 1980].

Now, suppose our loss function is not $\sum(\mu - \mu_i)^2$ but $\sum|\mu - \mu_i|$. Clearly the test statistic

$$F_1 = \sum_{i=1}^{I} n_i |\bar{X}_{i.} - \bar{X}_{..}|,$$

whose distribution, too, is easily determined by permutation means, will minimize the expected risk (Exercise 6.9).

In a sidebar, we've provided an outline of a computer program that uses a Monte Carlo to estimate the significance level (see Chapter 14.3). Our one programming trick is to pack all the observations into a single linear vector $\boldsymbol{X} = (X_{11}, \ldots, X_{1n_1}, X_{1(n_1+1)}, \ldots)$ and then to permute the observations within the vector. If we have k samples, we only need to select $k-1$ of them when we rearrange the data. The kth sample is determined automatically.

Note that we need to write a separate subprogram to compute the test statistic. If you're impatient, you can download such a program from http://mysite.verizon.net/res7sf1o/GoodStat.htm.

6.3.2 Pitman Correlation

The permutation version of the F-ratio test offers protection against any and all deviations from the null hypothesis of equality among treatment means. As a result, it may offer less protection against some specific alternative than

Program for Estimating Permutation Significance Levels

Monte, the number of Monte Carlo simulations; try 400
S_0, the value of the test statistic for the unpermuted observations
S, the value of the test statistic for the rearranged observations
$X[\ ]$, a one-dimensional vector that contains the observations
$n[\ ]$, a vector that contains the sample sizes
N, the total number of observations

Main program
    Get data
        Put all the observations into a single linear vector
    Compute the test statistic S_0
    Repeat Monte times:
        Rearrange the observations
        Recompute the test statistics S
        Compare S with S_0
    Print out the proportion of times S was equal to or larger than S_0

Rearrange
    Set s to the size of the combined sample
    Start: Choose a random integer k from 0 to $s-1$
        Swap $X[k]$ and $X[s-1]$:
            temp = $X[k]$;
            $X[k] = X[s-1]$;
            $X[s-1]$ = temp.
        Decrement s and repeat from start
    Stop after you've selected all but one of the samples.

Get data
    This user-written procedure gets all the data and packs it into a single long linear vector X.

Compute stat
    This user-written procedure computes the test statistic.

---

some other test function(s). When we have a specific alternative in mind, as is so often the case in biomedical research when we are testing for an ordered dose response, the F-ratio may not be the statistic of choice.

Frank, Trzos, and Good [1977] studied the increase in chromosome abnormalities and micronucleii as the dose of various known mutagens was increased. Their object was to develop an inexpensive but sensitive biochemical test for mutagenicity that would be able to detect even marginal effects. Thus they were more than willing to trade the global protection offered by the F-test for a statistical test that would be sensitive to ordered alternatives.

Fortunately, a most powerful unbiased test (and one that is also most powerful among tests that are invariant to changes in scale) has been known since the late 1930s. Pitman [1937] proposes a test for linear correlation using as test statistic

$$S = \sum_i f[i] \sum_{j=1}^{n_i} X_{ij},$$

where $f[i]$ is any monotone increasing function. The simplest choice is $f[i] = i$.

The permutation distributions of S' with $f[i] = ai + b$ and S with $f[i] = i$ are equivalent in the sense that if S_0 and S'_0 are the values of these test statistics corresponding to the same set of observations $\{X_i\}$, then $\Pr(S > S_0) = \Pr(S' > S'_0)$.

Theorem 6.3. *Suppose f is a monotone nondecreasing function. Then the distribution of the statistic $S = \sum_i f[i] \sum_{j=1}^{n_i} X_{ij}$ obtained from rearrangements provides an exact, unbiased test of the hypothesis H: $a_i = 0$ for all i against an ordered alternative K: $a_1 < a_2 < \cdots < a_I$.*

Proof. The rejection region R of the test consists of all rearrangements $\boldsymbol{X}^*$ of the sample $\boldsymbol{X}$ for which $S_I(\boldsymbol{X}^*)$ are greater than $S_I(\boldsymbol{X})$. Consider first those rearrangements $\boldsymbol{X}^{*\prime}$ consisting of a single pairwise exchange between any two groups. As f is a monotone nondecreasing function, then $\Pr\{S_I(\boldsymbol{X}^{*\prime}) \geq S_I(\boldsymbol{X})|H\} \geq \Pr\{S_I(\boldsymbol{X}^{*\prime})\} \geq S_I(\boldsymbol{X})|K\}$. But by definition, any rearrangement consists of such pairwise exchanges and the theorem follows. □

Let us apply the Pitman approach to the data collected by Frank et al. [1978] shown in Table 6.1. As the anticipated effect is proportional to the logarithm of the dose, we take $f[\text{dose}] = \log[\text{dose} + 1]$. (Adding 1 to the dose keeps this function from blowing up at a dose of zero.)

There are four dose groups; the original data for breaks may be written in the form

$$0\ 1\ 1\ 2 \quad 0\ 1\ 2\ 3\ 5 \quad 3\ 5\ 7\ 7 \quad 6\ 7\ 8\ 9\ 9.$$

Table 6.1. Micronuclei in polychromatophilic erythrocytes and chromosome alterations in the bone marrow of mice treated with CY.

Dose (mg/kg)	Number of Animals	Micronuclei per 200 Cells	Breaks per 25 Cells
0	4	0 0 0 0	0 1 1 2
5	5	1 1 1 4 5	0 1 2 3 5
20	4	0 0 0 4	3 5 7 7
80	5	2 3 5 11 20	6 7 8 9 9

6.3 k-Sample Comparisons—Other Loss Functions

As $\log[0 + 1] = 0$, the value of the Pitman statistic for the original data is $0 + 11^* \log[6] + 22^* \log[21] + 39^* \log[81] = 112.1$. The only larger values are associated with the small handful of rearrangements of the form

0 0 1 2	1 1 2 3 5	3 5 7 7	6 7 8 9 9
0 0 1 1	1 2 2 3 5	3 5 7 7	6 7 8 9 9
0 0 1 1	1 2 2 3 3	5 5 7 7	6 7 8 9 9
0 0 1 2	1 1 2 3 3	5 5 7 7	6 7 8 9 9
0 1 1 2	0 1 2 3 3	5 5 7 7	6 7 8 9 9
0 1 1 2	0 1 2 3 5	3 5 6 7	7 7 8 9 9
0 0 1 2	1 1 2 3 5	3 5 6 7	7 7 8 9 9
0 0 1 1	1 2 2 3 5	3 5 6 7	7 7 8 9 9
0 0 1 1	1 2 2 3 3	5 5 6 7	7 7 8 9 9
0 0 1 2	1 1 2 3 3	5 5 6 7	7 7 8 9 9
0 1 1 2	0 1 2 3 3	5 5 6 7	7 7 8 9 9

A statistically significant ordered dose response ($\alpha < 0.001$) has been detected. The micronucleii also exhibit a statistically significantly dose response when we calculate the permutation distribution of S with $f[i] = \log[\text{dose}_i + 1]$. To make the calculations, we took advantage of the computer program we used in the previous section; the only change was in the subroutine used to compute the test statistic.

A word of caution, if we use some function of the dose other than $f[\text{dose}] = \log[\text{dose}+1]$, we might not observe a statistically significant result. Our choice of a test statistic must always make practical as well as statistical sense (see Exercise 6.17).

6.3.3 Effect of Ties

Ties can complicate the determination of the significance level. Because of ties, each of the rearrangements noted in the preceding example might actually have resulted from several distinct reassignments of subjects to treatment groups and must be weighted accordingly. To illustrate this point, suppose we put tags on the 1's in the original sample:

0 1* 1# 2	0 1 2 3 5	3 5 7 7	6 7 8 9 9

The rearrangement

0 0 1 2	1 1 2 3 5	3 5 7 7	6 7 8 9 9

corresponds to the three reassignments

0 0 1 2	1* 1# 2 3 5	3 5 7 7	6 7 8 9 9
0 0 1* 2	1 1# 2 3 5	3 5 7 7	6 7 8 9 9
0 0 1# 2	1 1* 2 3 5	3 5 7 7	6 7 8 9

The 18 observations are divided into four dose groups containing 4, 5, 4, and 5 observations, respectively, so that there are $\binom{18}{4545}$ possible reassignments of observations to dose groups. Each reassignment has an equal probability of occurring so the probability of the rearrangement

$$0\ 0\ 1\ 2 \quad 1\ 1\ 2\ 3\ 5 \quad 3\ 5\ 7\ 7 \quad 6\ 7\ 8\ 9\ 9$$

is $3/\binom{18}{4545}$.

To determine the significance level when there are ties, weight each distinct rearrangement by its probability of occurrence. This weighting is done automatically if you use Monte Carlo sampling methods as is done in the computer program we provide in the previous section.

6.3.4 Cochran–Armitage Test

If our observations are binomial response variables, that is, they can take only the values 0 or 1, and if $f[i] = d_i$ is the magnitude of the ith dose and X_i denotes the number of responders in the ith dose group, then the Pitman correlation $\sum d_i X_i$ is more commonly known as the *Cochran–Armitage test for trend*. Bounds on the power can be obtained by the method of Mehta, Patel, and Senchaudhuri [1998].

6.3.5 Linear Estimation

Pitman correlation may be generalized by replacing the fixed function $f[i]$ by an estimate $\hat{\phi}$ derived by a linear estimation procedure such as least squares polynomial regression, kernel estimation, local regression, and smoothing splines [Raz, 1990].

Suppose the jth treatment group is defined by $\boldsymbol{X}_j$, a vector-valued design variable. ($\boldsymbol{X}_j$ might include settings for temperature, humidity, and phosphorous concentration.) Suppose, also, that we may represent the ith observation in the jth group by a regression model of the form $Y_{ji} = \mu(x_j) + e_{ji}$, $j = 1, \ldots, n$, where e_{ji} is an error variable with mean 0, and $\mu(x)$ is a smooth regression function (that is, for any x and ε sufficiently small, $\mu(x + \varepsilon)$ may be closely approximated by the first-order Taylor expansion $\mu(x) + b\varepsilon$).

The null hypothesis is that $\mu(x) = \mu$, a constant that does not depend on the design variable x. As always, we assume that the errors $\{e_{ji}\}$ are exchangeable so that all $n!$ assignments of the labels to the observations that preserve the sample sizes $\{n_j\}$ are equally likely.

Raz's test statistic is $Q = \sum \hat{\mu}[x_j]^2$, where $\hat{\mu}$ is an estimate of μ derived by a linear estimation procedure such as least-squares polynomial regression, kernel estimation, local regression, or smoothing splines.

This test may be performed using the permutation distribution of Q or, for large samples, a gamma distribution approximation.

6.3.6 A Unifying Theory

The permutation tests for Pitman correlation and the two-sample comparison of means are really special cases of a more general class of tests that take the form of a dot product of two vectors [Wald and Wolfowitz, 1944; De Cani, 1979]. Let $W = \{W_1, \ldots, W_N\}$ and $Z = \{Z_1, \ldots, Z_N\}$ be fixed sets of numbers, and let $z = \{z_1, \ldots, z_N\}$ be a random permutation of the elements of Z. Then we may use the dot product of the vectors Z and W, $T = \sum z_i w_i$, to test the hypothesis that the labeling is irrelevant. In the two-sample comparison, W is a vector of $m1$'s followed by $n0$'s. In the Pitman correlation, $W = \{f[1], \ldots, f[N]\}$ where f is a monotone function.

6.4 Four Ways to Control Variation

Although a test may be uniformly most powerful, and, in the case of a permutation test, its significance level "distribution-free," its power strongly depends on the underlying distribution.

Figure 6.1 depicts the effect of a change in the variance of the underlying population on the power of the permutation test for the difference in two means. As the variance increases, the power of a test decreases. *To get the most from your experiments, reduce the variance.*

There are four ways to account, correct for, or reduce variation:

1. control the environment;
2. block the experiment;
3. measure factors that cannot be controlled;
4. randomize.

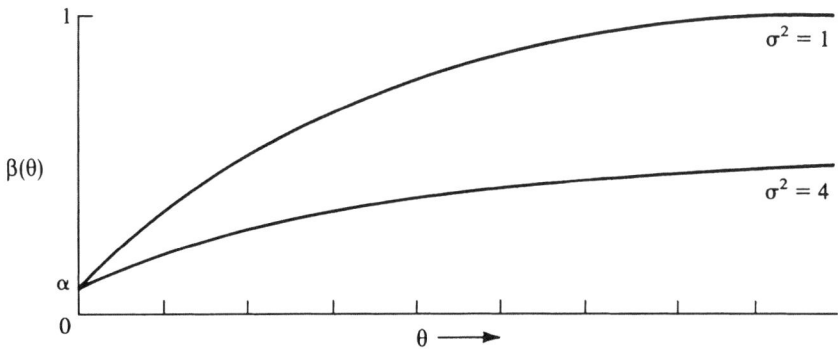

Fig. 6.1. Effect of the population variance on the power of a test of two means. $\theta = \theta_1 - \theta_2$.

6.4.1 Control the Environment

- Work if possible in a biosphere, where one can control temperature, humidity, and every other aspect of the experimental conditions.
- Make all observations at the same time of day (to correct for circadian rhythms).
- Make all observations in the same way, preferably by the same observer, and using the same measuring device.
- Use a more precise measuring device.

A major but often forgotten aspect of controlling the environment, particularly in nonhuman studies, is making sure that neither the subject (in animal studies) nor the experimenter (in all studies) is aware of which treatment the subject receives.

With humans it is almost self-evident that the subject will experience a placebo effect, providing, that is, she's not told that her treatment is merely a placebo. And she will be less inclined to adhere to the treatment regimen and to report side effects if she knows she is part of the control group.

The intelligence and sensitivity of animals should not be underestimated. The taste of a pill can be as significant to a rat as the pill's intended effect. And even a flatworm can tell which path in an unsterilized t-maze was previously followed.

Dogs, cats, horses, primates, and many birds can pick up subtle changes in their handler's behavior. This is the second major reason why the experimenter also should not be aware of which treatment is being administered. If he or she spends more time with the treated animals, conveying that "they're the important ones, after all," the treated animals may feel stressed, the controls neglected, and both groups react accordingly. Animals like humans are subject to the Hawthorne effect. Then, too, consciously or unconsciously, the experimenter is likely to pay closer attention to the treated subjects (even the inanimate ones) studying them more closely, looking for flaws. And because he or she is looking at them more closely, she is more likely to find flaws and record side effects. Triple blinding, wherein one individual treats subjects and a second observes, is strongly recommended.

6.4.2 Block the Experiment

We can reduce the variance if we subdivide the population under study into more homogeneous subpopulations and take separate samples from each. Suppose you were designing a survey on the effect of income level on the respondent's attitudes toward compulsory pregnancy. Obviously, the views of men and women differ markedly on this controversial topic. It would not be prudent to rely on randomization to even out the sex ratios in the various income groups.

The recommended solution is to block the experiment, to interview and to report on men and women separately. You would probably want to do the same type of blocking in a medical study. Similarly, in an agricultural study, you would want to distinguish among clay soils, sandy, and sandy–loam.

In short, whenever a population can be subdivided into distinguishable subpopulations, you can reduce the variance of your observations and increase the power of your statistical tests by blocking or stratifying your sample.

Suppose we have agreed to divide our sample into two blocks—one for men, one for women. If this were an experiment, rather than a survey, we would then assign subjects to treatments separately within each block.

In a study that involves two treatments and ten experimental subjects, four men and six women, we would first assign the men to treatment and then the women. We could assign the men in any of $\binom{4}{2} = 6$ ways and the women in any of $\binom{6}{3} = 20$ ways. That is, there are $6 \times 20 = 120$ possible random assignments in all.

When we come to analyze the results of our experiment, we use the permutation approach to ensure we analyze in the way the experiment was designed. Our test statistic is a natural extension of that used for the two-sample comparison of location parameters described in Chapter 3, $S = \sum_{b=1}^{B} \sum_{j=1}^{m_b} x_{bj}$, where B is the number of blocks, two in the present example, and the inner sum extends over the m_b treated observations within each block.

We compute the test statistic for the original data. Then, we rearrange the observations at random within each block, subject to the restriction that the number of observations within each treatment category remain constant.

In the example of two blocks, one for each sex, we compute S for each of the 120 possible rearrangements. If the value of S for the original data is among the largest values, then we reject the null hypothesis; otherwise, we accept it.

6.4.2.1 Using Ranks

Directly combining the results from different strata may not be appropriate in situations where different methods of measurement were used or different scales employed. Consider the following example:

An experimenter administered three drugs in random order to each of five recipients. He recorded their responses and now wishes to decide if there are significant differences among treatments. The problem is that the five subjects have quite different baselines. A partial solution would be to subtract the baseline value from each of that individual's observations, but who is to say that an individual with a high baseline value will respond in the same way as an individual with a low baseline reading?

Another solution would be to treat each individual as a block and then combine the results using the formula of the preceding section. But then we run the risk that the results from an individual with unusually large responses

100   6 Experimental Designs

Table 6.2a. Original observations

	A	B	C	D	E
Control	89.7	75	105	94	80
Treatment 1	86.2	74	95	98	79
Treatment 2	76.5	68	94	93	84

Table 6.2b. Ranks.

	A	B	C	D	E
Control	1	1	1	2	2
Treatment 1	2	2	2	1	3
Treatment 2	3	3	3	3	1

might mask the responses of the others. Or, suppose the measurements actually had been made by five different experimenters using five different measuring devices in five different laboratories. Would it really be appropriate to combine them?

The answer is *no*, for the five sets of observations measured on different scales are not exchangeable. By converting the data to ranks, separately for each case, we are able to put all the observations on a common scale, and then combine the results.

Using the block formula, we see that there are a total of $(3!)^5 = 7776$ possible rearrangements of labels within blocks (subjects) of which 2×5, or less than 1%, are as, or more, extreme than our original sets of ranks.

6.4.2.2 Matched Pairs

Blocking is applicable to any number of subgroups; in the extreme case, that in which every pair of observations forms a distinct subgroup, we have the case of matched pairs.

In a matched pairs experiment, we take blocking to its logical conclusion. Each subject in the treatment group is matched as closely as possible to a subject in the control group. For example, a 45-year-old black male hypertensive is given a pill to lower blood pressure, then a second similarly built 45-year-old black male hypertensive is given a placebo. One member of each pair is then assigned at random to the treatment group, and the other member is assigned to the controls.

Assuming we have been successful in our matching, we will end up with a series of independent pairs of observations (X_i, Y_i) where the members of each pair have been drawn from the distributions $F_i(x - \mu)$ and $F_i(x - \mu - \delta)$ respectively. Regardless of the form of this unknown distribution, the differences $Z_i = Y_i - X_i$ will be symmetrically distributed about the unknown parameter δ; hence,

$$\begin{aligned}
\Pr\{Z \leq z + \delta\} &= \Pr\{Y - X - \delta \leq z\} \\
&= \Pr\{(Y - \mu - \delta) - (X - \mu) \leq z\} \\
&= \Pr\{Y' - X' \leq z\} \\
&= \Pr\{X' - Y' \leq z\} \\
&= \Pr\{(X - \mu) - (Y - \mu - \delta) \leq z\} \\
&= \Pr\{X - Y \leq z - \delta\} \\
&= \Pr\{Z \geq -z + \delta\}.
\end{aligned}$$

But this is precisely the one-sample case we considered in Chapter 3 and the same readily computed tests are applicable with the following exception: If the observation on one member of a pair is missing, then we must discard the remaining observation.

For an almost most powerful test when one member of the pair is censored, see Chapter 11.5. For an application of a permutation test to the case where an experimental subject serves as her own control, see Shen and Quade [1986].

6.4.3 Measure Factors That Cannot Be Controlled

Some variables that affect the outcome of an experiment are under our control from the very beginning—e.g., light and fertilizer. But other equally influential variables, called covariates, we may only be capable of measuring, rather than controlling. An example of covariates in a biomedical experiment are blood chemistries. Various factors in the blood can affect an experimental outcome, and most blood factors will be affected by a treatment, but few are under our direct control.

In this section, you learn two methods for correcting for the effects of covariates. The first is for use when you know or suspect the nature of the functional relationship between the primary variables and the covariates. The second method is for use when the covariates take only a few discrete values and these values can be used to restrict the randomization.

6.4.3.1 Eliminate the Functional Relationship

Gail, Tan, and Piantadosi [1988] recommend eliminating the effects of covariates first and then applying permutation methods to the residuals. For example, suppose the observation Y depends both on the treatment $\tau_i (i = 1, \ldots, I)$ and on the p-dimensional vector of covariates $\boldsymbol{X} = (\boldsymbol{X}^1, \ldots, \boldsymbol{X}^p)$, that is $\boldsymbol{Y} = \boldsymbol{\mu} + \boldsymbol{\tau} + \boldsymbol{X}\boldsymbol{\beta} + \boldsymbol{e}$, where $\boldsymbol{Y}, \boldsymbol{\mu}, \boldsymbol{\tau}$, and $\boldsymbol{e}$ are $n \times 1$ vectors of observations, mean values, treatment effects, and errors, respectively, $\boldsymbol{X}$ is an $x \times p$ matrix of covariate values, and $\boldsymbol{\beta}$ is a $p \times 1$ vector of regression coefficients.

We might use least squares methods[3] to estimate the regression coefficients $\hat{\boldsymbol{\beta}}$ after which we would apply the permutation methods described in the preceding sections to the residuals $\boldsymbol{Z} = \boldsymbol{Y} - \boldsymbol{X}\hat{\boldsymbol{\beta}}$. Of course, we must assume that the residuals are exchangeable and both the concomitant variables (the X^i) and the regression coefficients are unaffected by the treatment [Kempthorne, 1952, p. 160].

---

[3] Note that the use of least squares methods of estimation makes sense only if the loss function is also least squares. This point is expanded upon in Good and Hardin [2003, p. 44].

A distribution-free multivariate analysis of covariance in which the effects of the treatments and the covariates are evaluated simultaneously is considered in Chapter 7.

6.4.3.2 Selecting Variables

Which covariates should be included in your model? Draper and Stoneman [1966] describe a permutation procedure for selecting covariates using a "forward stepping rule":

The first variable you select should have the largest squared sample correlation with the dependent variable y; thereafter, include the variable with the largest squared partial correlation with y given the variables that have already been selected. You may use any standard statistics package to obtain these correlations. Equivalently, you may select variables based on the maximum value of the square of the t-ratio for the regression coefficient of the entering variable, the so-called "F to enter." The problem lies in knowing when to stop, that is, in knowing when an additional variable contributes little beyond noise to the model.

Percentiles of the permutation distribution of the F-to-enter statistic can be used to test whether variables not yet added to the model would be of predictive value. Details for deriving the permutation distribution of this statistic defined in terms of Householder rotations of the permuted variable matrix are given in Forsythe et al. [1973].

6.4.3.3 Restricted Randomization

If the covariates take on only a few discrete values, e.g., smoker vs. nonsmoker, or status $= 0, 1, 2$, we may correct for their effects by restricting the rerandomizations to those permutations whose design matrices match the original design [Edgington, 1983].

Consider the artificial data set in Table 6.3 adapted from Rosenbaum [1984, p. 568]. To test the hypothesis that the treatment has no effect on the response, we would use the sum of the observations in the treatment group as our test

Table 6.3. Data for artificial example.

Subject	Treatment	Result	Covariate
A	1	6	1
B	1	2	0
C	0	5	1
D	0	4	1
E	0	3	1
G	0	1	0
H	0	0	0

statistic. The sum of 8 for the original observations is equaled or exceeded in 6 of the $\binom{7}{2} = 21$ possible rerandomizations. This result is not statistically significant.

Now let us take the covariate into consideration. One member of the original treatment group has a covariate value of 0, the other has a covariate value of 1. We limit our attention to the $12 = \binom{4}{1}\binom{3}{1}$ possible rerandomizations in which the members of the treatment group have similar covariate values. These consist of AB AG AH, CB CG CH, DB DG DH, EB EG EH. With only 1 of the 12 (that of AB) that we observed originally do we observe a result sum as large as 8. This result is statistically significant at the 10% level. Restricting the randomizations eliminates the masking effect of the covariate and reveals the statistically significant effect of the treatment. On the down side, the restriction has reduced the number of randomizations to the point that the p-value cannot be less than $1/12$.

If the covariate varies continuously, it may still be possible to apply the method of restricted randomizations by first subdividing the covariate's range into a few discrete categories. For example, if

$$x < -1, \text{ let } x' = 0,$$
$$-1 < x < 1, \text{ let } x' = 1,$$
$$1 < x, \text{ let } x' = 2.$$

Rosenbaum [1984] suggests that with larger samples one restrict the randomizations so that one attains a specific mean value of the covariate, rather than a specific set of values.

Subject to certain relatively weak assumptions, the method of restricted randomizations can also be applied to after-the-fact covariates (see Section 11.2).

6.4.4 Randomize

The fourth essential principle of experimental design is that to keep the errors exchangeable we need to randomly assign experimental units to treatment so that the innumerable factors that can neither be controlled nor observed directly are as likely to influence the outcome of one treatment as another.

Sometimes our biases are overt. A young surgeon wants to test a new technique but his supervisor will only permit him to try the new approach on patients his supervisor feels are hopeless. As often, our biases are subconscious or preconscious. And sometimes, they are the result of ignorance or overlooking the obvious.

Do you think the order in which animals are tested makes a difference? The first animals taken from a cage tend to be chosen because they are the most active (or, if they bite and the experimenter wants to keep his fingers, the most passive). Activity is related to corticosteroids, and the corticosteroid level affects virtually all physiological parameters.

If you are not convinced that the order in which subjects are tested makes a difference, let us suppose you are one of three candidates to be interviewed for a job. Would you prefer to be interviewed first, second, or third? You'll find the winning answer in a footnote.[4]

The use of randomization in design is illustrated in the next section. See, also, Maxwell and Cole [1991].

6.5 Latin Square

The random assignment of experimental units to treatment forestalls charges of bias and ensures residual errors are exchangeable, permitting the application of either parametric or permutation methods. Nevertheless, the luck of the draw may result in confounding undesirable effects with those that are of principal interest.

Suppose we are interested in testing the effects on plant growth of three brands of fertilizer. While it might be convenient to fertilize our plots as shown in Figure 6.2a, the result could be a systematic confounding of effects, particularly if, for example, there is a gradient in dissolved minerals from east to west across the field.

The layout adopted in Figure 6.2b, obtained with the aid of a computerized random number generator, reduces but does not eliminate the effects of this hypothetical gradient. Because this layout was selected at random, the exchangeability of the error terms and, hence, the exactness of the corresponding permutation test is assured. Unfortunately, the selection of a layout like Figure 6.2a or Figure 6.2b with their built-in bias is neither more nor less probable than any of the other $\binom{9}{333}$ possibilities.

What can we do to avoid such undesirable events? In the layout of Figure 6.2c or, equivalently, Figure 6.3, known as a Latin Square, each fertilizer level occurs once and once only in each row and in each column; if there is a systematic gradient of minerals in the soil, then this layout ensures that the gradient will have almost equal impact on each of the three treatment levels. It will have an almost equal impact even if the gradient extends from northeast to southwest rather than from east to west or north to south. We use the phrase "almost equal" because a gradient effect may still persist.

The Latin Square is one of the simplest examples of an experimental design in which the statistician takes advantage of some aspect of the model to reduce the overall sample size. A Latin Square is a three-factor experiment in

---

[4] You see, you do think order makes a difference or you wouldn't have looked. The only fair way to manage the interviews, and the only way to ensure that the errors are exchangeable, is to randomize the assignment of subjects to treatment.

Hi	Med	Lo		Hi	Med	Lo		Hi	Med	Lo
Hi	Med	Lo		Lo	Lo	Med		Lo	Hi	Med
Hi	Med	Lo		Hi	Hi	Med		Med	Lo	Hi
(a)				**(b)**				**(c)**		

Fig. 6.2. (a) Systematic assignment of fertilizer levels to plots; (b) random assignment of fertilizer levels to plots; (c) Latin Square assignment of fertilizer levels to plots.

which each combination of factors occurs once, and once only. We can use a Latin Square as in Figure 6.3 to assess the effects of soil composition on crop yield:

In this diagram, Factor 1—gypsum concentration, say—increasing from left to right, Factor 2 is increasing from top to bottom (or from north to south), and Factor 3, its varying levels denoted by the capital letters A, B, and C, occurs in combination with the other two in such a way that each combination of factors—row, column, and treatment—occurs once, and once only.

Because of this latter restriction, there are only 12 different ways in which we can assign the varying factor levels to form a 3 × 3 Latin Square. Among the other 11 designs are

	1	2	3
1	A	C	B
2	B	A	C
3	C	B	A

and

	1	2	3
1	C	B	A
2	B	A	C
3	A	C	B

We assume we begin our experiment by selecting one of these 12 designs at random and planting our seeds in accordance with the indicated conditions.

Because there is only a single replication of each factor combination in a Latin Square, we cannot estimate the interactions among the factors. Thus, the Latin Square is appropriate only if we feel confident in assuming that

Factor 1

		1	2	3
Factor 2	1	A	B	C
	2	B	C	A
	3	C	A	B

Fig. 6.3. A Latin Square.

the effects of the various factors are completely additive, that is, that the interaction terms are zero.

Our model for the Latin Square is

$$X_{kji} = \mu + q_k + r_j + s_i + \epsilon_{kji}.$$

Here μ stands for an overall population mean, while s_i, r_j, q_k denote the expected deviations from this mean occasioned by the use of the ith level of Factor 1, the jth level of Factor 2, and the kth level of Factor 3. The sum of each of these deviations over all possible levels is zero. The errors (residuals) ϵ_{kji} are assumed to be exchangeable with a zero expectation.

Consider the null hypothesis that Factor 3 does not affect the yield X, that is, that each of the q_k effects is zero. If we assume an ordered alternative $K: q_1 > q_2 > q_3$, our test statistic is the Pitman correlation with $f[i] = i$,

$$R' = \sum_{i=-1}^{+1} i\bar{X}_{i..} = \bar{X}_{C..} - \bar{X}_{A..}.$$

We evaluate this test statistic both for the observed design and for each of the 12 possible Latin Square designs that might have been employed in this particular experiment. We reject the hypothesis of no treatment effect only if the test statistic for the original observations is an extreme value.

For example, suppose we employed Design 1 and observed

21	28	17
14	27	19
13	18	23

Then $3y_{A..} = 58, 3y_{B..} = 65, 3y_{C..} = 57$ and our test statistic $R' = -1$. Had we employed Design 2, then the equations would be $3y_{A..} = 71, 3y_{B..} = 49$, $3y_{C..} = 65$ and our test statistic $R' = -6$. While with Design 3, they would be $3y_{A..} = 57, 3y_{B..} = 65, 3y_{C..} = 58$ and our test statistic $R' = +1$.

We see from the permutation distribution obtained in this manner that the value of our test statistic for the design actually employed in the experiment, $R' = -1$, is an *average* value, not an *extreme* one. We accept the null hypothesis and conclude that increasing the level of Factor 3 from A to B to C does not significantly increase the yield.

6.6 Very Large Samples

When the sample sizes are very large, from several dozen to several hundred observations per group, as they often are in clinical trials, the time required to compute a permutation distribution can be prohibitive even if we are taking advantage of one of the optimal computing algorithms described in Chapter 14. Fortunately, when sample sizes are large—and we refer here to the size of the smallest subsample corresponding to a specific factor combination, not to the size of the sample as a whole—we can make use of an asymptotic approximation in place of the exact permutation distribution. A series of papers by Hoeffding [1951], Box and Anderson [1955], and Kempthorne et al. [1961] support the replacement of the permutation distribution of the F-statistic by the tabulated distribution of the F-ratio. This approximation can often be improved on if we replace the observed values by their corresponding ranks or normal scores. Further discussion of these points is provided in Section 11.3.

6.7 Sequential Analysis

With very small samples, the permutation distribution is readily calculated. But with few observations, the power of the test may well be too small and we run the risk of overlooking a treatment effect that is of practical significance. A solution in some cases is to take our observations in stages, rejecting or accepting the null hypothesis at each stage only if the p-value of the data is very large or very small. Otherwise, we continue to take more observations.

6.7.1 A Vaccine Trial

Recently, I had the opportunity to participate in the conduct of a very large-scale clinical study of a new vaccine. I'd not been part of the design team, and when I read over the protocol, I was stunned to learn that the design called for inoculating and examining 100,000 patients! 50,000 with the experimental vaccine and 50,000 controls with a harmless saline solution.

Why so many? The disease at which the vaccine was aimed was relatively rare. In essence, we would be comparing two Poisson distributions. In accordance with the arguments provided in a previous chapter (Section 4.3.3), the power of our UMPU test would depend on the total number of those inoculated who came down with the disease, as well as the relative incidence of the disease in the control and treated populations. And this last, I soon learned, was 1.002, or something equally close to unity.

The big question, as far as I was concerned, was why this experiment was being done at all. Suppose 1 in 100,000 had a fatal reaction to the vaccine? Would the small decrease in the numbers getting the original nonfatal disease justify this death? And 1 in 100,000 is an average. A much higher percentage were disabled or killed by the Salk vaccine.

I joined the study at the halfway mark, 50,000 inoculated, 50,000 to go. The good news at that point was that no one, so far, had a bad reaction to the vaccine. The bad news was that very few—treated or control—had contracted the disease. The result, as noted in Section 4.3.3, was that the power of the test, conditioned as it was on the numbers who actually got

108   6 Experimental Designs

the disease, was close to the significance level. I did some calculations and found that if the incidence of the disease did not increase during the coming year, we'd be almost guaranteed to conclude that the vaccine was without value.

So why not quit while we were ahead? We would save the US taxpayer several hundred thousands dollars, to say nothing of the servicemen who would not have to undergo the risk of vaccination. But for both small-scale and large-scale political reasons we pressed on. The balance of this section will show you from the statistician's point of view that there is a much better way.

Let us plot the results of our experiment on a piece of graph paper as in Figure 6.4a. We start in the middle of the left hand side of the graph. Each time the disease is observed in the inoculated population we move one step to the right if the individual received the saline solution and one step upward if he or she received the vaccine. The situation after eight individuals came down with the disease is depicted in the figure.

We can turn this experiment into a sequential design if we agree to stop the inoculations as soon as the line runs off the graph paper. If it goes out

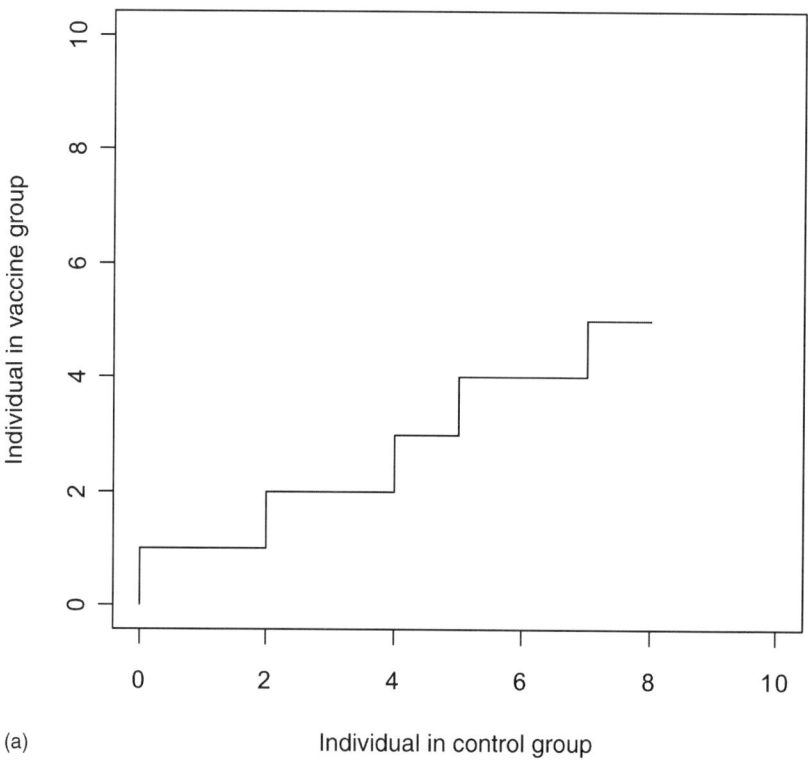

(a)

Fig. 6.4a. Plotting a sequential trial in progress.

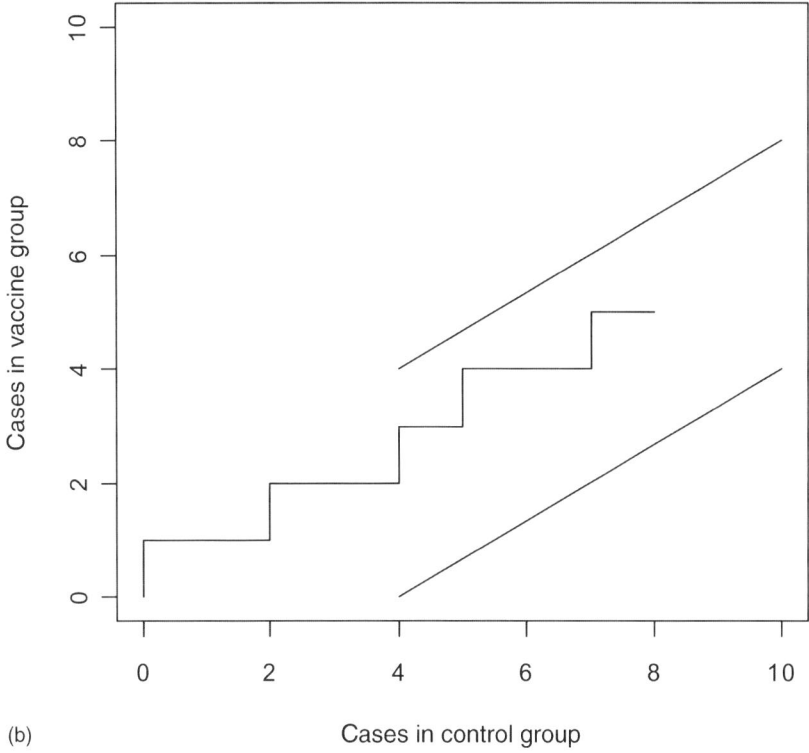

(b)

Fig. 6.4b. Sequential trial with stopping rule.

the bottom, we reject the null hypothesis; if it goes out the right side, we accept the null hypothesis, and if it goes out the top, we conclude the vaccine is dangerous.

A more reasonable set of boundaries is depicted in Figure 6.4b, where as we gather more data we are more likely to cross a boundary. To derive these boundaries, let $x_j = 0$ if the jth diseased individual is a control and $x_j = 1$ if the jth diseased individual is a treated subject. Let $p_i(x_n)$ denote the probability that the nth observation takes the value x_n; $i = 0$ when the hypothesis is true and $i = 1$ when it is false. In the case of our vaccine, $p_0(x_n) = 1/2$. Define $p_{in} = p_i(x_1) \cdots p_i(x_n)$. The boundaries are determined by the inequalities $A_0 < p_{1n}/p_{0n} < A_1$. The experiment is continued as long as the probability ratio satisfies the inequality.

It is obvious that we can chose the constants A_0 and A_1 so that we have guaranteed probabilities of rejecting the hypothesis p_0 when p_0 is true (the significance level) and of rejecting it when p_1 is true (the power). The important result is that our sequential probability ratio test will require fewer observations on the average than the equivalent fixed sample size test.

6.7.2 Determining the Boundary Values

In most instances, determining the exact values of A_0 and A_1 for specific values of α and β will be difficult. Fortunately, we can easily develop some close approximations.

Consider the subspace R_n of n observations $(x_1, \ldots, x_n)$ for which the sequential procedure stops with exactly $N = n$ observations, and subsequently rejects the hypothesis, that is $A_0 < p_{1k}/p_{0k} < A_1$ for $k < n$ and $A_1 < p_{1n}/p_{0n}$. Then

$$\alpha = \sum_{n=1}^{\infty} \int_{R_n} p_{on} \leq \frac{1}{A_1} \sum_{n=1}^{\infty} \int_{R_n} p_{1n} = \frac{\beta}{A_1}. \tag{6.8}$$

Similarly, if S_n denotes the subspace of observations in which $N = n$ and we accept the hypothesis, then

$$1 - \alpha = \sum_{n=1}^{\infty} \int_{S_n} p_{on} \geq \frac{1-\beta}{A_0}. \tag{6.9}$$

Assuming the probability that the procedure continues indefinitely is zero, we can use the inequalities (6.8) and (6.9) to approximate the boundaries, setting $A'_0 = (1-\beta)/(1-\alpha)$ and $A'_1 = \beta/\alpha$. The resultant error probabilities are $\alpha' \leq \alpha/\beta$ and $(1-\beta') \leq (1-\beta)/(1-\alpha)$.

In the case of the vaccine trials

$$\frac{p_{1n}}{p_{on}} = \frac{p_1^k(1-p_1)^{n-k}}{p_0^k(1-p_0)^{n-k}} = \left(\frac{p_1 q_0}{p_0 q_1}\right)^k \left(\frac{q_1}{q_0}\right)^n = \left(\frac{\lambda_T}{\lambda_C}\right)^k \left(\frac{2\lambda_C}{\lambda_T + \lambda_C}\right)^n,$$

where λ_T is the infection rate of those treated with the vaccine, λ_C is the infection rate of the controls, and k is the number of individuals treated with the vaccine who contracted the disease. Sequential sampling plans for such trials are widely available (see, for example, Hill [1962] and Girshick [1946]).

6.7.3 Power of a Sequential Analysis

If the vaccine is more effective than originally anticipated, we would hope the probability of detecting its value, the power of the test, would be greater than anticipated. In short, we would hope that $\beta(p) \leq \alpha$ for $p \leq 1/2$ and $\beta(p) \geq \beta$ for $p \geq p_1$. Fortunately, this is the case when testing the parameters of any member of the exponential family, like the binomial which has monotone likelihood ratio in a statistic.

Theorem 6.4. *Let $X_1, X_2, \ldots$ be independently distributed with probability density $p(x)$ and suppose the densities $p(x)$ have monotone likelihood ratio in $T(x)$. Then any sequential ratio test for testing θ_0 against $\theta_1 < \theta_0$ has a nondecreasing power function.*

Proof (after Lehmann, 1986). Let $Z_i = \log[p_1(x_i)/p_0(x_i)] = h[T_i]$ where h is non-decreasing. Our observations continue as long as the samples fall inside the band formed by the parallel straight lines $\sum_{i=1}^{n} h[t_i] = \log A_j, j = 0, 1$. Let $F[t|\theta]$ denote the cumulative distribution function T_i. By Lemma 3.2 of Section 3.6.2, if $\theta < \theta'$, then $F[t|\theta] \leq F[t|\theta']$ for all t. By Lemma 3.1, there exist a random variable V and functions f and f' such that $f(v) \leq f'(v)$ and the distributions of $f(V_i)$ and $f'(V_i)$ are F_θ and $F_{\theta'}$, respectively. The hypothesis is rejected if the path formed by the points $(1, h[t_1]), \ldots, (N, h[t_1]+\cdots+h[t_N])$ leaves the band. The probability of this event for $\theta, \beta(\theta)$ is the same when we replace T_i by $f(V_i)$, and for θ' when we replace T_i by $f'(V_i)$. Since $f(V_i) \leq f'(V_i)$ for all i, the path generated by the $f'(V_i)$ crosses the upper boundary whenever the path generated by the $f(V_i)$ does. So that $\beta(\theta) \leq \beta(\theta')$.

Note: Quality control practitioners generally reference the operating characteristic of a test, that is, the probability of making a Type II error for various values of the parameter, rather than its power function. □

6.7.4 Expected Sample Size

Let $Z_1, Z_2, \ldots$ be independent and identically distributed with $E|Z_n|$ finite. If a sequential rule is such that the total number of observations N is certain to be finite, then Wald's equation applies:

$$E(Z_1 + \cdots + Z_N) = \sum_{n=1}^{\infty} [\Pr\{N = n\} \sum_{i=1}^{n} E(Z_i|N = n)]$$

$$= \sum_{i=1}^{\infty} \sum_{n=i}^{\infty} \Pr\{N = n\} E(Z_i|N = n)$$

$$= \sum_{i=1}^{\infty} \Pr\{N \geq i\} E(Z_i|N \geq i)$$

$$= \sum_{i=1}^{\infty} \Pr\{N \geq i\} E(Z_i)$$

$$= E(N)E(Z).$$

Of course, the expected value of N depends on the parameter θ. Using the approximate boundary values established in Section 6.5.4, we see that if $E(Z|\theta) \neq 0$, $E(N|\theta)$ is approximately $\{\beta[\theta] \log A_1 + (1 - \beta[\theta]) \log A_1\}/E(Z|\theta)$.

Among all possible tests (sequential or not) for which the significance level is less than or equal to α when $\theta = \theta_0$ and the power is greater than or equal to β when $\theta = \theta_1$, and for which $E(N|\theta_0)$ and $E(N|\theta_1)$ are finite, the sequential probability ratio test with error probabilities α and $1 - \beta$ minimizes both $E(N|\theta_0)$ and $E(N|\theta_1)$ [Wald and Wolfowitz, 1948; Siegmund, 1985, pp. 19–22; Lehmann, 1986, pp. 104–110].

6.7.5 Curtailed Inspection

Typically, when a buyer such as a major wholesaler or distributor receives a large batch of items, a fixed number of the batch, say N, will be subject to detailed inspection and the entire batch rejected if c or more defectives are found. This describes a procedure called *curtailed inspection*.

Using the binomial distribution, it is easy to see that if we want to limit the probability of making a Type I error to 5% when p, the probability of an item being defective, is 2% or less, and $N=100$, we should use $c=6$ (Exercise 6.20). Then the probability of detection is 56% when p is 5% and 98% when p is 10%. Of course, if p is 10%, we are liable to exceed c well before all N items are inspected. This suggests that a sequential sampling plan, in which we stop inspecting as soon as c defectives are found, is called for instead.

Whether we used a fixed size sample or curtailed inspection, the probability of rejecting the hypothesis when it is true will be the same as the probability that the number of defectives equals or exceeds c. If a sample is rejected, the sample size ranges from c to N and the probabilities readily determined (Exercise 6.21).

We will also stop sampling and accept the batch if after inspecting $N-k+1$ items we have found no more than $c-k$ defectives, where $k=0,\ldots,c$. It is easy to see that the sample size is always less for curtailed inspection with a maximum somewhere near $p=c/N$. For small values of p, the average sample number (ASN) will be approximately $(N-c+1)/(1-p)$ and for large values of p, approximately c/p.

Combining the results of the preceding sections, we see that all sequential sampling schemes possess ASN curves with similar shapes, monotonically increasing until $\theta=\theta_0$, attaining a maximum between θ_0 and θ_1, and monotonically decreasing for $\theta > \theta_1$.

6.7.6 Restricted Sequential Sampling Schemes

Even though the *average* number of samples may be finite, the numbers required in any specific sequential trial may be prohibitive. Two types of restricted sequential sampling schemes are in common use:

- *Truncated schemes.* Trials end when a maximum of N samples are examined. This approach lends itself to comparing two treatments when either may be superior to the other. Referring once again to the graph of Figure 6.4b, we see that we can choose among Treatment 1 being superior to Treatment 2, or Treatment 2 being superior to Treatment 1, or, should the trials end at N samples without a prior decision, that neither treatment is superior.
- *One-sided stopping rules.* If testing is destructive, it's important to keep sampling to a minimum if the lot is ultimately accepted, but less so if the lot will be rejected. In clinical trials for safety, if the new treatment

has excessive side effects, we'll also want to bring the trials to a quick conclusion. Sampling schemes for such trials are described in Siegmund [1985, Chapter 3] and in the next section.

6.8 Sequentially Adaptive Treatment Allocation

6.8.1 Group Sequential Trials

So far in this text, we have passed lightly over the allocation of experimental units to treatment, writing, "Suppose there were n elements in the sample from the first population, and m from the second."

But how did we determine which n experimental units were to receive the treatment? If we were to make all the assignments at random then we would run the risk described in the section on the Latin Square, namely, that the first n animals taken from the cage would all be in the control group.

One extreme alternative, one assigning the experimental units alternately to each treatment so that we always have equal numbers in each treatment group, runs the risk that the experimenter will decode the system and his feelings color his interpretations. Standard practice therefore is to assign treatments in a groupwise fashion so that the subjects in each group receive treatment in the ratio n to m.

Such grouping can be done whether the sample size is fixed or determined sequentially. The results of the preceding sections apply whether the increment is a single subject or the entire group. (See, for example, Flehinger and Louis, 1971; Robbins and Siegmund, 1974; and Jennison and Turnbull, 2000.)

To avoid taking an excessive number of observations, the group size needs to be only a small fraction of the projected sample size. In our vaccination study, for example, randomization might be done on a day-by-day basis. Each morning, the sight of some 100 syringes will greet the paramedics, each bearing a coded label, already placed in random order on a dispensing tray.

6.8.2 Determining the Sampling Ratio

The question remains as to what fraction of the total sample should be allocated to each treatment group. Should all sample sizes be equal?

In the case of silicon implants, Dow Corning executives decided to assign all the subjects to the new treatment and none to the controls. That is, they marketed the product without bothering to do any experimentation. Later, when the treated silicon-bearing patients began to exhibit a variety of symptoms, accusations were made, lawsuits were filed, and huge judgments were awarded. Though later experiments—this time with controls, revealed that silicon implants posed no health hazard, Dow Corning had to file for bankruptcy.

My first mentor in the health field advocated that one use twice as many controls as subjects being treated in order to avoid having a promising treatment tarnished by a series of freak occurrences. In other words, "shit happens."

But what if we are treating a fatal disease—AIDS, for example—and the initial results suggest the new drug is certain to save lives. Should we continue to administer a worthless control substance to two-thirds of the patients?

The solution first proposed in a clinical setting by Zelen [1969] is to play the winner: "A success on a particular treatment generates a future trial on the same treatment with a different patient. A failure on a treatment generates a future trial on a different treatment with a different patient."

Strict adherence to such a rule would allow the experimenter to break the code, coloring his observations, so following Wei and Durham [1978], we let π_{in}, the probability of assigning the nth patient to treatment i (alternatively, the probability of assigning a patient in the mth group to treatment i), depend on the results of the first $n-1$ patients. If the new treatment is indeed the most favorable, then π_{in} converges to 1 and we terminate the trials, having exposed the least number of subjects to an inferior (and perhaps more dangerous) treatment.

Of course, in many practical situations, such as the treatment of AIDS, results aren't always immediately available. The probabilities π_{in} would then depend on the results that are available.

Wei and Durham propose the adaptive rule be represented by the following urn model when just two treatments are involved: Begin with A balls of each type in the urn. Draw balls from this urn with replacement to determine what treatment the next subject will receive. If the response to treatment k is a success, place B more balls of type k in the urn. If the treatment is a failure, place B more balls of the opposite type instead. Thus, draws from the urn gradually begin to favor the more effective treatment; the rapidity with which this occurs depends on the ratio of A and B.

6.8.3 Exact Random Allocation Tests

The play-the-winner rule was first studied in a nonclinical setting by Robbins [1952] as the one-armed bandit problem. A gambler has a choice between a fair bet at even odds or pulling the handle on a one-armed bandit. Which should he choose? The two-armed bandit problem arises when we need to decide between two treatments and either one could be preferable. (Which arm should the gambler pull?)

An obvious test statistic for comparing two treatments in the binomial case when the overall sample size is fixed is $S_n = \sum_{j=1}^{n} x_j Y_j$ where $x_j = 1$ if the jth trial results in success and $x_j = 0$ otherwise, and $Y_j = 1$ or 0 depending on which treatment the jth experimental subject is assigned. In other words, S_n is the number of successes for subjects in treatment group 1.

A network algorithm for use in obtaining the permutation distribution of S_n when employing a random urn scheme allocation is given in Chapter 14.6.1.

S_n is asymptotically normal. Using such an asymptotic approximation to the urn allocation scheme, Coad and Rosenberger [1999] show that the combination of a sequential stopping rule with adaptive treatment allocation reduces both the expected number of patients receiving the inferior treatment and the expected number of failures relative to a sequential trial with fixed equal allocation.

On the other hand, with small samples, less than 75 subjects, Stallard and Rosenberg [2002] find that the play-the-winner strategy is impractical. Using the *exact approach boundaries* developed by Lin et al. [1991], they find that (a) random allocation results in a reduction in power and (b) the additional patients required to maintain equivalent power increases the expected number of failures beyond those occurring with an equal allocation test.

6.9 Exercises

1. Let X be distributed according to a probability distribution $P_\theta, \theta \in \Omega$ and let G be a group of transformations of the sample space. If the distributions P_θ corresponding to different values of θ are distinct, show that a group of transformations G on the sample space induces a group of transformations G^* on the parameter space Ω.

2. Let $\beta_\alpha^*(\theta)$ be the supremum of the power function $\beta_\varphi(\theta)$ over all level-α tests of an hypothesis. Show (a) that if the problem of testing H is invariant under a group G, then $\beta_\alpha^*(\theta)$ is invariant under the induced group G^*; and (b) if a UMP invariant test exists it is most stringent.

3. Let $X_1, \ldots, X_n$ be independent, normally distributed as $N(\mu, \sigma^2)$. In Section 3.5.3, we developed a UMP unbiased test of the hypothesis $\sigma \geq \sigma_0$. This hypothesis remains invariant under the zero-point transformations $X_i' = X_i' + c$.

   a) Find a maximal invariant with respect to these transformations.
   b) Show that the family of normal distributions with fixed μ has monotone likelihood ratio in this statistic.
   c) Show that UMPI test coincides with the UMPU test.

4. Show that $f(k) = \sum_{\pi_j} \left(\prod_i (p_{i_j}'/q_{i_j}') \right) / \binom{n}{k}$ is an increasing function of k if $p_i/q_i > 1$.

5. Suppose it is known that $\mu_{ji} = a + b_j$ for $i = 1, \ldots, n_j; j = 1, \ldots, J$. Let $\boldsymbol{\mu}$ denote the vector $\{\mu_{1i}, \ldots, \mu_{2i}, \ldots, \mu_{Ji}, \ldots\}$. What is the dimension of the space to which $\boldsymbol{\mu}$ belongs?

6. Let $\boldsymbol{x} = (x_1, \ldots, x_n)$. Let G_1 be the set of all $n!$ permutations of $\boldsymbol{x}$. Let G_2 be the set of all continuous, strictly increasing transformations of the coordinates of $\boldsymbol{x}$ such that $x_i' = f(x_i), i = 1, \ldots, n$. Let G_3 be the group

of all orthogonal transformations of n space. Find the maximal invariants with respect to G_1, to G_2, and to G_3.

7. Let $X_1,\ldots,X_n$ be independent, normally distributed with means $\mu_1,\ldots,\mu_n$ and common variance σ^2. Define $\boldsymbol{Y} = \boldsymbol{CX}$ where C is an orthonormal matrix, that is, if c_i and c_j are any two distinct row vectors of C, $c_{i'}c_j = 0$ and $c_{i'}c_1 = 1$.

   a) Show that the $\{Yi\}$ are independent, normally distributed.
   b) Show that $\varphi^2 = \sum_{i=1}^{k} \mu_i^2/\sigma^2$ is a maximal invariant with respect to the four transformations introduced in Section 6.1.2.

8. Find variance-equalizing transformations for the Poisson and binomial distributed variables.

9. What would be the optimal statistic in the k-sample problem for testing the null hypothesis against the alternative that $\mu_1 > \mu_2$?

10. Let $X_{ij} = \mu + \alpha_i + e_{ij}, i = 1,\ldots,I, j = 1,\ldots,n_i$, where $\sum \alpha_i = 0$ and the $\{e_{ij}\}$ are exchangeable random elements. Let $F_1 = \sum_{i=1}^{I} n_i |\bar{X}_{i.} - \bar{X}_{..}|$. Show that a permutation test based on F_1 is unbiased for testing the hypothesis $\mu_i = \mu$ for all i against an alternative $\mu_i \neq \mu_j$ for some $i < j$. (Hint. First show that it is true for $I = 2$.)

11. Use both the F-ratio and Pitman correlation to analyze the data for micronucleii in Table 6.2. Explain the difference in results.

12. The following vaginal virus titres were observed in mice by H.E. Renis of the Upjohn Company 144 hours after inoculation with herpes virus type II (see Good, 1979, for complete details).

    Saline controls 10,000, 3000, 2600, 2400, 1500
    Treated with antibiotic 9000, 1700, 1100, 360, 1

    Is this a one-sample, two-sample, k-sample, or matched pairs study? Does treatment have an effect?

    Most authorities would suggest using a logarithmic transformation before analyzing these data. Repeat your analysis after taking the logarithm of each of the observations. Is there any difference? Compare your results and interpretations with those of Good [1979].

13. Using the logarithm of the viral titre, determine an approximate 90% confidence interval for the treatment effect. (Hint: Keep subtracting a constant from the logarithms of the observations on saline controls until you can no longer detect a treatment difference.)

14. Suppose you make a series of I independent pairs of observations $\{(x_i, y_i); i = 1,\ldots,I\}$. y_i might be tensile strength and x_i the percentage of some trace metal. You know from your previous work that each of the y_i has a symmetric distribution.

    a) How would you test the hypothesis that for all i, the median of y_i is x_i? (Hint: See Section 3.1.2.)
    b) Do you need to assume that the distributions of the $\{y_i\}$ all have the same shape, i.e., that they are all normal or all double exponential? Are the $\{y_i\}$ exchangeable? Are the differences $\{x_i - y_i\}$ exchangeable?

15. Rewrite the computer program in Section 6.2.1 so it will yield the permutation distributions of the three k-sample statistics, F_1, F_2, and R. Would you still accept/reject the hypothesis if you used F_2 or R in place of F_1?
16. Design an experiment.
    a) List all the factors that might influence the outcome of your experiment.
    b) Write a model in terms of these factors.
    c) Which factors are under your control?
    d) Which of these factors will you use to restrict the scope of the experiment?
    e) Which of these factors will you use to block?
    f) Which of the remaining factors will you neglect initially, that is, lump into your error term?
    g) How will you deal with each of the remaining covariates? By correction? By blocking after the fact?
    h) How many subjects will you observe in each subcategory?
    i) Is the subject the correct experimental unit?
    j) Write out two of the possible assignments of subjects to treatment.
    k) How many possible assignments are there in all?
17. Without thinking through the implications, you analyze your data from matched pairs as if you had two independent samples and obtain a significant result. Consequently, you decide not to waste time analyzing the data correctly. Is your decision right or wrong? (Hint: Were the results within each matched pair correlated? What if one but not both of the observations in a matched pair were missing?)
18. In Chapter 3, we proposed the use of the sum of the cross-products $\sum x_i y_i$ for testing the hypothesis that X and Y are independent against the alternative that $F[X|Y = b] \leq F[X|Y = a]$ if $a < b$. Show that this test is UMP invariant with respect to the group of all transformations $X_i' = aX_i + b, Y_i' = cY_i + d$ for which $a > 0$ and $c > 0$.
19. In order to derive Wald's equation, why must the expected value of $E|Z_n|$ be finite? Why does $\sum_{i=1}^{\infty} \Pr\{N \geq i\} E(Z_i) = E(N)E(Z)$.
20. Set up an inspection plan for detecting bad batches. The probability should be no greater than 10% of rejecting a batch with no more than 3% defective and no less than 90% of rejecting a batch with 10% or more defective. Complete the computations for the curtailed inspection plan and determine what advantage, if any, results from the use of curtailed rather than total inspection when $p = 5\%, 10\%$, and 20%.
21. a) Set up a fixed sample inspection plan for comparing two binomials. The hypothesis is that $p_0 = p_1 = 0.5$ versus the alternative that $p_1 = 0.6$. $\beta(0.5) = 10\%$ and $\beta(0.6) = 70\%$. What is the sample size? What is the rejection region? What is $\beta(0.7)$?
    b) Set up a sequential sampling plan for the same problem. What is $\beta(0.7)$? What is the expected sample size when $p = 0.5, 0.6$, and 0.7?

22. Use the requirement of invariance under a group of transformations to find a set of confidence bounds on an unknown continuous cumulative distribution function. Note that the resultant bounds can be used to test the hypothesis of *goodness-of-fit*, that is, that the unknown distribution F is actually that of some known distribution F_0. Is there a UMP test? See Tallis [1983].
23. An alternative to using a one-sample test in the case of matched pairs is to use a two-sample comparison restricting rearrangements to the exchange of values within a pair. In either instance, n pairs yield 2^n possible rearrangements. Which approach leads to the more powerful test?

7
Multifactor Designs

The analysis of randomized blocks can be generalized to very complex experimental designs with multiple control variables and confounded effects. In this chapter, we consider the evaluation of main effects and interactions via synchronized rearrangements and the analysis of variance. We also study the analysis of covariance and the analysis of unbalanced designs via a combination of bootstrap and permutation methods.

7.1 Multifactor Models

What distinguishes the complex experimental design from the simple one-sample, two-sample, and k-sample experiments we have considered so far is the presence of multiple control factors. For example, we may want to assess the simultaneous effects on crop yield of hours of sunlight and rainfall. We determine to observe the crop yield X_{ijm} for I different levels of sunlight, $i = 1, \ldots, I$, and J different levels of rainfall, $j = 1, \ldots, J$, and to make M observations at each factor combination, $m = 1, \ldots, M$. We adopt as our model relating the dependent variable crop yield (the *effect*) to the independent variables of sunlight and rainfall (the *causes*)

$$X_{ijm} = \mu + s_i + r_j + (sr)_{ij} + \epsilon_{ijm}. \tag{7.1}$$

In this model, terms with a single subscript like s_i, the effect of sunlight, are called *main effects*. Terms with multiple subscripts like $(sr)_{ij}$, the residual and nonadditive effect of sunlight and rainfall, are called *interactions*. The $\{\epsilon_{ijm}\}$ represent that portion of crop yield that cannot be explained by the independent variables alone; these are variously termed the *residuals*, the *errors*, or the *model errors*. To ensure the residuals are exchangeable so that permutation methods can be applied, the experimental units must be assigned at random to treatment.

If we wanted to assess the simultaneous effect on crop yield of three factors simultaneously—sunlight, rainfall, and fertilizer, say—we would observe the crop yield X_{ijkm} for I different levels of sunlight, J different levels of rainfall, and K different levels of fertilizer, $k = 1, \ldots, K$, and make n_{ijk} observations at each factor combination. Our model would then be

$$X_{ijkm} = \mu + s_i + r_j + f_k + (sr)_{ij} + (sf)_{ik} + (rf)_{jk} + (srf)_{ijk} + \epsilon_{ijkm}. \quad (7.2)$$

In this model we have three main effects, s_i, r_j and f_k, three first-order interactions, $(sr)_{ij}, (sf)_{ik}$, and $(rf)_{jk}$, a single second-order interaction, $(srf)_{ijk}$, and the error term, ϵ_{ijkm}.

Including the additive constant μ in the model allows us to define all main effects and interactions so they sum to zero; thus $\sum s_i = 0, \sum r_j = 0$, $\sum_i (sr)_{ij} = 0$ for $j = 1, \ldots, J$, $\sum_j (sr)_{ij} = 0$ for $i = 1, \ldots, I$, and so forth. Under the hypothesis of no nonzero interactions, the expected effect of the joint presence of the two factors s_i and r_j is the sum $s_i + r_j$. Under the hypothesis of "no effect of sunlight on crop yield," each of the main effects are equal, that is $s_1 = \cdots = s_I = 0$. Under one alternative to this hypothesis, the different terms s_i represent deviations from a zero average.

Clearly, when we have multiple factors, we must also have multiple test statistics. In the preceding example, we require three separate tests and test statistics for the three main effects of sunlight, rainfall, and fertilizer, plus four other statistical tests for the three first-order and the one second-order interactions. Will we be able to find statistics that measure a single intended effect without confounding it with a second unrelated effect? Will the several p-values be independent of one another?

7.2 Analysis of Variance

The analysis of variance (ANOVA) relies on the decomposition (that is, the analysis) of the sum of squares of a set of observations about their grand mean into a series of sums. For example in the two-way ANOVA,

$$\sum_i \sum_j \sum_k (X_{ijk} - \bar{X}_{...})^2 = \sum_i n_{i.}(\bar{X}_{i..} - \bar{X}_{...})^2 + \sum_j n_{.j}(\bar{X}_{.j.} - \bar{X}_{...})^2$$
$$+ \sum_i \sum_j n_{ij}(\bar{X}_{ij.} - \bar{X}_{i..} - \bar{X}_{.j.} + \bar{X}_{...})^2$$
$$+ \sum_i \sum_j \sum_k (X_{ijk} - \bar{X}_{ij.})^2.$$

Associated with the model of Equation (7.1) are a set of univariate linear hypotheses concerning the main effects $\{s_i\}$ and $\{r_j\}$ and the interactions $(sr)_{ij}$.

7.2 Analysis of Variance

The methods of Section 6.2.1 lead us to adopting as test statistics

$$W_I = \frac{nJ\sum_i (\bar{X}_{i..} - \bar{X}_{...})/(I-1)}{\sum_i \sum_j \sum_k (X_{ijk} - \bar{X}_{ij.})^2/IJ(n-1)},$$

$$W_J = \frac{nI\sum_i (\bar{X}_{.j.} - \bar{X}_{...})/(J-1)}{\sum_i \sum_j \sum_k (X_{ijk} - \bar{X}_{ij.})^2/IJ(n-1)},$$

and

$$W_{IJ} = \frac{\sum_i \sum_j n_{ij}(\bar{X}_{ij.} - \bar{X}_{i..} - \bar{X}_{.j.} + \bar{X}_{...})^2 (I-1)(J-1)}{\sum_i \sum_j \sum_k (X_{ijk} - \bar{X}_{ij.})^2/IJ(n-1)}.$$

To show that these three statistics lead to independent tests, we need to trace our steps through Section 6.2.1 in reverse to show that we may write

$$Y = CX,$$

such that the first $IJ(n-1)$ row vectors of C span Ω, the first $I-1$ row vectors of C span the subspace under the hypothesis that all the effects $\{s_i\}$ are zero, the next $J-1$ row vectors of C span the subspace under the hypothesis that all the effects $\{r_j\}$ are zero, and the next $(I-1)(J-1)$ span the subspace under the hypothesis that all the interactions $(sr)_{ij}$ are zero (Exercise 7.2).

If the design is balanced, that is, if $n_{ij} = n$ for all i and j, and if the error terms are independent, identically normally distributed, then W_I, W_J, W_{IJ} will be independently distributed and have the F-distribution with $I-1$, $IJ(n-1)$; $J-1$, $IJ(n-1)$, and $(I-1)(J-1)$, $IJ(n-1)$ degrees of freedom, respectively; see, for example, Lehmann [1986; pp 392–396].

As shown in Section 6.2.1, tests based on these statistics will be UMP among tests that are invariant with respect to transformations in zero-point and scale against alternatives in which one or more of the effects are nonzero. If the observations X_i are independent and drawn from distributions which differ only by a shift, that is, $F[x - \mu_i]$ where F is an arbitrary distribution with finite variance σ^2, then for large samples the size and power of these tests is the same as it would be if the observations had been independent and normally distributed. This is because under the stated conditions, the within-cells sum of squares in the denominator of the test statistic divided by its degrees of freedom tends in probability to σ^2 as the sample size grows larger, and the between-samples sum of squares in the numerator divided by its degrees of freedom and by σ^2 tends to a chi-square distribution.

The determining quantity in the convergence of the numerator is not the aggregate sample size N, but the number n of observations associated with each combination of factors (see, for example, Lehmann, 1986, p. 376). In the example of the two-way ANOVA, $n = N/(IJ)$. In many studies (clinical trials are one example), although the number of observations N may be quite large, n is small because of the large number of factors involved.

122   7 Multifactor Designs

In a recent series of simulations aimed at studying the small-sample properties of the analysis of variance when applied to multifactor designs, the results were different for balanced and unbalanced designs.

When the designs were balanced and the observations drawn from a mixture of normal distributions, even with sample sizes as small as 3 or 4 observations per cell, the significance levels provided by the assumption of normality for the analysis of variance were accurate to within the level of precision provided by 10,000 simulations.

When working with unbalanced designs, the simulated distributions could be divided into two categories: those in which the various distributions in the mixture were sufficiently close so the result resembled a skewed normal as in Figure 7.1; and those in which the modes were clearly differentiated, as in Figure 7.2.

In the case that gave rise to Figure 7.1, a mixture of three distributions, $N(0, 1)$, $N(2, 4)$, and $N(0, 3)$ in the proportions of 50%, 40%, and 10%, respectively, the interaction term in the 2×3 design was significant in just 499 cases out of 10,000 at the 5% level. The result was unaffected when a row effect of 1 was introduced, demonstrating that the tests provided by the analysis of variance for the interaction and row effects remain independent even when the data are not drawn from a normal distribution.

Repeating the analysis, but with an unbalanced design in which only 3 observations were in cell 23, the interaction term was significant in 491 cases out of 10,000 at the 5% level.

With only 3 observations in each of cells 22 and 23, and no row effect, the interaction term was found to be significant in 496 cases out of 10,000 at the

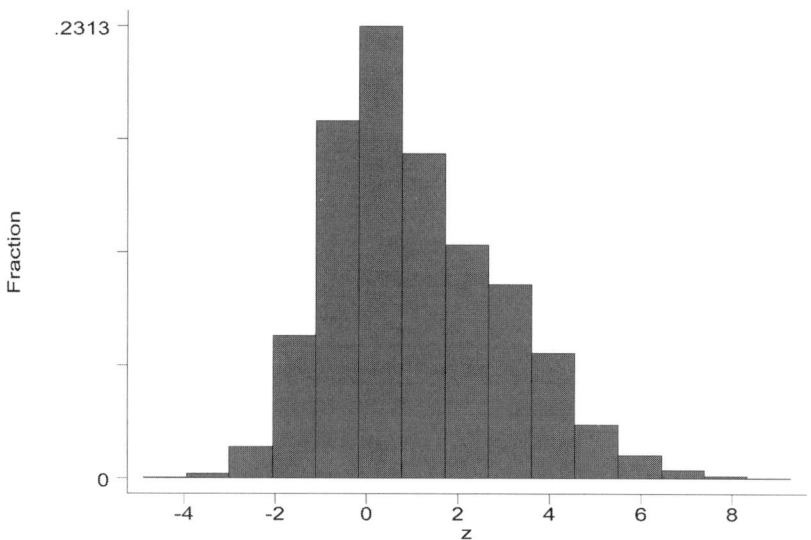

Fig. 7.1. A mixture of three normal distributions. 10,000 observations.

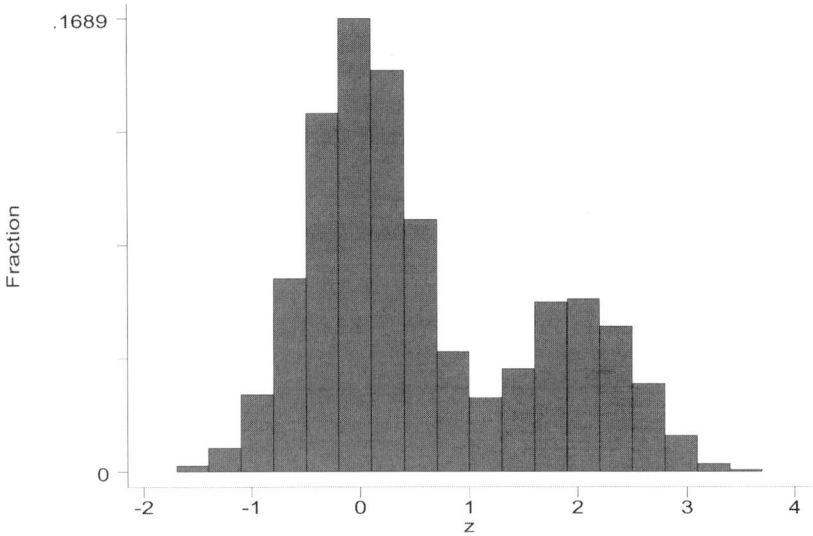

Fig. 7.2. A mixture of two well-separated normal distributions. 10,000 observations.

5% level, the row effect in 532 cases, and the column effect in 538 cases, all three within the limits of precision about 5% provided by the simulation.

The distribution depicted in Figure 7.1 has mean 1.097 and standard deviation 1.871. The probability of detecting a row effect of 1 using ANOVA was the same, approximately 24%, whether the data were drawn from an $N(1.097, 1.871^2)$ distribution or from the mixture of distributions depicted in the figure. It would appear that the power as well as the significance level of ANOVA is robust when distributions are of this type.

In the case that gave rise to Figure 7.2, a mixture of two normal distributions, $N(0.5, 1)$ and $N(2, 1)$ in the proportions of 70% and 30%, the results were quite different for the balanced and unbalanced 2×3 designs with 24 and 23 observations, respectively. As seen in Table 7.1, the omission of even a single observation dramatically alters the significance levels. When the stated significance level was 5%, the actual frequency of Type I error was 7%. When

Table 7.1. ANOVA results of 10,000 simulations with a mixture of normal distributions in a 2×3 design.

	5%		1%	
	Balanced	Unbalanced	Balanced	Unbalanced
Rows		523		92
Columns	525	708	129	213
$R \times C$	516	759	109	243

the stated significance level was 1%, the actual frequency of Type I error was more than twice that expected.

Note that in this instance, as with any $2 \times K$ experimental design, the p-value for rows is left unaffected as it is based on the square of the relatively robust t-statistic.

As in the k-sample, one-factor case analyzed in the preceding chapter, the analysis of variance is optimal only for a mean square loss function. A further limitation in the multifactor $2 \times K \times \cdots$ case is that only two-sided tests for rows are possible even though there are only two rows to be compared.

7.3 Permutation Methods: Main Effects

In a k-way analysis with equal sample sizes M in each category, we can assess the main effects by permutation means using essentially the same statistics we would use for randomized blocks. Take sunlight in the three-factor example of Equation (7.2). If we have only two levels of sunlight, then, as shown in Section 3.6.2, our test statistic for the effect of sunlight is

$$t = \sum_j \sum_k \sum_m X_{1jkm}. \tag{7.3}$$

If we have more than two levels of sunlight, and a mean square loss function, our test statistic is

$$F_2 = \sum_i \sum_j \sum_k \left(\sum_m X_{ijkm}/n_{ijk} \right)^2. \tag{7.4}$$

If we have a linear loss function, our test statistic would be

$$F_1 = \sum_i \sum_j \sum_k \left| \sum_m X_{ijkm}/n_{ijk} \right| \tag{7.5}$$

The statistics F_2 and F_1 offer protection against a broad variety of shift alternatives including

$$K1 : s1 = s2 > s3 = \cdots$$
$$K2 : s1 > s2 > s3 = \cdots$$
$$K3 : s1 < s2 > s3 = \cdots$$

As a result, they may not provide a most powerful test for any single one of these alternatives. If we believe the effect to be monotone increasing, then, in line with the thinking detailed in Section 6.3.2, we would use the Pitman correlation statistic

$$R = \sum_i \sum_j \sum_k g[i] \sum_m X_{ijkm}/n_{ijk} \tag{7.6}$$

To obtain the rearrangement distributions of the test statistics t, F_2, F_1, and R, we permute the observations independently in each of the JK blocks determined by a specific combination of rainfall and fertilizer. Exchanging observations within a category corresponding to a specific level of sunlight leaves the statistics t, F_2, F_1, and R unchanged. We can concentrate on exchanges between categories and within blocks as described in Section 6.4.2.

We compute the test statistic ($t, F_1,$ or R) for each rearrangement, rejecting the hypothesis that sunlight has no effect on crop yield only if the value of t (or F_1 or R) that we obtain using the original arrangement of the observations lies among the α most extreme of these values.

A third alternative to F_1 and F_2 is

$$F_3 = \sum_i \sum_j \sum_k \frac{n_{i..}(n_{i..}-1)(\bar{X}_{i...} - \bar{X}_{....})^2}{\sum_m (X_{ijkm} - \bar{X}_{ijk.})^2} \quad (7.7)$$

due to James [1951], which Hall [1989] recommends for use with the bootstrap when we cannot be certain that the observations in the various categories all have the same variance. In simulation studies with k-sample permutation tests and variances that differed by an order of magnitude, we found F_3 was inferior to F_2.

A final alternative to the statistics F_1 and F_2 is the standard F-ratio statistic

$$F = \frac{\sum_i n_{i..}(\bar{X}_{i...} - \bar{X}_{....})^2/(I-1)}{\sum_i \sum_j \sum_k \sum_m (X_{ijkm} - \bar{X}_{ijk.})^2/(N-IJK)}. \quad (7.8)$$

But F reduces to F_2 on removing factors that are invariant under exchanges between blocks; determining its rearrangement distribution would entail unnecessary and redundant computations.

7.3.1 An Example

In this section, we apply the permutation method to determine the main effects of sunlight and fertilizer on crop yield using the data from the two-factor experiment depicted in Table 7.2a. As there are only two levels of sunlight in this experiment, we use t, Equation (7.3), to test for the main effect. For the original observations, $t = 23 + 55 + 75 = 153$. One possible rearrangement is shown in Table 7.2b, in which we have interchanged the two observations marked with an asterisk, the 5 and 7. The new value of t is 154.

As can be seen by a continuing series of straightforward hand calculations, the test statistic t for the main effect of sunlight is as small or smaller than it is for the original observations in only 8 out of the $\binom{6}{3}^3 = 8000$ possible rearrangements. For example, it is smaller when we swap the 9 of the Hi–Lo group for the 10 of the Lo–Lo group (the two observations marked with the

126 7 Multifactor Designs

Table 7.2a. Effect of sunlight and fertilizer on crop yield.

	Fertilizer		
Sunlight	LO	MED	HI
LO	5	15	21
	10	22	29
	8	18	25
HI	6	25	55
	9	32	60
	12	40	48

Table 7.2b. Effect of sunlight and fertilizer. Data rearranged.

	LO	MED	HI
LO	6*	15	21
	10#	22	29
	8	18	25
+03 HI	5*	25	55
	9#	32	60
	12	40	48

pound sign). As a result, we conclude that the effect of sunlight is statistically significant.

The computations for the main effect of fertilizer are more complicated—we must examine $\binom{9}{3\ 3\ 3}^2$ rearrangements, and compute the statistic F_1 for each. We use F_1 rather than R because of the possibility that too much fertilizer—the "HI" level—might actually suppress growth. Only a computer can do this many calculations quickly and correctly, so we adapted our program from Section 6.3 to pack the data into a two-dimensional array in which each row corresponds to a block. The estimated significance level is 0.001 and we conclude that this main effect, too, is statistically significant.

7.4 Permutation Methods: Interactions

In the preceding analysis of main effects, we assumed the effect of sunlight was the same regardless of the levels of the other factors. If it is not, if sunlight and fertilizer interact in a nonadditive fashion, then we shall have to test for and report the effects of sunlight separately for each level of fertilizer.

To test the hypothesis of no interaction, we first eliminate row and column effects by subtracting the row and column means from the original observations. That is, we set $X'_{ijk} = X_{ijk} - \bar{X}_{i..} - \bar{X}_{.j.} + \bar{X}_{....}$, where, by adding the

Table 7.3. Effect of sunlight and fertilizer on crop yield. Testing for nonadditive interaction.

Sunlight	Fertilizer		
	LO	MED	HI
LO	4.1	−2.1	−11.2
	9.1	4.1	−3.2
	7.1	0.1	−7.2
HI	−9.8	−7.7	7.8
	−7.8	−0.7	12.8
	−3.8	7.2	0.8

grand mean $\bar{X}_{...}$, we ensure the overall sum will be zero. In the example of the effect of sunlight and fertilizer on crop yield we are left with the residuals shown in Table 7.3.

The pattern of plus and minus signs in this table of residuals suggests that fertilizer and sunlight affect crop yield in a superadditive fashion. Note the minus signs associated with the mismatched combinations of a high level of sunlight and a low level of fertilizer and a low level of sunlight with a high level of fertilizer. To encapsulate our intuition in numeric form, we sum the deviates within each cell, square the sum, and then sum the squares to form the test statistic

$$W_{IJ} = \sum_i \sum_j \left(\sum_k X'_{ijk} \right)^2.$$

We compute this test statistic for each rerandomization of the 18 deviates into 6 subsamples. In most cases, the values of the test statistic are close to zero as the entries in each cell cancel. The value of the test statistic for our original data, $I = 2127.8$, stands out as an exceptional value, but how are we to interpret it?

Recall that $X'_{ijk} = X_{ijk} - \bar{X}_{i..} - \bar{X}_{.j.} + \bar{X}_{...}$ or, in terms of our original linear model, that $X'_{ijk} = \epsilon_{ijk} - \bar{\epsilon}_{i..} - \bar{\epsilon}_{.j.} + \bar{\epsilon}_{...}$. But this means that two residuals in the same row, such as X'_{i11} and X'_{i23}, will be correlated while two observations from different rows and columns will not. The residuals are not exchangeable, the arguments of Section 3.6.2 do not apply, and a test based on the distribution of W_{IJ} with respect to *all* possible permutations will not be exact.

7.5 Synchronized Rearrangements

Recently, the class of experimental designs that are analyzable by permutation means to yield exact significance levels was extended to the two-factor case through the use of synchronized rearrangements [Salmaso, 2003; Pesarin, 2001]. Using the concept of weak exchangeability, Good [2003] obtained similar results, albeit employing a different test statistic. Here we show how the results of these authors can be easily extended to general multifactor designs via the group-theoretic concept of similarities. We also show in contrast to the results of Pesarin [2001] that we remain free to select the test statistic best suited to the problem at hand.

First, we show how the class of exact tests can be extended via weak exchangeability. We define synchronous rearrangements in terms of exchanges. We define a semimetric on the set of finite sums of design observations and show how similarities yield exact unbiased tests for experimental designs of any number of factors.

7.5.1 Exchangeable and Weakly Exchangeable Variables

A *rearrangement* is a permutation of a set of values in which elements are exchanged among predesignated subsets. Thus the group of permutations P of a combined sample can be divided into equivalence classes corresponding to distinct rearrangements among the samples.

A set of observations is said to be *exchangeable*, if their joint distribution is invariant under *all* rearrangements of their subscripts. A set of observations is said to be *weakly exchangeable* if their joint distribution is invariant with respect to *some nonempty subset* R of the rearrangements of their subscripts, $R \subseteq P$ [Good, 2002].

As was described in Chapter 1, obtaining a test of a hypothesis via rearrangements requires four steps:

1. A test statistic is chosen. Normally, this would be one that, based on some set of predetermined criteria, best discriminates between the primary hypothesis and the alternative hypothesis.
2. The value of this statistic is determined for the set of observations as they were originally, that is, prior to any rearrangement of their labels.
3. A rearrangement distribution is generated by computing the value of the test statistic for each rearrangement $\Pi \in R$.
4. The value of the statistic obtained at step 2 is compared with the set of possible values generated at step 3. If the original value of the test statistic lies in the tail(s) of the rearrangement distribution favoring the alternative hypothesis, the primary hypothesis is rejected.

If a set of observations is weakly exchangeable with respect to R under a null hypothesis H, one can obtain an exact test of H by computing the rearrangement distribution of a statistic S over R. (See Exercise 7.7.)

7.5.2 Two Factors

Let the set of observations $\{X_{ijk}\}$ in a two-factor experimental design be thought of in terms of a rectangular lattice L with K colored, patterned balls at each vertex. All the balls in the same column have the same color initially, a color that is distinct from the color of the balls in any other column. All the balls in the same row have the same pattern initially, a pattern that is distinct from the pattern of the balls in any other row. See, for example, Figure 7.3a.

Let $\boldsymbol{P}$ denote the set of rearrangements that preserve the number of balls at each row and column of the lattice. $\boldsymbol{P}$ is a group.

Let $\boldsymbol{P_R}$ denote the set of rearrangements of balls among rows which (a) preserve the number of balls at each row and column of the lattice, and (b) result in the numbers of each color within each column being the same in each row. The set of all rearrangements generated by the set $\boldsymbol{P_R}$ is a subgroup of $\boldsymbol{P}$.

In a $2 \times J$ balanced design with K observations in each category, there would be exactly 1 rearrangement (the identity) with no exchanges in $\boldsymbol{P_R}$ and 1 with all K elements exchanged in each column between rows. There would be $\binom{K}{1}^{2J}$ rearrangements with a single exchange of elements in each column between rows for a total of $\sum_{k=1}^{K} \binom{K}{k}^{2J}$ rearrangements in $\boldsymbol{P_R}$ in all.

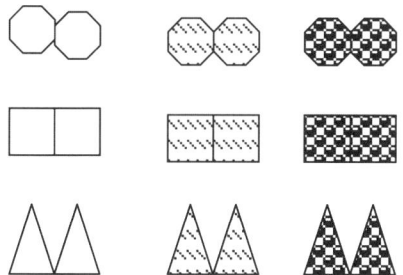

Fig. 7.3a. Part of a two-factor experimental design.

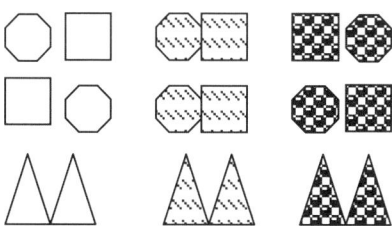

Fig. 7.3b. The same design after a synchronized exchange of elements between the first and second rows. $\boldsymbol{P}$ is in $\boldsymbol{P_R}$.

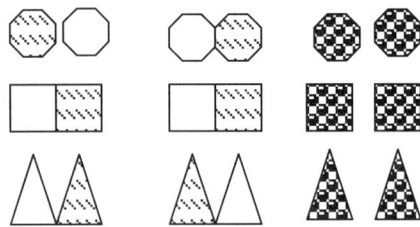

Fig. 7.3c. The same design after a synchronized exchange of elements between the first and second columns. P is in P_C.

Let P_C denote the set of rearrangements of balls among columns which (a) preserve the number of balls at each row and column of the lattice, and (b) result in the numbers of each pattern within each row being the same in each column. The set of all rearrangements generated by the set P_C is a subgroup of P.

In a $2 \times J$ balanced design with K observations in each category, there would be exactly 1 rearrangement (the identity) with no exchanges in P_C and 1 with all K elements exchanged in each row between columns. There would be $\binom{K}{1}^{2J}$ rearrangements with a single exchange of elements in each row between columns for a total of $\sum_{k=1}^{K} \binom{K}{k}^{2J}$ rearrangements in P_C in all.

Let P_{RC} denote the set of exchanges of balls that preserve the number of balls at each row and column of the lattice, and result in (a) an exchange of balls between both rows and columns (or no exchange at all), (b) the numbers of each color within each column being the same in each row, and (c) the numbers of each pattern within each row being the same in each column.

As demonstrated in Figures 7.3d and 7.3e, the rearrangements in P_{RC} are the result of some but not all successive exchanges involving an element of P_R and an element of P_C.

$$P_{RC} \cap P_R = P_{RC} \cap P_C = P_R \cap P_C = I.$$

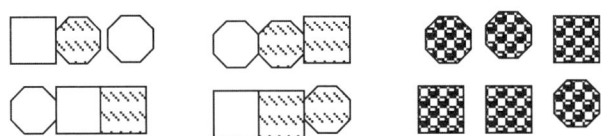

Fig. 7.3d. In this design synchronized exchanges of elements have taken place between the first and second rows and the first and second columns. All requirements (a), (b), and (c) are satisfied so that this rearrangement is in P_{RC}.

7.5 Synchronized Rearrangements 131

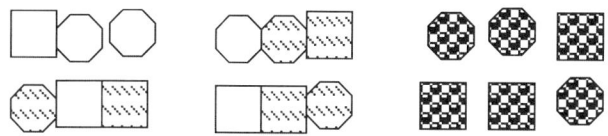

Fig. 7.3e. In this design successive synchronized exchanges of elements have taken place. First, between the first and second rows and then, between the first and second columns. All requirements (a), (b), and (c) are *not* satisfied so that this rearrangement is *not* in $\boldsymbol{P_{RC}}$.

Define $X_{ijk} = Z_{ijk} + \Delta_{ij}$ where

$$\Delta_{ij} = \mu + \alpha_i + \beta_j + \gamma_{ij}, \quad \sum \alpha_i = \sum \beta_j = \sum_i \gamma_{ij} = \sum_j \gamma_{ij} = 0,$$

and $E(Z_{ijk}) = 0$. Let

$$p[\Delta; X] = \prod_i \prod_j \prod_k f[x_{ijk} - \Delta_{ij}]$$

where f is a density function that is continuous almost everywhere, that is, f is the density function of each of the identically distributed Z_{ijk}.

Without loss of generality, we may assume $\mu = 0$ or, equivalently, we may work with the set of observations $\{X'_{ijk}\}$ obtained by subtracting μ from each element of $\{X_{ijk}\}$. Suppose, now, the hypothesis H_1: $\alpha_i = 0$ for all i holds. Then the joint distribution of the vector $(x_{i1k'}, x_{i2k''}, \ldots, x_{ijk^*})$ obtained by taking an arbitrary element from each column of the ith row is identical with the joint distribution of $(z - \beta_1 - \gamma_{i1}, z - \beta_2 - \gamma_{i2}, \ldots, z - \beta_J - \gamma_{iJ})$ where f is the probability density of z. The probability density of the sum of these latter elements is identical with the probability density of $nz - \sum_{j=1}^{J} \beta_j - \sum_{j=1}^{J} \gamma_{ij} = nz$; that is, $f(z/n)$.

Under H_1,

- f is the probability density of the mean of each of the rows of X.
- Applying any of the elements of $\boldsymbol{P_R}$ leaves this density unchanged.
- Applying any of the elements of $\boldsymbol{P_R}$ leaves the density of the test statistic $F_2 = \sum_i \left(\sum_j \sum_k x_{ijk} \right)^2$ unchanged.[1]

Similarly, to test H_2, we may use the rearrangement distribution over $\boldsymbol{P_C}$ of any of the statistics $F_2 = \sum_j (\sum_i \sum_k x_{ijk})^2$, $F_1 = \sum_j |\sum_i \sum_k x_{ijk}|$, or $R_2 = \sum_j g[j] \sum_i \sum_k x_{ijk}$, where $g[j]$ is a monotone function of j. Note that we may obtain one-sided, as well as two-sided, tests when there are only two rows or two columns, by focusing, for example, on the statistic $(\sum_i \sum_k x_{i2k})^2$.

If $q \in \boldsymbol{P_R}$ and $s \in \boldsymbol{P_C}$, then under H_3, $S_{ij} = \sum_k x_{ijk} - x_{i..}/I - x_{.j.}/J$ is[2] invariant with respect to $p = qt \in \boldsymbol{P_{RC}}$, and, by induction, applying any of

---

[1] That is, as well as the densities of any of the other test statistics for ordered and unordered effects proposed in Chapter 6.
[2] Here and in similar expressions, the use of a dot as in $x_{.i.}$ denotes summation over the missing subscript(s).

the elements of $\boldsymbol{P_{RC}}$ leaves the density of the test statistic $S = \sum_i \sum_j (S_{ij})^2$ unchanged. As only the identity $\boldsymbol{I}$ is common to the corresponding rearrangement sets, the permutation tests[3] of the three hypotheses are independent of one another.

Alternatively, to test H_3 in a 2^2 factorial design, it is easy to see that the statistic $S = x_{11.} - x_{12.} - x_{21.} + x_{22.}$ is invariant with respect to $\boldsymbol{P_{RC}}$ providing H_3 is true, and that under an alternative the statistic S for the original nonrearranged observations is more likely to be an extreme value of the rearrangement distribution over $\boldsymbol{P_{RC}}$.

As an example, suppose we are studying the effects of two factors each at two levels and have taken only a single replication. Then $\boldsymbol{P_R}$, $\boldsymbol{P_C}$, and $\boldsymbol{P_{RC}}$ each contain only two rearrangements. If we have made K replications, then $\boldsymbol{P_R}$, $\boldsymbol{P_C}$, and $\boldsymbol{P_{RC}}$ will each contain $K^4 + 2$ of the $\binom{4K}{K\ K\ K}$ possible rearrangements.

Suppose there are differences in the time and location at which each replication was made (see Section 6.4.2). The set $\boldsymbol{P_R}$ would be limited to row-by-row exchanges where all the elements exchanged come from the same replication. In our example of a 2^2 factorial, there would be $K(K-1)$ possible single exchanges, $K(K-1)(K-1)(K-2)$ possible double exchanges, and so forth. The test statistics would be the same as in our original example.

7.5.3 Three or More Factors

We define an M-factor *experimental design* as an M-dimensional lattice $\boldsymbol{L}$ such that with each element of the lattice $\boldsymbol{s} = \{i_1, i_2, \ldots, i_M\}$, $i_1 = 1, \ldots, I_1$; $\ldots, i_M = 1, \ldots, I_M$, is associated a set of independent random variables $X_k(\boldsymbol{s}), k = 1, \ldots, n_s$ distributed as $F[x - \Delta_s]$ where the following conditions hold:

$$\Delta_s = EX_k(\boldsymbol{s}) = \mu + \sum_{m=1}^{M} \alpha_{i_m} + \sum_{m=1}^{M} \sum_{j=1}^{m-1} \beta_{i_j i_m} + \cdots ; \qquad (7.9)$$

$$\sum_{i_m=1}^{I_m} \alpha_{i_m} = 0 \quad \text{for all } m; \qquad (7.10)$$

$$\sum_{i_j=1}^{I_j} \beta_{i_j i_m} = \sum_{i_m=1}^{I_m} \beta_{i_j i_m} = 0 \quad \text{for all } m \text{ and } j; \qquad (7.11)$$

and so forth.

If $n_s = n$ for all s, we term the design *balanced*.

---

[3] Strictly speaking, one ought write rearrangement test, but "permutation test" is already in common usage.

A well-known result from group theory is that every rearrangement can be constructed from a succession of pairwise exchanges. Thus, one way to convert $\{A, B, C\}$ to $\{B, C, A\}$ is to apply the changes $(1,3)(1,2)$ in that order. The pairwise exchanges we will be concerned with are those in which an observation in one cell of an experimental design is swapped with an observation in another cell at the same location in the design with the exception of its lth coordinate. For example, in a two-factor design, such an exchange might be between two cells in the same row but different columns. We let $(l; l_1, l_2; k_1, k_2; s)$ denote a pairwise exchange in which we swap the k_1th observation at $\{i_1, \ldots, i_{l-1}, l1, \ldots, i_M\}$ with the k_2th observation at $\{i_1, \ldots, i_{l-1}, l_2, \ldots, i_M\}$.

We let $(t : l; l_1, l_2; s)$ denote an exchange in which t such pairwise exchanges take place simultaneously between the same pair of cells. In a balanced design, for each fixed pair of cells $(l; l_1, l_2)$ there are exactly $\binom{n}{t}^2$ such exchanges.

Let r_l denote the vector $\{i_1, \ldots, i_{l-1}, i_{l+1}, \ldots, i_M\}$ and reorder the coordinates of s so that $s = \{l, r_l\}$. A synchronized pairwise rearrangement used for testing main effects, for example,

$$H_l\colon \alpha_{l_j} = 0 \text{ for } l_j = 1, \ldots, I_l$$

has the form

$$(t\colon l; l_1, l_2) = \prod_{r_l}(t : l; l_1, l_2; \{l, r_l\}),$$

that is, any exchanges of observations between levels $l1$ and $l2$ of the lth factor take place in synchrony at all levels of all the remaining factors.

A synchronized pairwise rearrangement used for testing the interaction between factors l and j has the form $(t\colon l; l_1, l_2)(t\colon j; j_1, j_2)$ in which an initial synchronized pairwise exchange made between rows j_1 and j_2 at each combination of all other factors in the design is followed by a synchronized pairwise exchange between columns l_1 and l_2.

Synchronized rearrangements are composed of combinations of synchronized pairwise rearrangements involving distinct pairs of rows and columns.

7.5.4 Similarities

Let X be the vector space formed from finite combinations of the random variables $\{X_j(s)\}$ that form an experimental design. We define a semimetric ρ on X such that if $X, Y \in X$ with distribution functions F_X, F_Y, then $\rho(X, Y) = \sup_z |F_X[z] - F_Y[z]|$. Note that if $\rho(X, Y) = 0$ then $X = Y$ except on a set of probability zero.

In line with group-theoretic convention (see, for example, Yale, 1968), we term a rearrangement Π a *similarity* if, for all points $W, X, Y,$ and Z in X, $\rho(X, Y) = \rho(W, Z)$ if and only if $\rho(\Pi X, \Pi Y) = \rho(\Pi W, \Pi Z)$.

In what follows, dot notation will be used to represent a sum. Thus, $X. = \sum X_i$ and $X_{\cdot j \cdot} = \sum_i \sum_k X_{ijk}$.

Theorem 7.1. *Let $\{X_j(s)\}$ be a set of independent observations in a complete balanced experimental design with factors in S such that $X_j(s) = \Delta_s + \epsilon_j(s)$, where Δ_s satisfies conditions (1), (2), and (3) of Section 7.5.3. and the $\{\epsilon_j(s)\}$ are a set of independent identically distributed random variables with mean zero, then if π is a synchronous rearrangement with respect to the factor i_j, then π is a similarity under the null hypothesis $\alpha_{i_j} = 0$ for $i_j = 1, \ldots, I_j$.*

Proof. Without loss of generality, suppose we designate as "rows" the factor for which we wish to test a main hypothesis and "columns" a second factor which we wish to test for a first-order interaction with "rows." To simplify the notation, we assume that the factor subscripts are reordered so that the subscripts denoting rows and columns are in the initial positions.

Let Π_{mp} denote a pairwise synchronous exchange between rows m and p as in Figure 7.4a and b, (that is $\Pi_{mp} = (1:1; m, p)$ in our previous notation). Of course, only two dimensions and three of the rows and columns are illustrated in our diagram. Similar synchronous exchanges have taken place at all levels of the remaining factors.

If $i \neq m$ and $i \neq p$, then $E\Pi_{mp}X_{i\ldots} = EX_{i\ldots}$, as the remaining factors in condition (1) vanish in accordance with conditions (2) and (3). If $i = m$, then $E\Pi_{mp}X_{i\ldots} = EX_{i\ldots} - I(\alpha_{Im} - \alpha_{Ip}) - \beta_{1Ji\cdot}$. But if H_I is true, $\beta_{IJi\cdot} = 0$, so that $E\Pi_{mp}X_{i\ldots} = EX_{i\ldots}$. Similarly for $i = p$ when H_1 is true. The $\epsilon_j(s)$ are independent and identically distributed; the pairwise exchange involved the swap of equal numbers of independent, identically distributed variables. So Π_{mp} is a similarity. Since all synchronous exchanges are made up of similar pairwise synchronous exchanges, all are similarities as was to be proved. □

Corollary 7.1. *The results appear to extend to unbalanced designs as long as there is at least one observation per cell (see Section 7.6). But if the design is unbalanced, main effects may already be confounded with interactions. Symmetric rearrangements will merely preserve the imbalance.*

The proof of the corollary is immediate, as the proof of the main theorem did not require that the design be balanced.

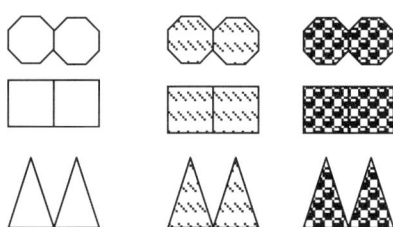

Fig. 7.4a. Part of an M-factor design. M-2 of the dimensions are hidden in this representation.

7.5 Synchronized Rearrangements

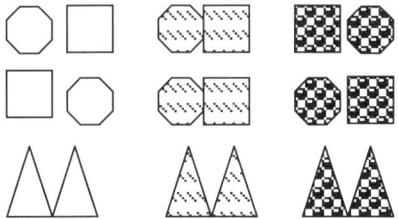

Fig. 7.4b. The same design after an exchange of row elements. Similar synchronous exchanges have taken place in the hidden dimensions.

Imbalance has further practical implications, as it results in fewer opportunities for an exact test without randomization on the boundary and a reduction in power. Consider a 2×2 experimental design with 10 observations in 3 of the cells and 1 in the fourth. Thus, we can exchange at most one pair of observations between cells. Only 1001 synchronous rearrangements are available to test for a row effect—one for the identity plus $1^\star \binom{10}{1}^3$. 1000 distinct synchronous rearrangements plus the identity are available to test for a column effect and 1000 for interaction. The total of 3001 distinct synchronous rearrangements are a miniscule fraction of the $\binom{31}{10 \ \ 10 \ \ 10}$ possible rearrangements for this design.

At least two observations per cell are essential if we are to be guaranteed independent tests of all factors.

Typically, the loss function associated with a testing problem will be symmetric about zero and monotone nondecreasing on the positive half-line.

Corollary 7.2. *Suppose g is a monotone nondecreasing function such as $g[i] = i$ or $g[i] = log[i+1]$. Then the distribution of the statistic $S_I = \sum_{i=1}^{I_i} g[i] X_{..i..}$ obtained from synchronous rearrangements can be used to obtain an exact, unbiased test of a hypothesis concerning a main effect in an M-factor design, such as H_I: $\alpha_{Ij} = 0$ for all j against an ordered alternative, such as K_I: $\alpha_{I1} < \alpha_{I2} < \cdots < \alpha_{IJ}$.*

The proof parallels that of Theorem 6.2.

7.5.5 Test for Interaction

Theorem 7.2. *Let $\{X_k(s)\}$ be a set of independent observations in a complete balanced experimental design with factors in S such that $X_k(s) = \Delta_s + \epsilon_k(s)$, where Δ_s satisfies conditions (1), (2), and (3) and the $\{\epsilon_k(s)\}$ are a set of independent, identically distributed random variables with mean zero. Then the distribution of the statistic $T_{IJ} = \sum_{1 \leq i < i' \leq I} \sum_{1 \leq j < j' \leq J} (X_{..i..j..} +$*

136 7 Multifactor Designs

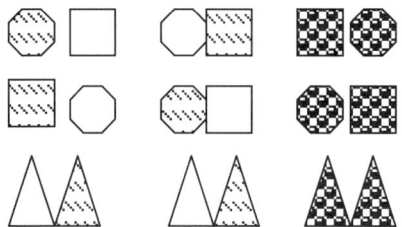

Fig. 7.5a. The same design as in Figure 7.4b after a further exchange of column elements. Similar synchronous exchanges have taken place in the hidden dimensions. Note that each combination of shape and pattern corresponds to a distinct first-order interaction term.

$X_{..i'..j'..} - X_{..i'..j..} - X_{..i..j'..})^2$ obtained from synchronous rearrangements can be used to obtain an exact, unbiased test of a hypothesis concerning a first-order interaction in an N-factor design such as H_{IJ}: $\beta_{ij} = 0$ for $i = 1, \ldots, I; j = 1, \ldots, J$.

Proof. Let Π_{nmlk} denote a synchronous pairwise exchange between the cells of the nth and mth rows and the lth and kth columns of the design, that is, reading from right to left, $\Pi_{nmlk} = (1: J; l, k) \, (1: I; m, n)$ in our previous notation). As can be seen from Figure 7.5a, the second exchange affects not only the four cells $(l, m), (l, n), (k, m)$, and (k, n) in the $I \times J$ plane, but also the cells in the adjacent rows and columns of the plane. We also see that it suffices to show that the $T_{ij} = X_{ij...} + X_{.i'..j'..} - X_{..i'..j...} - X_{..i..j'...}$ remain invariant with respect to Π_{nmlk} when the hypothesis H_{IJ} is true for three of the $2 \times 2 \times \cdots$ subdesigns depicted.

From the condition (1), we see that X_{ij} may be written as the sum of a deterministic portion, which we write as Δ_{ij} and a stochastic portion that we denote by E_{ij}.

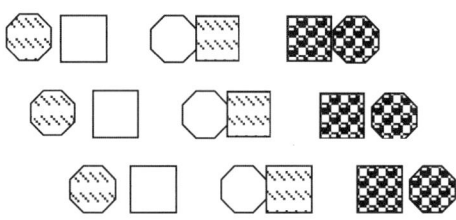

Fig. 7.5b. The same design as Figure 7.5a from a new perspective. This new perspective confirms that similar synchronous exchanges also have taken place in the hidden dimensions. The first row and column of Figure 7.5a are displayed at the bottom of this figure. The two additional lines correspond to different levels of a third factor.

Noting that under $H_{12}, \beta_{ij} = 0$ for all i and j, we see that for the $2 \times 2 \times \cdots$ subdesign in the upper right hand corner of Figure 7.5a, the deterministic portion D_{ij} of $\Pi_{nmlk} T_{ij}$, is equal to

$$\Delta_{..i..j..} - \alpha_{1n} + \alpha_{1m} - \alpha_{2l} + \alpha_{2k}$$
$$+ \Delta_{..i'..j'..} - \alpha_{1m} + \alpha_{1n} - \alpha_{2k} + \alpha_{2l}$$
$$- \Delta_{..i'j..} + \alpha_{1m} - \alpha_{1n} + \alpha_{2l} - \alpha_{2k}$$
$$- \Delta_{..i..j'..} + \alpha_{1n} - \alpha_{1m} + \alpha_{2k} - \alpha_{2l}$$

which is equivalent to the deterministic portion of T_{ij}.

The stochastic portion of T_{ij} also remains unchanged by the rearrangement as it involves a swap of equal numbers of independent, identically distributed variables.

For the $2 \times 2 \times \cdots$ subdesign in the lower right hand corner of Figure 7.5a,

$$\prod_{nmlk} D_{ij} = \Delta_{..i..j..} - \alpha_{1m} + \alpha_{1n} - \alpha_{2k} + \alpha_{2l}$$
$$+ \Delta_{..i'..j'..}$$
$$- \Delta_{..i'j..} - \alpha_{2l} + \alpha_{2k}$$
$$- \Delta_{..i..j'..} - \alpha_{1n} + \alpha_{1m}$$
$$= D_{ij}.$$

For the $2 \times 2 \times \cdots$ subdesign formed from the first and second rows and first and third columns of Figure 7.5a,

$$\prod_{nmlk} D_{ij} = \Delta_{..i..j..} - \alpha_{1n} + \alpha_{1m} - \alpha_{2l} + \alpha_{2k}$$
$$+ \Delta_{..i'..j'..} - \alpha_{1m} + \alpha_{1n}$$
$$- \Delta_{..i'j..} + \alpha_{1m} - \alpha_{1n} + \alpha_{2l} - \alpha_{2k}$$
$$- \Delta_{..i..j'..} + \alpha_{1n} - \alpha_{1m}$$
$$= D_{ij}.$$

All other altered 2×2 subdesigns are similar to one of these three. As synchronous rearrangements are made up of synchronous pairwise exchanges, the theorem follows. □

7.6 Unbalanced Designs

As noted in Section 7.2, the analysis of variance may not be applicable to severely unbalanced designs. Imbalance in the design will result in the confounding of main effects with interactions. Neither is the permutation test a panacea. When one or more cells in a design have no or only a few observations, it may be impossible to find sufficient synchronous rearrangements to perform a meaningful test. As we shall show in the next few sections, the bootstrap can help us here.

7.6.1 Missing Combinations

If an entire factor combination is missing, we may not be able to estimate or test any of the effects. One very concrete example is an unbalanced design I encountered in the 1970s:

Makinodan et al. [1976] studied the effects of age on the mediation of the immune response. They measured the anti-SBRC response of spleen cells derived from C57BL mice of various ages. In one set of trials, the cells were derived entirely from the spleens of young mice, in a second, they came from the spleens of old mice, and in a third they came from mixtures of the two.

Let $X_{i,j,k}$ denote the response of the kth sample taken from a population of type i,j ($i = 0 = j$: controls; $i = 1, j = 0$: cells from young animals only; $i = 0, j = 1$: cells from old animals only; $i = 1 = j$: mixture of cells from old and young animals). We assume that for lymphocytes taken from the spleens of young animals,

$$X_{2,1,k} = \mu + \alpha + \epsilon_{2,1,k};$$

for the spleens of old animals,

$$X_{1,2,k} = \mu - \alpha + \epsilon_{1,2,k};$$

and for a mixture of p spleens from young animals and $(1-p)$ spleens from old animals, where $0 \leq p \leq 1$,

$$X_{2,2,k} = p(\mu + \alpha) + (1-p)(\mu - \alpha) - \gamma + \epsilon_{2,2,k}$$

where the errors $\epsilon_{2,2,k}$ are independent values.

Makinodan knew in advance of his experiment that $\alpha > 0$. He also knew that the distributions of the errors $\epsilon_{i,j,k}$ would be different for the different populations. We can assume only that these errors are independent of one another and that their medians are zero.

Makinodan wanted to test the hypothesis $\gamma = 0$ as there are immediate biological interpretations for the three alternatives: (1) from $\gamma = 0$ one may infer independent action of the two cell populations; (2) $\gamma < 0$ means excess lymphocytes in young populations; and (3) $\gamma > 0$ suggests the presence of suppressor cells in the spleens of older animals.

The standard parametric (ANOVA) approach won't work because of the empty cell. We can still approach the problem parametrically. Let $S = \bar{X}_{2,2} - p\bar{X}_{1,2} - (1-p)\bar{X}_{2,1}$. Then $T = S/\hat{V}(S)$, where the denominator is an estimate of the variance of S, will have approximately Student's t-distribution. If the samples consist of only three or four observations and the observations do not come from a normal distribution, as was the case with Makinodan's data, then Student's t-distribution is a poor approximation, and a nonparametric approach is called for.

There are no synchronous rearrangements, so this method is ruled out, too. Fortunately, another resampling method, the bootstrap, can provide a solution.

Here is the *bootstrap procedure*:

Draw an observation at random and with replacement from the set $\{x_{2,1,k}\}$; label it $x^*_{2,1,j}$. Continue until you have drawn the same number of observations as were in the original sample taken from young animals Similarly, draw the bootstrap observations $x^*_{1,2,j}$ and $x^*_{2,2,j}$ from the sets $\{x_{1,2,k}\}$ and $\{x_{2,2,k}\}$.

Let
$$z_j = p\bar{X}^*_{1,2} + (1-p)\bar{X}^*_{2,1} - \bar{X}^*_{2,2}.$$

Repeat this resampling procedure a thousand or more times, obtaining a bootstrap estimate z_j of the interaction each time you resample. Use the resultant set of bootstrap estimates $\{z_j\}$ to obtain a confidence interval for γ. If $z = 0$ belongs to this confidence interval, accept the hypothesis of additivity; otherwise reject.

One word of caution: Unlike a permutation test, a bootstrap is exact only for very large samples. The probability of a Type I error may be greater than the significance level you specify.

Bootstrap Analysis of an Unbalanced Design

Mean DPFC response. Effect of pooled old BC3FL spleen cells on the anti-SRBC response of indicator pooled BC3FL spleen cells. Data extracted from Makinodan et al. [1976]. Bootstrap analysis:

	Young cells	Old cells	1/2 + 1/2
	5640	1150	7100
	5120	2520	11020
	5780	900	13065
	4430	50	
	7230		

Bootstrap sample 1:	5640 + 900 − 11020	−4480
Bootstrap sample 2:	5780 + 1150 − 11020	−4090
Bootstrap sample 3:	7230 + 1150 − 7100	1280
⋮	... ...	...
Bootstrap sample 600:	5780 + 2520 − 7100	1200

7.6.2 The Boot-Perm Test

The preceding was an extreme example of an unbalanced design. More often, we will have a few observations in each category. In either set of circumstances, we may proceed as follows:

Bootstrap from the original data to create a balanced design, sampling with replacement separately from each category, so that the ab's are selected from the ab sample, the aB's from the aB sample and so forth. Analyze the resultant balanced design using a permutation test. Bootstrap 10 times. If you reject every time or accept every time, draw the corresponding conclusion. Otherwise bootstrap 100 times and check again. If you still have ambiguity, then either a highly significant interaction is present or the differences and main effects are not significant. Only by taking additional observations to obtain a more balanced sample can you make an informed decision.

7.7 Which Test Should You Use?

For a balanced experimental design, the analysis of variance (ANOVA) provides a set of independent uniformly most powerful invariant tests that are remarkably robust to deviations from normality.

With data that are distinctly non-normal, an attempt should be made in line with the guidelines provided in Chapter 11.3 to find a transformation such that the transformed data is close to normal. If this is not possible, then a test using synchronous rearrangements should be employed.

The use of synchronous rearrangements is also recommended for the analysis of unbalanced designs providing there are an adequate number of synchronous rearrangements despite the imbalance.

The preceding section contained an extreme example of an unbalanced design; more often we have a few observations in each category. In these circumstances, we could use synchronized rearrangements or we could proceed as follows:

Bootstrap from the original data preserving categories so the observations in the ijkth cell of the bootstrap design are selected with or without replacement from the ijkth cell of the original design. Analyze the resultant balanced design using the analysis of variance. Repeat 100 times. If the results are consistent so that you reject or accept the hypothesis in nearly every instance, draw the corresponding conclusion. If the results are mixed, varying from bootstrap sample to bootstrap sample, then you probably have a highly significant interaction and must draw additional observations from the original population before proceeding further.

7.8 Exercises

1. Find the expected value of the numerator and denominator of W_I in Section 7.2.
2. Show that if a vector Y is distributed as normal $N(0, \sigma^2 I)$ and if $Z = CY$ where C is an orthogonal matrix, then Z is distributed as $N(0, \sigma^2 I)$, also.
3. For model (7.1), express the test statistics W_1, W_J, W_{IJ} in terms of the variables Y in the canonical form $Y = CX$ to show that the numerators of these expressions lie in separate subspaces.

4. *Confidence interval.* Derive a 90% confidence interval for the main effect of sunlight using the crop yield data in Table 7.1. First, restate the model so as to make clear what it is you are estimating:

$$X_{ikl} = \mu + s_i + f_k + sf_{ik} + \epsilon_{ikl}, \text{ with } s_1 = -\delta \text{ and } s_2 = \delta.$$

Recall that we rejected the null hypothesis that $\delta = 0$. Suppose you add $d = 1$ to each of the observations in the low sunlight group and subtract $d = 1$ from each of the observations in the high sunlight group. Would you still reject the null hypothesis at the 90% level? If your answer is "yes" then $d = 1$ does not belong to the 90% confidence interval for δ. If your answer is "no" then $d = 1$ does belong. Experiment (be systematic) until you find a value δ_o such that you accept the null hypothesis whenever $d > \delta_o$.

5. a) Is the Still–White test for interaction asymptotically exact?
   b) If we were to generate random relabelings of the original observations, and then compute the Still–White statistic, would the resulting distribution provide an exact test?

6. Suppose a set of observations is weakly exchangeable with respect to two nonempty subsets of rearrangements R_1 and R_2.
   a) Show that they are also weakly exchangeable with respect to the union of R_1 and R_2.
   b) Would a test based on the rearrangement distribution of the observations on the union of R_1 and R_2 always be as or more powerful than a test based on the rearrangement distribution of the observations on R_1 alone?

7. The following designs were among those used to validate the GoodStats program. For each design, specify the total number of synchronized rearrangements, and the p-values associated with row effects, column effects and interaction.

   Design a: 2   1   0   0
               0   0   0   0
   Design b: 2   0   0   0
               0   0   0   1
   Design c: 2   1   0   0   0   0
               0   0   0   0   0   0.

8. Show that P is the group generated by the union of P_R, P_C and P_{RC}.
9. Show that while a synchronized rearrangement used for testing the interaction between factors l and j has the form $(t: l; l1, l2)$ $(t: j; j1, j2)$, the converse is not necessarily true.
10. In an $I \times J$ two-factor experimental design, will the statistics S and T produce equivalent results using a) analysis of variance, b) synchronized rearrangements, where $S = \sum_{1 \leq i < i' \leq I} \sum_{1 \leq j < j' \leq J} g[X_{ij} + X_{i'j'} - X_{i'j} - X_{ij'}]$ and $T = \sum_{i=1}^{I} \sum_{j=1}^{J} (\bar{X}_{ij.} - \bar{X}_{i..} - \bar{X}_{.j.} + \bar{X}_{...})^2$.
11. Prove Corollary 7.2.

12. Extend Theorem 7.2 to the case of second-order interactions in a three-factor design.
13. Are the following two bases for permutation-based inference equivalent?
    (i) The rearrangement distribution is derived by considering the outcomes associated with all possible assignments of treatments to experimental units.
    (ii) The rearrangement distribution is derived by considering the outcomes associated with all possible rearrangements of the existing labels on the experimental units.
    (Hint: Consider an experiment in which we allocate treatments to n experimental units by a succession of coin flips.)
14. Establish properties that will determine without explicit representation whether the product of two rearrangements pq will belong to $\boldsymbol{P_{RC}}$ when p is in $\boldsymbol{P_R}$ and q is in $\boldsymbol{P_C}$.

8
Categorical Data

In many experiments and in almost all surveys, many if not all the results fall into discrete categories rather than being measurable on a continuous scale: e.g., male vs. female, African–American vs. Hispanic vs. Asian vs. white, in favor vs. against vs. undecided. The corresponding hypotheses concern proportions: ("African–Americans are as likely to be Democrats as they are to be Republicans." "The dominant genotype 'spotted shell' occurs with three times the frequency of the recessive." "The data from all 14 treatment sites may be combined as the effects of treatment are identical at each site." In this chapter you will learn methods for testing these hypotheses. The techniques you will learn also are applicable when your measurements are ordinal but not metric, as with preference ratings: e.g. "Do you prefer the acting of Toni Colette to Nicole Kidman? Strongly prefer? Slightly prefer? Indifferent?"

8.1 Fisher's Exact Test

Suppose, upon examining the cancer registry in a hospital, we uncover the data that we put in the form of a 2×2 contingency table, Table 8.1.

The 9 denotes the number of males who survived cancer, the 1 denotes the number of males who died from the disease, and so forth. The four marginal totals or *marginals* are 10, 14, 13, and 11. The total number of men in the study is 10, while 14 denotes the total number of women, and so forth.

We see in this table an apparent difference in the survival rates for men and women: Only 1 of 10 men died following treatment, but 10 of the 14 women failed to survive. Is this difference statistically significant?

The answer is "yes" if the data represent a random sample of cancer patients. Let's see why, using the same line of reasoning that Fisher advanced at the annual Christmas meeting of the Royal Statistical Society in 1934. After Fisher's talk was concluded, incidentally, a seconding speaker compared Fisher's talk to "the braying of the Golden Ass." I hope the reader will take

Table 8.1.

	Survived	Died	Total
Men	9	1	10
Women	4	10	14
Total	13	11	24

more kindly to my own explanation. Fisher's test maximizes the minimum power [Tocher, 1950].

The marginals in this table are fixed because, indisputably, there are 11 dead bodies among the 24 persons in the study and 14 women. Suppose that before completing the table, we lost the subject identification labels so that we could no longer identify which subject belonged in which category. Imagine you are given two sets of 24 labels. The first set has 14 labels with the word "woman" and 10 labels with the word "man." The second set of labels has 11 labels with the word "dead" and 13 labels with the word "alive." Under the null hypothesis, you are allowed to distribute the labels to subjects independently of one another, one label from each of the two sets per subject.

The following two tables are the result of this relabeling procedure. The first of these tables could make a strong case for the superior fitness of the male, stronger even than our original observations. In the second table, the survival rates for men and women are more alike than they were in our original table.

There are a total of $N = \sum_{x=0}^{10} \binom{13}{x}\binom{11}{10-x} = \binom{24}{10}$ ways you could hand out the labels. $\binom{14}{10}\binom{10}{1}$ of the assignments result in tables that are as extreme as our original table (that is, in which 9 of the men survive), and $\binom{14}{11}\binom{10}{0}$ in tables that are more extreme (all 10 of the men survive). This is a very small fraction of the total, so we conclude that a difference in survival rates of the two sexes as extreme as the difference we observed in our

Table 8.2a.

	Survived	Died	Total
Men	10	0	10
Women	3	11	14
Total	13	11	24

Table 8.2b.

	Survived	Died	Total
Men	8	2	10
Women	5	9	14
Total	13	11	24

Table 8.3.

	Category 1	Category 2	Total
Category A	x	$m - x$	m
Category B	$t - x$		n
Total	t		$m + n$

original table is very unlikely to have occurred by chance alone. We reject the hypothesis that the survival rates for the two sexes are the same and accept the alternative hypothesis that, in this instance at least, males are more likely to profit from treatment.

8.1.1 Hypergeometric Distribution

How did we determine the total number of possible tables? The component terms are taken from the *hypergeometric* distribution:

$$\sum_{x=0}^{t} \binom{m}{x}\binom{n}{t-x} / \binom{m+n}{t} \tag{8.1}$$

where n, m, t, and x occur as indicated in the 2×2 contingency table, Table 8.3.

If the proportion of category 1 is the same for both categories A and B, then all tables with the same marginals are equally likely and $\sum_{k=0}^{t-x} \binom{m}{t-k}\binom{n}{k}$ tables are more extreme.

8.1.2 One-Tailed and Two-Tailed Tests

In the preceding example we tested the hypothesis that survival rates do not depend on sex against the alternative that men diagnosed with cancer are likely to live longer than women similarly diagnosed. We rejected the null hypothesis because only a small fraction of the possible tables were as or more extreme than the one we observed initially. This is an example of a one-tailed test. But is it the appropriate test? Is this really the alternative hypothesis we would have proposed if we had not already seen the data? Wouldn't we have been just as likely to reject the null hypothesis that men and women profit the same from treatment if we had observed Table 8.4?

Of course, we would! In determining the significance level in the present example, we must add together the total number of tables that lie in either of the two extremes or tails of the permutation distribution.

The critical values and significance levels are quite different for one-tailed and two-tailed tests and, all too often, the wrong test has been employed in published work. McKinney et al. [1989] reviewed some 70 plus articles that

Table 8.4.

	Survived	Died	Total
Men	0	10	10
Women	13	1	14
Total	13	11	24

appeared in six medical journals. In over half of these articles, Fisher's exact test was applied improperly. Either a one-tailed test had been used when a two-tailed test was called for, or the authors of the paper simply hadn't bothered to state which test they had used.

Of course, unless you are submitting the results of your analysis to a regulatory agency, no one will know whether you originally intended a one-tailed test or a two-tailed test and subsequently changed your mind. No one will know whether your hypothesis was conceived before you started or only after you'd examined the data. All you have to do is lie about the condition of the data. Just recognize that if you test an after-the-fact hypothesis without identifying it as such, you are guilty of scientific fraud.

When you design an experiment, decide at that time whether you wish to test your hypothesis against a two-sided or a one-sided alternative: A two-sided alternative dictates a two-tailed test; a one-sided alternative dictates a one-tailed test.

As an example, suppose we decide to do a follow-on study of the cancer registry to confirm our original finding that men diagnosed as having tumors live significantly longer than women similarly diagnosed. In this follow-on study, we have a one-sided alternative. Thus, we would analyze the results using a one-tailed test rather than the two-tailed test we applied in the original study.

8.1.3 The Two-Tailed Test

Unfortunately, it is not as obvious which tables should be included in the second tail. Is Table 8.4 as extreme as Table 8.2 in the sense that it favors an alternative more than the null hypothesis? One solution is simply to double the p-value we obtained for a one-tailed test. Alternately, we can define and use a test statistic as a basis of comparison. One commonly used measure is the χ^2 (chi-square) statistic defined for the 2×2 contingency table after eliminating terms that are invariant under permutations as $[x - tm/(m+n)]^2$. For Table 8.1, this statistic is 12.84, for Table 8.4, it is 29.34.

We leave it to you to do the computations to show that Table 8.5 is more extreme than Table 8.1, but Table 8.6 is not.

8.1.4 Determining the p-Value

A problem with any of the methods we've used so far is that they produce only discrete significance values. We're very unlikely to observe 0.05 exactly;

Table 8.5.

	Survived	Died	Total
Men	1	9	10
Women	12	2	14
Total	13	11	24

Table 8.6.

	Survived	Died	Total
Men	2	8	10
Women	11	3	14
Total	13	11	24

Table 8.7.

	Survived	Died	Total
Men	7	3	10
Women	6	8	14
Total	13	11	24

if the number of observations is small, p-values may jump from 0.040 (as in Table 8.2a) to 0.185 (as in Table 8.7) as a result of a single additional case.

What are the appropriate criterion for rejection of the null hypothesis? Limiting ourselves to 4% of the tables means we err on the conservative side. Rejecting for 18% means we are settling for an excessively high frequency of Type I errors. There are at least six solutions:

1) Deliberately err on the conservative side, that is, reject when $p \leq C(n, m, t)$ where $C(n, m, t)$ is the smallest integer for which the proportion of tables with the same marginals (n, m, t) is less than or equal to the significance level. See Boschloo [1970] and McDonald, Davis, and Miliken [1977] for some slight improvements on this approach.
2) Randomize on the boundary. If you get a p-value of 0.185 and the next closest value would have been 0.040, let the computer choose a random number between 0 and 1 for you. If this number is less than $(0.05 - 0.04)/(0.185 - 0.040)$, reject the hypothesis at the 5% level, accept it otherwise. Unless you don't care to leave your decisions to chance.
3) Use the mid-p-value. Let p be equal to half the probability of the table you actually observe plus all of the probability of more extreme results. See Lancaster [1961].
4) Present your audience with the data and the p-value you calculated; let them make up their own minds whether it is a significant result or not. See Section 3.2.6.
5) Make use of a back-up statistic; see Section 8.5.1.2.

6) Conduct a sensitivity analysis, Dupont [1986]. Add a single additional case to one of the cells (say the cell that has the most observations already, so your addition will have the least percentage impact). Does the significance value change appreciably? (Again, leave it to your audience to decide what is an "appreciable" change.) This approach is particularly compelling if you are presenting statistical evidence in a courtroom, as it turns impersonal percentages into individuals; see Good [2001b].

8.1.5 What is the Alternative?

In Chapter 3, we noted that every test requires both a primary hypothesis and an alternative hypothesis. The primary hypothesis for the 2×2 contingency table is that the labels for alternate categories are independent of one another; but what is the alternative? We need a model.

Let us assume we have taken two independent samples represented in the two rows, and that each sample consists of independent identically distributed observations. In the first row, m independent binomial trials resulted in x observations in the first row, the successes. In the second row, the n independent binomial trials resulted in $t - x$ successes. The joint probability may be written as

$$\binom{m}{x} p_1^x (1-p_1)^{m-x} \binom{n}{t-x} p_2^{t-x} (1-p_2)^{n-(t-x)}$$

or, equivalently,

$$\binom{m}{x}\binom{n}{t-x}(1-p_1)^m(1-p_2)^n \exp\left[x\log[-\theta] + t\log\frac{p_2}{(1-p_2)}\right],$$

where θ is the odds ratio

$$\frac{p_2/(1-p_2)}{p_1/(1-p_1)}.$$

Our null hypothesis of identical distributions in the two rows is that $p_1 = p_2$ or $\theta = 1$, while a one-sided alternative might be that $\theta > 1$.

8.1.6 Increasing the Power

Providing we are willing to randomize on the boundary as described in Section 8.1.4, Fisher's exact test based on the conditional distribution of x given m, n, and t is UMP among all unbiased tests for comparing two binomial populations [Lehmann, 1986, pp 151–162].

It is UMP under any of the following four world views:

i) Binomial sampling—one set of marginals in the contingency table is random; the other set and the total number of observations $n + m$ are fixed.

ii) Independent Poisson processes—all marginals and the total number of observations are random.
iii) Multinomial sampling—all marginals are random and the total number of observations is fixed.
iv) An experiment in which a fixed number $n + m$ of subjects is randomly assigned on an individual basis to one of two different treatments—all marginals are fixed.

The power of Fisher's test depends strongly on the composition of the sample. A balanced sample, with equal numbers in each category, is the most desirable. If the sample is too unbalanced, for example, if 100 of the observations have the row attribute and only 1 does not, it may not be possible to determine if the column attribute is independent of the row attribute.

The question arises whether one ought to take samples from (a) the population at large, or (b) by selecting samples from subjects with and without the row attribute, or (c) selecting samples from subjects with and without the column attribute. If (b) or (c), should the samples be of equal size?

For very large sample sizes (Exercise 8.1), one can use a normal approximation to show that more powerful tests can be obtained by selecting samples of equal size. If you have some prior knowledge about the frequency of the two attributes, then select samples on the basis of that attribute whose probability is closest to $1/2$.

Studies of the power of Fisher's exact test against various alternatives were conducted by Haber [1987] and Irony and Pereira [1986]. It is easy to see that

$$p\{f_{11} = x\} = \frac{\binom{m}{x}\binom{n}{t-x}\theta^x}{\sum_u \binom{m}{u}\binom{n}{t-u}\theta^u},$$

where θ is the odds ratio; this is the *noncentral hypergeometric distribution* [Fisher, 1934; Cornfield, 1956].

8.1.7 Ongoing Controversy

Fisher's original presentation of his "exact test" was marked by acrimony and dissent that has continued to the present day. Fisher's exact test agrees asymptotically with the chi-square test based on one degree of freedom, a fact that is no longer in dispute today (see Kendall and Stuart, 1979, page 586; Mehta and Patel, 1986). But in 1934, Fisher's listeners raged over whether there should be three or four degrees of freedom in a 2×2 contingency table or just one degree as Fisher asserted.

To understand, and hopefully disagree with, the objections, read the discussions following Fisher [1935], as well as Box [1978]. Also, see Exercises 8.2, 8.3, and 8.4.

Today, the controversy has taken a somewhat different form. Most analysts choose to view the contingency table as we have here, conditional on the margins. But others view the table as resulting from two separate and independent binomial samples, the so-called *unconditional case*. The column sums still are considered fixed, each corresponding to a distinct population, but the row sums now are viewed as random variables. Instead of having only $\left(\begin{smallmatrix} f_{..} \\ \min_{ij} f_{ij} \end{smallmatrix}\right)$ possibilities, thus limiting the increments in which significance levels may be achieved as noted in Section 8.1.4, all tables that satisfy $\sum_i f_{1j} = nj$ for $j = 1, \ldots, C$ are considered.

In the conditional approach advanced by Fisher, the sum of successes from the two populations is sufficient under the null hypothesis for the unknown probability of success (Exercise 8.2). In the unconditional case, the obvious choices for test statistic (see, for example, those advanced by Barnard, 1945 and Chan, 1998) do depend on this unknown parameter. The solution is to take as the *p*-value the supremum of the exact *p*-values over some range of possible values for the parameter [Barnard, 1945 and Berger and Boos, 1994].

The unconditional approach appears to us less desirable as it requires we include in the rejection region tables whose marginals did not occur, but see Barnard [1945, 1949, 1979, 1989], Greenland [1991], Haber [1987], Storer and Kim [1990], and Suissa and Shuster [1984, 1985].

8.2 Odds Ratio

In most instances, we won't be satisfied with merely rejecting the null hypothesis but will want to make some more powerful statement like "men are twice as likely as women to get a good-paying job" or "women under 30 are twice as likely as men over 40 to receive as academic appointment."

In the discrimination case of Fisher vs. Transco Services of Milwaukee [1992], the plaintiffs claimed that Transco was ten times as likely to fire older employees. Can we support this claim with statistics? The Transco data are provided in Table 8.8.

Let p_1 denote the probability of firing a young person, and π_2 the probability of firing an older person. We want to go beyond testing the null hypothesis $p_1 = p_2$ to determine a confidence interval for the odds ratio $\theta = (p_2/(1-p_2))/(p_1/(1-p_1))$.

Table 8.8. Transco employment.

Outcome	Young	Old
Fired	1	10
Retained	24	17

> Statistic based on the observed 2 × 2 table:
>
> Binomial proportion for column <young>: pi_1 = 0.04000
> Binomial proportion for column <old   >: pi_2 = 0.3704
>
> $$\text{Odds Ratio} = \frac{(\text{pi\_2})/(1-\text{pi\_2})}{(\text{pi\_1})/(1-\text{pi\_1})} = 14.12$$
>
> Results:
> p-value(2-sided)   95.00% Confidence Interval
> 0.007145   (1.649,   636.5)

Fig. 8.1. StatXact-4 Output Datafile: C:\SX3WIN\EXAMPLES\TRANSCO.CY3 ODDS RATIO OF TWO BINOMIAL PROPORTIONS.

One can obtain confidence intervals for the odds ratio by iterative methods as described in Section 3.2, see also Cornfield [1956], Mantel and Hankey [1971], and Thomas [1971]. Baptista and Pike [1977] describe an approach that sometimes gives shorter confidence intervals. We turn for aid to StatXact^{TM}, a statistical package whose emphasis is the analysis of categorical and ordinal data. Choosing "statistics", "two binomials", and "CI odds ratios" from successive StatXact menus, we obtain the results of Figure 8.1.

Based on these results, we can tell the judge that older workers were fired at a rate at least 1.6 times the rate at which younger workers were discharged.

8.2.1 Stratified 2 × 2's

In trying to develop a cure for a relatively rare disease, we face the problem of having to gather data from a multitude of test centers, each with its own set of procedures and its own way of executing them. Before we can combine the data, we must be sure the odds ratios across the test centers are approximately the same. Consider the set of results in Table 8.9, obtained by the Sandoz drug company and reproduced with permission from the StatXact-3 manual. One of the sites, number 15, stands out from the rest. But is the difference statistically significant?[1]

Zelen [1971] proposed a test based on the number of 2 × 2 tables with the same marginals that are as likely or less likely than the table that was actually observed. With 22 contingency tables, the number of computations needed to examine all rearrangements is in the billions. Fortunately, StatXact utilizes

---

[1] Similar problems were encountered in a study in which test subjects might use one of several different "identical" machines. We couldn't combine the results from the different machines or the different technicians who operated them until we performed an initial test of their equivalence.

Table 8.9. Sandoz drug data.

Test Site	New Drug Response	#	Control Drug Response	#
1	0	15	0	15
2	0	39	6	32
3	1	20	3	18
4	1	14	2	15
5	1	20	2	19
6	0	12	2	10
7	3	49	10	42
8	0	19	2	17
9	1	14	0	15
10	2	26	2	27
11	0	19	2	18
12	0	12	1	11
13	0	24	5	19
14	2	10	2	11
15	0	14	11	3
16	0	53	4	48
17	0	20	0	20
18	0	21	0	21
19	1	50	1	48
20	0	13	1	13
21	0	13	1	13
22	0	21	0	21

several time-saving algorithms, including the one introduced in Mehta, Patel, and Senchaudhuri [1988] to obtain a Monte Carlo estimate of the significance level. We pull down menus Statistics, Stratified 2 × 2 Tables, Homogeneity of Odds Ratios to obtain the results in Figure 8.2.

The estimated p-value of .013, just a fraction greater than 1%, tells us it would be unwise to combine the results from the different sites.

The output of the StatXact program provides us with one more important finding. Displayed above the Monte Carlo estimate of the exact p-value, 0.01237 is the asymptotic or large-sample approximation based on the chi-square distribution. Its value, 0.0785, is many times larger than the correct value; relying on this so-called approximation would have led us to a completely different and erroneous conclusion.

8.3 Exact Significance Levels

The preceding result is not an isolated one. Asymptotic approximations such as the use of the chi-square distribution for the small-sample distribution of Pearson's chi-square statistic are to be avoided in the analysis of contingency

```
[18  2 × 2 informative tables]

Observed Statistics:
   BD: Breslow and Day Statistic = 25.78
   ZE: Zelen Statistic           = 9.481e − 009

Asymptotic p-value: (based on chi-square distribution with 17 df)
   Pr{BD.GE. 25.78} = 0.0785

Monte Carlo estimate of p-value:
   Pr{ZE.GE. 9.481e-009}    = 0.0127
   99.00% Confidence Interval = (0.0119,    0.0135)

Elapsed Time is 0:16:15.37 (10000 tables sampled; starting seed 85190)
```

Fig. 8.2. StatXact-4 Output Datafile: C:\SX3WIN\EXAMPLES\SANDOZ.CY3 TEST FOR HOMOGENEITY OF ODDS RATIOS.

tables except for tables with a large number of observations in each and every cell. Table 8.10 contains data on oral lesions observed in three regions of India derived from Gupta et al. [1980]. We want to test the hypothesis that the location of oral lesions is unrelated to geographical region. Possible test statistics include Freeman–Halton p (see Section 8.4), p_χ, and p_L. This latter statistic is based on the log-likelihood ratio $\sum\sum f_{ij}\log(f_{ij}f_{..}/f_{i.}f_{.j})$.

We may calculate the exact significance levels of these test statistics by deriving their permutation distributions or use asymptotic approximations obtained from tables of the chi-square statistic. Table 8.11 taken from the StatXact-3 manual compares the various approaches.

The exact significance level varies from 1% to 3.5%, depending on which test statistic we select. Tabulated p-values based on large-sample approximations

Table 8.10. Oral lesions in three regions of India.

Site of Lesion	Kerala	Gujarat	Andh
Labial Mucosa	0	1	0
Buccal Mucosa	8	1	8
Commissure	0	1	0
Gingiva	0	1	0
Hard Palate	0	1	0
Soft Palate	0	1	0
Tongue	0	1	0
Floor of Mouth	1	0	1
Alveolar Ridge	1	0	1

Table 8.11. Three tests of independence.

Statistic	χ^2	F–H	LR
Exact p-value	.0269	.0101	.0356
Tabulated p-value	.1400	.2331	.1060

vary from 11% to 23%. Using the Freeman–Halton statistic, the permutation test tells us the differences among regions are significant at the 1% level; the large-sample approximation says *no*, they are *insignificant* even at the 20% level. *The permutation test is correct.* The large-sample approximation is grossly in error. With so many near-zero entries in the original contingency table, the chi-square large-sample approximation is not appropriate.[2]

8.4 Unordered $r \times c$ Contingency Tables

With a computer at hand, the principal issue in the analysis of a contingency table with r rows ($r > 2$) and c columns ($c > 2$) is deciding on an appropriate test statistic. Halter [1969] showed that we can find the probabilities of any individual $r \times c$ contingency table through a straightforward generalization of the hypergeometric distribution given in Equation (8.1). An $r \times c$ contingency table consists of a set of frequencies $\{f_{ij}, 1 \le i \le r; 1 \le j \le c\}$ with row marginals $\{f_{i.}, 1 \le i \le r\}$ and column marginals $\{f_{.j}, 1 \le j \le c\}$. Suppose once again we have mixed up the labels. To make matters worse, this time every item/subject is to be assigned both a row and a column label from the $r + c$ stacks of labels of which $f_{1.}$ are labeled row 1, $f_{2.}$ are labeled row 2, and so forth.

Let P denote the probability with which a specific table assembled at random will have these exact frequencies. $P = Q/R$ with[3]

$$Q = \prod_{i=1}^{r} f_{i.}! \prod_{j=1}^{c} f_{.j}! f_{..}!$$

and

$$R = \prod_{i=1}^{r} \prod_{j=1}^{c} f_{ij}!$$

An obvious extension of Fisher's exact test is the Freeman and Halton [1951] test based on the proportion p of tables for which P is greater than or equal to P_0 for the original table.

---

[2] See also, Mudholkar and Hutson [1997].
[3] $\prod_{i=1}^{n} f_i! = f_1! f_2! \cdots f_n!$

8.4 Unordered $r \times c$ Contingency Tables

While the extension itself may be obvious, it's not as obvious that this extension offers any protection against the alternatives of interest. Just because one table is less likely than another under the null hypothesis does not mean it is going to be more likely under the alternatives of interest to us. Consider the 1×3 contingency table $\boxed{f_1 | f_2 | f_3}$, which corresponds to the multinomial with probabilities $p_1 + p_2 + p_3 = 1$, the table whose entries are 1, 2, 3 argues more in favor of the null hypothesis $p_1 = p_2 = p_3$ than of the ordered alternative $p_1 > p_2 > p_3$.

The classic statistic for independence in a contingency table with r rows and c columns is

$$\chi^2 = \sum_{i=1}^{r}\sum_{j=1}^{c} (f_{ij} - Ef_{ij})^2 / Ef_{ij},$$

where Ef_{ij} is the number of observations in the ijth category one would expect on theoretical grounds.

With very large samples this statistic has the chi-square distribution with $(r-1)(c-1)$ degrees of freedom. But in most practical applications, the chi-square distribution is only an approximation and notoriously inexact for small and unevenly distributed samples.

The permutation statistic based on the proportion p_χ of tables for which χ^2 is greater than or equal to χ_0^2 for the original table provides an exact test and possesses all the advantages of the original chi-square. The distinction between the two approaches, as we observed in Chapter 2, is that with the original chi-square we look up the significance level in a table, while with the permutation statistic, we derive the significance level from the permutation distribution. With large samples, the two approaches are equivalent, as the permutation distribution converges to the tabulated distribution (see Chapter 14 of Bishop, Fienberg, and Holland [1975]).

An alternative is the likelihood ratio test based on the statistic

$$p_L = 2\sum_{i=1}^{r}\sum_{j=1}^{c} f_{ij} \log\left[\frac{f_{ij}}{f_{i.}f_{.j}/f_{..}}\right].$$

If you wish to compare the strength of the association between the row and column variables across different $r \times c$ tables having different row and column dimensions, use one of the contingency coefficient's described by Liebetrau [1983]. One example is *Cramer's V*, which ranges between 0 and 1, with 0 signifying no association and total dependence:

$$V = \sqrt{\frac{\chi^2}{f_{..}(\min[r,c]-1)}}.$$

All these permutation tests have one of the original chi-square test's disadvantages: while they offer global protection against a wide variety of alternatives, they offer no particular protection against any single one of them. The statistics p, p_χ, and p_L treat row and column categories symmetrically, and no attempt is made to distinguish between cause and effect. To address this deficiency, Goodman and Kruskal [1954] introduce an asymmetric measure of association for nominal scale variables called $tau(\tau)$, which measures the proportional reduction in error obtained when one variable, the "cause," or independent variable, is used to predict the other, the "effect," or dependent variable.

Assuming the independent variable determines the row,

$$\tau = \frac{\sum_j f_{mj} - f_{m.}}{f_{..} - f_{m.}},$$

where $f_{mj} = \max_i f_{ij}$, and $f_{m.} = \max_i f_{i.}$.

$0 \leq \tau \leq 1$. $\tau = 0$ when the variables are independent; $\tau = 1$ when, for each category of the independent variables, all observations fall into exactly one category of the dependent. These points are illustrated in the following 2×3 tables:

3	6	9
6	12	18

$\tau = 0$

18	0	0
0	36	0

$\tau = 1$

3	6	9
12	18	6

$\tau = 0.166$

A permutation test of independence is based on the proportion of tables p_τ for which $\tau \geq \tau_0$.

An alternative is the uncertainty coefficient derived from the likelihood ratio statistic

$$U_{R|C} = \frac{\sum_{i=1}^r \sum_{j=1}^c f_{ij} \log[Ef_{ij}]}{\sum_{i=1}^r f_{i.} \log[f_{i.}/f_{..}]}.$$

Like tau, this statistic measures the reduction in error and ranges between 0 and 1 as the association ranges from complete dependence to complete dependence of the row, column variables.

8.4.1 Agreement Between Observers

Suppose two observers assign the same sample set to various categories so the results can be put in the form of an $r \times r$ table. An example would be two teachers assigning letter grades to the same set of students. A permutation test based on *Cohen's kappa* (κ) as described in Agresti [1990] and Berry and

Mielke [1988], allows us to measure the degree of agreement between the two observers:

$$\kappa = \frac{f_{..}\sum_{i=1}^{r}(f_{ii} - f_{i.}f_{.i})}{f_{..}^2 - \sum_{i=1}^{r} f_{i.}f_{.i}}.$$

Note that $0 < k \leq 1$.

Which Test?

The data are in categories.
The categories can't be ordered.
(a) If there are exactly two rows and two columns
    Use Fisher's Exact Test
If there are more than two rows and at least two columns, and
(i) you want to test whether the relative frequencies are the same in each row and in each column:
    Use the Freedman–Halton Test or use chi-square;
(ii) you want to test whether the column frequencies depend on the row:
    Use tau or the uncertainty coefficient.
(b) If the number of rows is equal to the number of columns, and you want to test whether the row and column classifications are in agreement:
    Use Kappa.

8.4.2 What Should We Randomize?

Table 8.12a summarizes Clarke's [1960, 1962] observations on the relation between habitat and the relative frequencies of different varieties of *C. nemoralis* snail. It is tempting to analyze this table using the methods of the preceding section, but before we can analyze a data set, we need to understand *how* it was collected. In this instance, observers went to a series of locations in southern England. At each location, they noted the type of habitat—beechwoods, grasslands, and so forth—and the frequencies of each of 12 different varieties of snail. The original findings are summarized in Table 8.12b reproduced from Manly [1983]. Note that each row in this table corresponds to a single multivariate observation.

Manly computed the chi-square statistic for the original data as summarized in Table 8.12a. Then, using the information in Table 8.12b, he randomly reassigned the location labels to different habitats, preserving the number of locations at each habitat. For example, in one of the rearrangements, Clipper Down Wood, Boarstall Wood, Hatford, and Charlbury Hill—and only these four locations—were designated as "fens." He formed a summary table similar to 8.12a for each rearrangement and computed the chi-square statistic for that table. He found the original value of the chi-square statistic 1756.9 was greater than any of the values he observed in each of 500 random reassignments and

158   8 Categorical Data

Table 8.12a. Summary of Clarke's [1960, 1962] data on *C. nemoralis*.

Habitat	N1	N2	N3	N4	N5	N6	N7	N8	N9	N10	N11	N12
Beechwoods	9	1	34	26	0	46	8	59	126	6	40	115
Other deciduous	10	1	1	0	0	85	8	13	44	2	1	12
Fens	73	3	8	4	6	89	1	23	21	11	0	22
Hedgerows	76	15	32	19	36	98	3	12	8	14	1	18
Grasslands	49	29	75	7	28	23	17	60	12	14	14	24

concluded that habitat type has a significant effect on the distribution of the various body types of the *C. nemoralis* snail.

Manly's analysis combines multivariate and categorical techniques. It makes optimal use of all the data because it takes into account how the data were collected. Could Manly have used Table 8.12b alone to analyze the data? He could not, because this table lacks essential information about interdependencies among the various types of snail.

8.4.3 Underlying Assumptions

The assumptions that underlie the analysis of an $r \times c$ contingency table are the same as those that underlie the analysis of the k- or r-sample problem. To see this, note that a contingency table is merely a way of summarizing a set of N bivariate observations. We may convert from this table to r distinct samples by using the first, or row, observation as the sample or treatment label and the second, or column, observations as the "value." Keeping the marginals fixed while we rearrange the labels ensures that the r sample sizes and the N individual values remain unchanged.

As in the r-sample problem, the labels must be exchangeable under the null hypothesis. This entails two assumptions: First, that the row and column scores are mutually independent, and second, that the observations themselves are independent of one another. We as statisticians can only test the first of these assumptions. We rely on the investigator to ensure that the latter assumption is satisfied. (See Question 4 at the end of this chapter.)

8.4.4 Symmetric Contingency Tables

Suppose now that we wish to compare two methods of assignment or two assigners. The resulting two-way contingency table in which both row and column variables have the same categories is termed *symmetric*. Although in many applications most of the observations lie on the main diagonal and the off-diagonal counts are small, it is the off-diagonal counts that are of greatest interest.

8.4 Unordered $r \times c$ Contingency Tables   159

Table 8.12b. Clarke's [1960, 1962] data* on *C. nemoralis*.

Habitat Type	Location	N1	N2	N3	N4	N5	N6	N7	N8	N9	N10	N11	N12
Beechwood	Clipper Down Wood	1	0	0	0	0	8	0	1	12	1	0	0
	Hackpen Wood	0	0	5	4	0	0	0	5	20	0	1	1
	Kingstone Coombes	0	0	0	2	0	4	1	0	0	0	0	2
	Danks Down Wood	0	0	2	0	0	9	0	15	21	0	1	27
	Fawley Bottom Wood	0	1	0	0	0	5	3	0	2	3	0	0
	Maidensgrove Wood	0	0	0	0	0	3	2	0	5	2	0	0
	Aston Rowant Wood	0	0	0	0	0	6	1	0	23	0	0	0
	Rockley Wood	0	0	10	15	0	0	0	4	20	0	0	21
	Manton Wood	0	0	3	1	0	0	1	6	2	0	3	9
	Knoll Down A	3	0	0	0	0	8	0	9	2	0	35	47
	Knoll Down B	0	0	7	4	0	0	0	0	0	0	0	8
	Roundway Wood	5	0	7	0	0	3	0	19	20	0	0	0
Other deciduous	Boarstall Wood	0	0	0	0	0	13	0	9	28	1	0	0
woods	Rockley Copse	9	1	1	0	0	63	8	4	10	0	0	8
	Elsfield Covert	1	0	0	0	0	6	0	0	4	0	0	0
	Uffington Wood 2	0	0	0	0	0	3	0	0	2	1	1	4
Fens	Shippon	54	1	3	3	1	54	0	8	13	7	0	20
	Headington Wick	5	1	3	0	2	14	1	13	4	2	0	0
	Cothill Fen	2	1	1	0	1	3	0	0	1	1	0	0
	Shippon Fen 2	12	0	1	1	2	18	0	2	3	1	0	2
Hedgerows and	Hatford	1	1	0	15	0	2	0	1	3	2	0	4
rough herbage	Shepherd's Rest 1	16	7	9	0	19	11	1	0	0	6	0	0
	Shepherd's Rest 2	13	4	4	0	9	0	1	0	0	1	0	0
	Standford in Vale	5	0	0	0	0	5	0	1	4	0	0	0
	Wootton	2	0	3	0	0	7	0	1	0	0	0	0
	Chisledon	6	2	0	2	4	9	0	0	0	1	0	1
	Faringdon	18	0	8	0	1	34	0	5	0	0	1	4
	The Ham	8	0	2	1	1	1	0	1	0	0	0	9
	Wanborough Plain	2	0	0	0	0	24	0	0	1	3	0	0
	Watchfield	3	1	0	0	0	2	1	0	0	1	0	0
	Hill Barn Tumulus	1	0	5	0	0	0	0	3	0	0	0	0
	Littie Hinton	1	0	1	1	2	3	0	0	0	0	0	0
Grasslands	Charlbury Hill	2	0	5	1	0	1	0	4	7	0	0	5
	White Horse 1	4	10	4	0	3	3	3	7	0	1	2	1
	White Horse 2	6	6	10	0	0	0	0	0	0	0	0	0
	White Horse 3	7	2	12	0	7	7	4	5	0	2	0	0
	White Horse 4	7	0	2	0	2	0	1	1	0	0	0	0
	Dragons Hill 1	2	4	5	0	0	3	4	19	0	5	2	4
	Dragons Hill 2	1	1	6	0	0	0	1	4	0	0	2	2
	Dragons Hill 3	1	2	3	0	2	2	3	12	0	1	0	4
	West Down 1	0	1	4	3	1	0	0	0	0	0	7	2
	West Down 2	0	0	5	3	0	0	0	0	1	1	0	5
	Sparsholt Down	13	1	15	0	6	0	0	0	0	0	0	0
	Little Hinton	5	0	1	0	5	5	0	1	3	1	0	0

(*continued*)

Table 8.12b. Continued.

Habitat Type	Location	N1	N2	N3	N4	N5	N6	N7	N8	N9	N10	N11	N12
White Horse	5	0	2	2	0	0	1	0	1	0	1	1	0
Dragons Hill	4	1	0	2	0	2	1	1	6	1	2	0	1

*The morph types are similar to those for *hortensis*, with up to five bands present. They are: N1, yellow fully banded (Y12345); N2, yellow part-banded (N00345); N3, yellow mid-banded (Y00300); N4, yellow unbanded (Y00000); N5, other yellows; N6, pink fully banded (P12345); N7, pink part-banded (P00345); N8, pink mid-banded (P00300); N9, pink unbanded (P00000); N10, other pinks; N11, brown banded; N12, brown unbanded.

Note: From "Analysis of polymorphic variation in different types of habitat," B.F.J. Manly, which appeared in *Biometrics*; 1983; **16**: 13–27. Reprinted with permission from the Biometric Society.

A saturated log-linear model for an $r \times r$ symmetric table with multinomial distributed cell counts f_{ij} is

$$E(f_{ij}) = \lambda + \alpha_i + \beta_j + \gamma_{ij} \quad i,j = 1, \ldots, r.$$

The standard test for independence is of the hypothesis $\gamma_{ij} = 0$ for all i, j. But we may also be interested in tests of quasi-independence $\gamma_{ij} = 0$ for $i = j$ and quasi-symmetry $\gamma_{ij} = \gamma_{ji}$ for all i, j. As McDonald, DeRoure, and Michaelides [1998] note, while an exact goodness-of-fit test of independence uses the conditional distribution of the cell counts given *just* the marginals, an exact goodness-of-fit test of quasi-independence uses the conditional distribution of the cell counts given *both* the marginals *and* the diagonal counts. These same authors provide rapid computation algorithms for the needed conditional distributions expanding on an earlier article by McDonald and Smith [1995].

8.5 Ordered Contingency Tables

When data are measured on a continuous basis with multiple decimal places, such as 1.1213, 1.130, 1.141, ties are a relatively infrequent occurrence. But when we ask someone to provide a self-rating on a discrete ordinal scale, 1 through 5, for example, ties are inevitable, the rule, not the exception, and the methods of this chapter may be more appropriate for analyzing such ordinal data than those of Chapter 3.

8.5.1 Ordered 2 × c Tables

Our analysis of a $2 \times c$ ordered contingency table is straightforward and parallels the approach used in Section 6.3.2 for a k-sample comparison, once we have determined what value to assign each of the ordered categories. We illustrate this with data gathered by Graubard and Korn [1987], shown in Table 8.13.

8.5 Ordered Contingency Tables

Table 8.13. Data gathered by Graubard and Korn [1987].

	Maternal Alcohol Consumption (drinks/day).					
Malformation	0	<1	1–2	3–5	≥6	Total
Absent	17066	14464	788	126	37	32481
Present	48	38	5	1	1	93
Total	17114	14502	793	127	38	32574

Recall that our test statistic is $\sum g[j] f_{ij}$ where $g[j]$ is any monotone increasing function, and f_{ij} is the number of observations in the ith row and jth column of the table. Among the leading choices for a scoring method are:

i) the category number: 1 for the 1st category, 2 for the second and so forth;
ii) the midrank scores;
iii) scores determined by the user, the choice we made in Section 6.3.2 when we tested the micronucleii data for the presence of a dose-related trend.

Consider the following 1×2 contingency table

	Alcohol Consumption		Total
Drinks/day	0	1–2	
Frequency	3	5	8

The category or equidistant scores are 1 and 2. The ranks of the 8 observations are 1 through 3, and 4 through 8, so that the mid-rank score of those in the first category is 2, and in the second 6. Our user-chosen scores, corresponding to alcohol consumption, are 0 and 1.5.

Analyzing the data in Table 8.13, we obtain p-values that range from the insignificant, 0.29 for mid-rank scores, to marginally significant, 0.10 for the equidistant scores, to highly significant, 0.01 for our user-chosen scores. A user-chosen score based on the user's knowledge of underlying cause and effect is always recommended, as it will be the most effective at distinguishing between hypothesis and alternative.

The chi-square approximation yields values ranging from 0.017 for the classic Pearson chi-square statistic to 0.19 when the likelihood ratio is employed (Agresti [1992]).

8.5.1.1 Alternative Hypotheses

A variety of one-sided alternative hypotheses may be appropriate when columns are ordered; see, for example, Cohen and Sackrowitz [2000]. Let p_{ij} represent the (unknown) probability of an event resulting in an entry in the ith row and jth column of the table, $i = 1,\ldots,r$, $j = 1,\ldots,c$. Let $\pi_{ij} = p_{ij}p_{(i+1)(j+1)}/p_{(I+1)j}p_{i(j+1)}$ denote the local odds ratio. In an $r \times c$ table, one possible alternative is K: $\pi_{ij} > 0$ for $i = 1,\ldots,r-1; j = 1,\ldots,c-1$. A second possible alternative in the case of a $2 \times c$ table is that of K_S: $\sum_{j=k}^{C} p_{2j}/p_{2.} \geq \sum_{j=k}^{C} p_{1j}/p_{1.}$, for $k = 2,\ldots,c$ with strict inequality for some k. Note that $K \subseteq K_S$.

8.5.1.2 Back-up Statistics

As noted in Section 8.1.4, a major limitation of permutation methods with small samples is that there will be only a limited number of rearrangements. In consequence, we either have to accept a larger probability of making a Type I error or settle for a smaller acceptance region than may be desirable. One possible solution, first proposed by Streitberg and Roehmel [1990], is to make use of two statistics, a primary statistic to make an initial coarse division into acceptance, rejection, and boundary regions, and a second back-up statistic to resolve ties on the boundary.

For testing H against K in an $r \times c$ table, Cohen and Sackrowitz [1992] used $\sum_{i=1}^{R} \sum_{j=1}^{C} \sum_{k=1}^{i} \sum_{l=1}^{j} f_{kl}$ as their primary statistic and Freeman and Halton's statistic (Section 8.4) as their back-up.

For testing against H against K_S in an $2 \times c$ table, the mid-rank statistic proposed by Graubard and Korn served as the primary statistic for Streitberg and Roehmel [1990], while their back-up statistic was $\sum_{j=1}^{C} x_{2j}v_j$ where

$$v_j = \begin{cases} 2\bar{r}_j - 1 & \text{if } \bar{r}_j \leq (f_{..} + 1)/2, \\ 2(f_{..} - \bar{r}_j + 1) & \text{if } \bar{r}_j > (f_{..} + 1)/2, \end{cases}$$

and $\bar{r}_j$ is the average rank for the jth category.

8.5.1.3 Directed Chi-Square

On example of a statistic that takes on larger values when the data are drawn from a one-sided ordered alternative in K_S is the directed chi-square statistic proposed by Cohen and Sackrowitz [2000].

$$\chi_D^2 = \inf_{u \in A} \sum_{j=1}^{C} u_j^2 f_{.j}, \text{ where}$$

$$A = \left\{ u : \sum_{k=1}^{j} u_k f_{.k} \geq \sum_{k=1}^{j} f_{1k}; j = 1,\ldots,C-1; \sum_{k=1}^{C} u_k f_{.k} = f_{.1} \right\}$$

These authors claim that when used as the primary statistic with the mid-rank statistic as a back-up almost all ties may be eliminated. To verify, enter cell frequencies at http://stat.rutgers.edu/~madigan/dvp.html.

8.5.2 More Than Two Rows and Two Columns

Two cases need to be considered: The first when the columns but not the rows of the table may be ordered (the other variable being purely categorical), and the second when both columns and rows can be ordered.

8.5.2.1 Singly Ordered Tables

Several tests have been proposed (see Agresti [1992], Haberman [1974], and Soms [1985]). The test statistic takes the form $F_2 = \sum (T_i - \bar{T})^2$ where $T_i = \sum g_j f_{ij}$. As in the case of the $2 \times c$ table, our problem is in deciding on the appropriate scores (g_j). Among the proposals are ranks, normal scores, and Savage scores.[4]

8.5.2.2 Doubly Ordered Tables

Our log-linear model is that

$$\log[E(n_{ij})] = \mu + \lambda_i^X + \lambda_j^Y + \lambda_{ij}.$$

In an $r \times c$ contingency table conditioned on fixed marginal totals, Cornfield [1956] showed that the outcome depends only on the $(r-1)(c-1)$ odds ratios

$$\phi_{ij} = \frac{\pi_{ij} \pi_{i+1,j+1}}{\pi_{i,j+1} \pi_{i+1,j}},$$

where π_{ij} is the probability of an individual being classified row i and column j.

In a 2×2 table, conditional probabilities depend on a single odds ratio, and hence, one- and two-tailed tests of association are easily defined. In an $r \times c$ table, there are potentially $n = 2(r-1)(c-1)$ sets of extreme values, two for each odds ratio. Hence, an omnibus test for no association, e.g., χ^2, might have as many as 2^n tails.

Following Patefield [1982], we consider tests of the null hypothesis of no association between row and column categories $H: \phi_{ij} = 1$ for all i, j against the alternative of a positive trend $K: \phi_{ij} \geq 1$ for all i, j.

The two principal test statistics considered by Patefield, are

$$\lambda_3 = \sum \sum f_{ij} r_i c_j,$$

where $\{r_i\}$ and $\{c_j\}$ are user-chosen row and column scores,[5] and

$$\lambda_2 = \sup \sum \sum f_{ij} x_i y_j,$$

---

[4] See Chapter 11.3 for an explanation of these terms.
[5] This statistic is actually just another form of Mantel's U, perhaps the most widely used of all multivariate statistics, See Chapter 9.

where the supremum is taken over all sets $\{x_i\}$ and $\{y_j\}$, satisfying the conditions $\sum f_{i.}x_i = 0$, $\sum f_{.j}y_j = 0$, $\sum f_i x_i^2 = n_{..}$, $\sum f_{.j}y_j^2 = f_{..}$; and

$$x_1 \leq x_2 \cdots \leq x_r; \quad y_1 \leq y_2 \cdots \leq y_c.$$

Patefield finds that λ_2 has higher power than the linear-by-linear association test λ_3 when some but not all of the odds ratios ϕ_{ij} are close to unity, whereas λ_3 has higher power than λ_2 when the odds ratios all have about the same value.

The log-likelihood ratio behaves like λ_2; the Goodman and Kruskal test of association behaves like λ_3.

Other possible statistics, including one based on the difference between the numbers of concordant and discordant pairs, are considered by Agresti and Wackerly [1977].

8.6 Covariates

The presence of a covariate adds a third dimension to a contingency table. We consider two approaches to the analysis of higher-dimension tables: Bross' method and blocking.

8.6.1 Bross' Method

Bross [1964] studies the effects of treatment on the survival of premature infants. His results are summarized in Table 8.14. These results, though suggestive, are not statistically significant.

Bross notes that survival is very much a function of a third, concomitant variable—the birth weight of the child. A low birth weight indicates greater prematurity and, hence, greater odds against a child's survival. An analysis of treatment is out of the question unless, somehow, he can correct for the effects of birth weight.

A solution we studied in earlier chapters is to set up an experiment in which we study the effects of treatment in pairs that have been matched on the basis of birth weight. But Bross' study of the premature was not an experiment; he could only observe, not control birth weight.

Table 8.14. Effect of treatment of survival of the premature.

	Dead	Recovered	Totals
Placebo	6	5	11
Treatment	2	12	14
Totals	8	17	25

Table 8.15. Effect of treatment and birth weight on survival of the premature.

Weight	Treatment	Outcome	NI TR/PL	I PL/TR
1.08	TR	D		
1.13	TR	R	3	
1.14	placebo	D		
1.20	TR	R	2	
1.30	TR	R	2	
1.40	placebo	D		
1.59	TR	D		
1.69	TR	R	1	
1.88	placebo	D		

Table 8.15 depicts his first nine observations, ordered by birth weight. The last two columns of this table deserve explanation. The column headed *NI* records the number of cases in which a child of lower birth weight treated with ukinase recovered when an untreated child of higher birth weight died. Such a result is to be expected under the alternative of a positive treatment effect, though it would occur only occasionally by chance under the null hypothesis.

The column headed *I* records the number of cases in which an untreated child of lower birth weight recovered when a child of higher birth weight treated with ukinase died. Such an event or inversion would be highly unlikely under the alternative.

As his test statistic, Bross adopts

$$S = \frac{(NI - I)^2}{NI + I}.$$

Note that total of $NI = 8$, $I = 0$, and $S = 8$ for the original observations. Bross computes S for each of the $\binom{9}{3}$ possible rearrangements of the treatment labels—and only the label were changed, the pairing of birth weight with outcome was preserved. None of the other rearrangements yield as large a value of S as the original observations. Bross concludes that the treatment has a statistically significant effect on survival of the premature.

8.6.2 Blocking

Another way to correct for the effects of a covariate is to divide the observations into blocks so that the value of the covariate is approximately constant within each block. Under the assumption that the odds ratio is the same for each block, Mehta, Patel, and Gray [1985] provide a method for combining the results from several 2×2 contingency tables. To test the assumption of a constant odds ratio, $\theta_1 = \theta_2 = \cdots = \theta_B$, use Zelen's test (see Section 8.2.1).

On the other hand, it may be that an apparent association between the variables determining the rows and columns of a contingency table is actually the result of an association with a third factor. By separating the data into blocks based on the values of this third factor, we may test this latter assumption, $\theta_1 = \theta_2 = \cdots = \theta_n = 1$. Lehmann [1986, pp 162–166] showed that a UMPU test exists and is given by rejecting if $T = \sum_{k=1}^{B} f_{11k}$ is an extreme value relative to that of other tables with the same fixed marginals.

A two-sided test can be obtained by doubling the exact one-sided p-value, or by specifying that $(T - E(T))$ be less than or equal to the value actually observed.

Birch [1964] showed that this result can be extended to B $2 \times c$ contingency tables whose c columns are ordered, using the linear rank statistic given by $T = \sum_{k=1}^{B} \sum_{j=1}^{c} w_j f_{1jk}$, where the weights or scores w_j are selected as described in Section 8.5.1.[6]

Agresti [1992] showed this result may be extended still further to tests of the conditional independence of the row and column variables in an $r \times c$ table given a third blocking variable. If we can assume that the $(r-1)(c-1)$ odds ratios are identical for all values of the blocking factor, our test statistic is

$$dV^{-1}d,$$

where d is the matrix with elements $d_{ij} = \sum_k [f_{ijk} - f_{i.k} f_{.jk}/f_{..k}]$, $i = 1, \ldots, R-1$; $j = 1, \ldots, C-1$; and V is the null covariance matrix of d.

If we cannot assume the odds ratios are identical under the alternative, then we may still test for conditional independence using the statistic $\sum \chi_k^2$, where χ_k^2 is the chi-square statistic for testing independence of rows and columns within the kth level of the blocking factor.

8.7 Exercises

1. Use a normal approximation to show that more powerful tests in a 2×2 contingency table can be obtained by selecting samples of equal size on the basis of that attribute whose expected frequency is closest to 1/2.
2. Would you use the same test for comparing two or more binomials as you would when all the margins are random and only the total number of observations is fixed? Suppose that X is $B(p_1, m)$ and Y is $B(p_2, n)$. Show that Fisher's exact test based on the conditional distribution of X given the number of successes in the two samples is UMP among all unbiased tests of the hypothesis $p_1 = p_2$ against the alternatives $p_1 < p_2$.

---

[6] We need to assume the absence of a joint dependence among the three variables.

3. In many epidemiological studies, each entry in the table is the result of an independent Poisson process and the total number of observations is itself a random variable. In a 2×2 table with categories ab, aB, Ab, and AB, one might want to test the hypothesis that the ratio $\lambda_{ab}/\lambda_{Ab} \leq \lambda_{aB}/\lambda_{AB}$. Would you use the same test here as you would for comparing two binomials?
4. Show that once you have selected $(r-1)(c-1)$ of the entries in a contingency table with r rows and c columns the remainder of the entries are determined.
5. A recent report in the New England Journal of Medicine concerned a group of patients with a severe bacterial infection of their blood stream who received a single intravenous dose of a genetically altered antibody.

   a) In comparing death rates of the treated and untreated groups, should we use a one-tailed or a two-tailed test?
   The report stated that those in the treated group had a 30% death rate compared with a 49% death rate for a group of untreated patients.
   b) How large a sample size would you require using Fisher's exact test to show that such a percentage difference was statistically significant at the 5% level?

6. Suppose you observed the following table:

10	90
20	90

   Determine the p-value as many different ways as you can. Conduct a sensitivity test by determining the p-values for the table

10	91
20	89

   For a discussion of your results, see Dupont [1986].
7. How would you go about obtaining a confidence interval for $\pi_1 - \pi_2$? π_1/π_2? See Santner and Snell [1980].
8. Referring to Table 8.9, if Sandoz excluded site 15 from their calculations, could they safely combine the data from the remaining sites?
9. Suppose we have K pairs of binomials (in Table 8.9, we have 22) and we feel safe to assume that all have the same odds ratio. Show that a UMP unbiased test of the hypothesis that the odds ratio is 1 against the alternative that it is greater than 1 is based on the sum over all the tables of the number of successes of the first variable.
10. Will encouraging your child promote his or her intellectual development? A sample of 100 children and their mothers were observed and the children's IQs tested at 6 and 12 years. Before examining the data,

a) Do you plan to perform a one-tailed or two-tailed test? Results were as follows:

	Mothers Encourage Schoolwork		
	Rarely	Sometimes	Never
IQ increased	8	15	27
IQ decreased	30	9	11

b) What is the significance level of your test?

11. Holmes and Williams [1954] studied tonsil size in children to verify a possible association with the virus *S. pyrogenes*. Do you feel there is an association? How many rows and columns are in the following contingency table? Which, if any, of the variables is ordered?

Tonsil Size by Noncarrier and Carrier of *S. Pyrogenes*			
	Not Enlarged	Enlarged	Greatly Enlarged
Noncarrier	497	560	269
Carrier	19	29	24

9
Multivariate Analysis

The value of an analysis based on simultaneous observations on several variables—for example, height, weight, blood pressure, and cholesterol level—is that it can be used to detect subtle changes that might not be detectable, except with very large, prohibitively expensive samples, were you to consider only one variable at a time.

In this chapter we consider four approaches to the analysis of multivariate data: via the nonparametric combination of univariate tests; by parametric means, utilizing canonical forms and the properties of the multivariate normal; by permutation means utilizing essentially the same statistics as are used in the parametric approach but obtaining reference values from a permutation distribution; and by means of a nonparametric runs test.

We also consider methods for analyzing repeated measures.

9.1 Nonparametric Combination of Univariate Tests

To obtain a test that will take full advantage of the multivariate approach, we follow in the footsteps of Pesarin [1990, 2001] and harness several of the ideas we've developed previously—univariate statistics for optimally exposing differences among groups of metric, ordinal, or categorical observations, the use of ranks to place diverse observations on a single common scale, the Fisher omnibus statistic, and the permutation test.

As is the case with all the methods considered in this chapter, all the observations on a single experimental unit are maintained as a single indivisible vector. Labels are applied to and exchanged among these vectors, and not among the individual observations.

Let X denote the original $n \times K$ matrix of multivariate observations. Corresponding to X is a $1 \times K$ vector $T_0 = T(X)$ of univariate statistics. In a clinical trial, for example, some of the observations might relate to the occurrence or nonoccurrence of certain side effects, some might be the values of certain blood chemistries, and others might relate to quality of life. The corresponding univariate statistics might include Pearson's chi-square, several t-statistics, and several Pitman correlations.

Permuting the labels on the observation vectors yields a new matrix $\boldsymbol{X'}$ and a new vector $\boldsymbol{T'} = \boldsymbol{T}(\boldsymbol{X'})$ of univariate statistics. To obtain a single summary statistic encompassing the information provided by all the observations, we proceed as follows:

1. Generate a large number N of permutations of $\boldsymbol{X}$ and thus obtain N vectors of univariate test statistics $\boldsymbol{T}_i$, $i = 1, \ldots, N$.
2. Rank the $N+1$ values of each single-variable test statistic separately. The rank should be related to the extent to which the statistic favors the alternative. For example, if large values of T_{ik} are to be expected when the principal hypothesis concerning the kth variable is false, then,

$$R_{ik} = R(T_{ik}) = \sum_h I[T_{hk} \leq T_{ik}], \quad \text{for } i = 0, \ldots, N,$$

where the indicator function $I[\mathrm{E}]$ takes values 1 or 0 according to whether the event E is true or false.
3. Combine the ranks of the K individual univariate tests using Fisher's omnibus statistic

$$U_i = -\sum_{k=1}^{K} \log\left[\frac{N + 0.5 - R_{ik}}{N+1}\right]; i = 1, \ldots, N.$$

4. Determine from the individual permutation distributions the marginal significance level of each of the single-variable statistics for the original nonpermuted observations,

$$p_k = \frac{0.5 + \sum_{m=1}^{N} I[T_{mk} \geq T_{0k}]}{N+1}, \quad k = 1, \ldots, K.$$

5. Combine these values into a single statistic

$$U_0 = -\sum_{k=1}^{K} \log[p_k] = -\sum_{k=1}^{K} \log\left[\frac{N + 0.5 - R_{0k}}{N+1}\right];$$

note that R_{0k} can take any value in the range 0 to N.
6. Determine the significance level of the combined test,

$$p = \frac{0.5 + \sum_{m=1}^{N} I[U_m \geq U_0]}{N+1}.$$

Liptak's or Tippett's combining functions, described in Section 5.2.1, can be employed in preference to Fisher's.

The range of application of this method is very general. If H_i, A_i denote the hypothesis, alternative associated with the ith variable, respectively, then the multivariate hypothesis you are testing is H_1 and H_2 and ... H_J and the alternative is A_1 or A_2 or ... A_K. Individual alternatives may be omnibus in nature or restricted. The individual variables may be independent or interdependent and may be metric or nonmetric, discrete, or continuous. Their distributions need be known only in so far as this knowledge would influence the choice of univariate test statistics.

Software to perform the combination method is available from www.methodologica.it.

9.2 Parametric Approach

9.2.1 Canonical Form

The ideas espoused in Section 6.1.2 can be extended to multivariate observations $(X_1, \ldots, X_K)$ that are distributed in accordance with the multivariate normal probability density

$$\frac{\sqrt{|D|}}{(2\pi)^{K/2}} \exp\left[-\frac{1}{2}\sum_{i=1}^{K}\sum_{j=1}^{K} d_{ij}(x_i - \mu_i)(x_j - \mu_j)\right].$$

where the matrix $D = (d_{ij})$ is positive definite, and $|D|$ denotes its determinant;

$$E(X_j) = \mu_j; E(X_j - \mu_j)(X_j - \mu_j) = \sigma_{ij}; (\sigma_{ij}) = D^{-1}.$$

Our arguments parallel those of Section 6.1.2. Let $\boldsymbol{X}_1, \ldots, \boldsymbol{X}_n$ denote independent multivariate normal vectors; $\boldsymbol{X}_i = (X_{i1}, \ldots, X_{iK}), i = 1, \ldots, n$ with $E(X_{ik}) = \mu_{ik}$ for $k = 1, \ldots, K$ and common covariance matrix D^{-1}. The column vector of means $\boldsymbol{\mu}_j = (\mu_{1j}, \ldots, \mu_{nj})$ is known to lie in a given s-dimensional subspace Ω where $s < n$ and the hypothesis to be tested is that $\boldsymbol{\mu}$ lies in an $(s-r)$-dimensional subspace of Ω.

Again, the hypothesis can be given a particularly simple form by making an orthogonal transformation to vectors $\boldsymbol{Y}_1, \ldots, \boldsymbol{Y}_n$

$$\boldsymbol{Y} = \boldsymbol{CX},$$

such that the first s row vectors of $\boldsymbol{C}$ span Ω, and the first r row vectors of $\boldsymbol{C}$ span the subspace under the hypothesis.

The rows of $\boldsymbol{Y}$ are independent multivariate normal vectors (Exercise 9.3). Change the notation so that Y's, U's, and X's denote the first r, the next $s-r$ and the last $n-s$ row vectors in $\boldsymbol{Y}$:

$$\begin{pmatrix} Y \\ U \\ Z \end{pmatrix} = \boldsymbol{CX}$$

Thus, the expected values of the Z's are zero and the hypothesis to be tested is that the expected values of the Y's are zero, also.

172 9 Multivariate Analysis

To find an optimal statistic for testing the hypothesis, we shall use our objective of impartiality to reduce the potential choices to that of a maximal invariant.

Adding an arbitrary constant to each of the variables in $\mathbf{U}$ leaves the testing problem invariant, thus eliminating these variables from further consideration.

It is easy to see that the orthogonal transformations B and G, $Y^* = BY$ and $Z^* = GZ$, also leave the problem invariant and that the maximal invariants with respect to the two groups of such transformations are $V = Y^T Y$ and $S = Z^T Z$ (Exercise 9.4).

Assume that $n-s$, the number of degrees of freedom we have for estimating the components of the covariance matrix, is greater than the number of covariates k (see Exercise 9.5). Consider the group of nonsingular $p \times p$ matrices M, such that $Y^* = YM$ and $Z^* = ZM$. These transformations leave the testing problem invariant and induce the transformations $V^* = M^T V M$ and $S^* = M^T S M$ with respect to which the roots of the determinant equation $|V - \lambda S| = 0$ are maximal invariants (Exercise 9.6).

9.2.2 Hotelling's T^2

The number of nonzero roots of the determinant equation $|V - \lambda S| = 0$ is $\min(k, p)$. Consider the case $p=1$, $k > 1$. The equivalent equation $|VS^{-1} - \lambda I| = 0$ reduces to

$$(-\lambda)^k + W(-\lambda)^{k-1} = 0,$$

where W is the trace of VS^{-1}.

$$W = \sum_i \sum_j S^{ij} Y_i Y_j,$$

where S^{ij} denotes the ijth element of S^{-1}.

In the one-sample case, we draw $n > k$ observations from a k-variate normal distribution with unknown mean $\boldsymbol{\mu} = (\mu_1, \ldots, \mu_K)$ and wish to test the hypothesis that $\boldsymbol{\mu} = \boldsymbol{\mu}^*$ against the omnibus alternative $\boldsymbol{\mu} \neq \boldsymbol{\mu}^*$. It is easy to show (Exercise 9.7) that $Y_i Y_j = (X_i - \mu_i^*)(X_j - \mu_j^*)/n$ and

$$S_{ij} = \sum_{h=1}^{n} (X_{hi} - \bar{X}_{.i})(X_{hj} - \bar{X}_{.j}),$$

where S_{ij} is the ijth element of S. The statistic $(n-1)W$ is known as *Hotelling's T^2*, and we reject the hypothesis if it is large. The statistic $(n-k)W/k$ has the F-distribution with k and $n-k$ degrees of freedom. The associated confidence sets

$$n(n-1) \sum_{i=1}^{n} \sum_{h=1}^{n} (\mu_i - \bar{X}_{.i}) S^{ij} (\mu_j - \bar{X}_{.j}) \leq C$$

are ellipsoids centered at the sample means.

The two-sample comparison with samples of size m and n is only slightly more complicated. We assume that the multivariate observations $(\boldsymbol{X}_{11},\ldots,\boldsymbol{X}_{1m},\boldsymbol{X}_{21},\ldots,\boldsymbol{X}_{2n})$ are independent, multivariate normal with a common covariance matrix and expectations $\boldsymbol{\mu}_1$ and $\boldsymbol{\mu}_2$, and that we have sufficient observations, $n + m - 2 > k$ that we can estimate the covariance matrix. We wish to test the hypothesis $\boldsymbol{\mu}_1 = \boldsymbol{\mu}_2$ against the omnibus alternative $\boldsymbol{\mu}_1 \neq \boldsymbol{\mu}_2$, so that $s = 2$ and $p = 1$. Then

$$Y_i Y_j = m(\bar{X}_{1.i} - \bar{X}_{..i})(\bar{X}_{1.j} - \bar{X}_{..j}) + n(\bar{X}_{2.i} - \bar{X}_{..i})(\bar{X}_{2.j} - \bar{X}_{..j})$$

and

$$S_{ij} = \sum_{h=1}^{m}(\bar{X}_{1hi} - \bar{X}_{1.i})(\bar{X}_{1hj} - \bar{X}_{1.j}) + \sum_{h=1}^{n}(\bar{X}_{2hi} - \bar{X}_{2.i})(\bar{X}_{2hj} - \bar{X}_{2.j})$$

Hotelling's T^2 is given by $(n+m)(n+m-2)(\bar{\boldsymbol{X}}_1 - \bar{\boldsymbol{X}}_2)^T S^{-1}(\bar{\boldsymbol{X}}_1 - \bar{\boldsymbol{X}}_2)$ and we reject the hypothesis when it is large.

9.2.3 Multivariate Analysis of Variance (MANOVA)

When there are three or more samples, the number of vector constraints imposed by the hypothesis exceeds one (see Exercise 9.6), and a UMP invariant test no longer exists. A number of tests based on the roots of the determinant equation $|V - \lambda S| = 0$ have been proposed; for example, the Lawley–Hotelling trace test which rejects for large values of $\sum \lambda_i$. Significance levels for these tests can be determined by parametric means (standard in most statistics software) or by the permutation methods described in the next section.

Tests against specific restricted alternatives are difficult to obtain in the parametric setting and a nonparametric approach is recommended; see, for example, Shorack [1967], Robertson, Wright, and Dykstra [1988], El Barmi and Dykstra [1995], and Dardanoni and Forcina [1998].

9.3 Permutation Methods

Cut-off values for any of the statistics considered in the previous section may be determined by reference to its permutation distribution.

In the multivariate version of the permutation methodology, each vector of observations on an individual subject is treated as a single indivisible entity. When we relabel, we relabel on a subject-by-subject basis so that all observations on a single subject receive the same new label. If the original vector of observations on subject i in sample j consists of k distinct observations on k different variables and we give this vector a new label j^*, then the individual observations remain together as a unit, each with the new label.

To test the hypothesis that the midvalues of two distributions are the same, we could use Hotelling's T^2 given by

$$(n+m)(n+m-2)(\bar{X}_1 - \bar{X}_2)^T S^{-1}(\bar{X}_1 - \bar{X}_2),$$

but then we would be forced to recompute the covariance matrix S and its inverse for each new rearrangement. To reduce the number of computations, Wald and Wolfowitz [1943] suggest a slightly different statistic, T', that is, a monotonic function of T (see Exercise 3.13). Let

$$U_j = N^{-1} \sum_{k=1}^{2} \sum_{i=1}^{n_k} X_{kij}$$

$$c_{ij} = \sum_{k=1}^{2} \sum_{m=1}^{n_k} (X_{kmi} - U_i)(X_{kmj} - U_j).$$

Let C be the matrix with components c_{ij}. Then $T'^2 = (\bar{X}_{1.} - \bar{X}_{2.})^T C^{-1} (\bar{X}_{1.} - \bar{X}_{2.})$

As with all permutation tests we proceed in three steps:

1. Compute the test statistic for the original observations;
2. compute the test statistic for all relabelings;
3. determine the percentage of relabelings that lead to values of the test statistic that are as, or more, extreme than the original value.

Regrettably no commercial software is presently available. We need to program and implement three procedures:

a) one to rearrange the stored data;
b) one to compute the T^2 statistic;
c) one to compute the significance level.

Only the first of these procedures, devoted to rearranging the data, represents a significant change from the simple calculations performed in the univariate case. In a multivariate analysis, we can't afford to manipulate the actual data; a simple swap could mean the exchange of 9 or 10 or even 100 different variables; so we rearrange a vector of indices that point to the data instead. Here is a fragment of C code that does just that:

```
float Data [length, variates];
int index[length];
....
rearrange (index, length);
....
for (j = 0; j < n control; j++) Mean [k] += Data[index[j], k];
```

Program for Computing Multivariate Permutation Statistics

#define length   119
#define control   60
#define variates   9

Set aside space for a multivariate array data [length, variates]; and a vector of sample sizes index[length];

Main program
   load (data)
   compute stat0 (data, index);
   repeat Nsim times;
      rearrange data;
      compute stat (data, index);
      record whether stat >= stat0;
   print out the significance level of the test;

Load
   packs the data into a long matrix, each row of which corresponds to k observations on a single subject; the first n rows are the control group; the last m rows are the treatment group. (A second use of this subroutine will be to eliminate variables and subjects that will not be included in the analysis, e.g., to eliminate all records that include missing data, and to define and select specific subgroups.)

Rearrange
   randomly rearranges the rows of the data array; the elements in each row are left in the same order.

Compute
   calculate the mean of each variable for each sample and store the results in a 2 by n array N;
   calculate n by n array V of covariances for the combined sample and invert V;
   matrix mult (mean, W, *W);
   matrix mult (W, mean);
   return T'^2;

9.3.1 Which Test—Parametric or Permutation?

All the tests described in this and the preceding section require that all multivariate observations be independent of one another,[1] that they be metric, and that the covariance matrix be the same for all observations regardless of the values of the underlying parameters. Hotelling's T^2 is applicable only to shift alternatives ($F[x] = G[x - \delta]$). A further requirement is that there be a sufficient number of observations so that the covariance matrix can be estimated, that is, $n > K$.[2]

---

[1] Strictly speaking, permutation methods only require that the multivariate vectors be exchangeable.

[2] Use of the bootstrap would only result in the creation of a singular noninvertible matrix. See Dempster [1958] for a possible approach.

As described in Section 9.1, permutation solutions based on the nonparametric combination of univariate tests only require that the multivariate distributions be the same under the null hypothesis. The observations may be metric, ordinal, categorical, or a mixture thereof. The alternatives and testing method may vary from variable to variable and the alternatives are not confined to shifts of means. As univariate permutation tests in general do not require estimation of standard deviations, they are applicable even when $n < K$.

Nonetheless, based on power considerations, Hotelling's T^2 is the appropriate statistic in the one-sample and two-sample cases, providing:

1. All the observations are metric.
2. A two-sided test against shift alternatives is desired.
3. The metric data have a distribution close to that of the multivariate normal.
4. The samples are large.

Under these conditions, the distribution of the statistic converges to a chi-square distribution with K degrees of freedom independent of the distribution of the observations provided that the latter have finite second moments.

The stated significance level of the parametric version of Hotelling's T^2 cannot be relied on for small samples if the data are not normally distributed (Davis [1982]; Srivastava and Awan [1982]). As always, the corresponding permutation test yields an exact significance level even if the errors are not normally distributed, providing that the errors are exchangeable from sample to sample. Under the assumption of multivariate normality, the power of the permutation version of Hotelling's T^2 converges with increasing sample size to the power of the most powerful parametric test that is invariant under transformations of scale.

Much of the theoretical work on permutation tests using Hotelling's T^2 has focused on the properties of the *unconditional permutation test* in which the original observations are replaced by ranks. Details of the asymptotic properties and power of the unconditional test are given in Barton and David [1961], Chatterjee and Sen [1964, 1966], and Gill and Siotani [1987]. The effect of missing observations on the significance level and power of the test is studied by Servy and Sen [1987]. But see the cautionary comments on the use of ranks made by Blair, Sawilowsky, and Higgins [1987].

9.3.2 Interpreting the Results

The significance of T^2 or some equivalent multivariate statistic still leaves unanswered the question of which variables have led to the rejection of the multivariate hypothesis. As noted in Chapter 5, the presence of one or more falsely significant univariate results is not unexpected. My own preference, reflecting my studies under Jerzy Neyman, is to search for a mechanistic, cause-and-effect model that will explain the findings. Some of the many problems associated with multiple regression are documented in Good and Hardin [2003], chapters 8–10.

9.4 Alternative Statistics

9.4.1 Maximum-*t*

Hotelling's T^2 is designed to test the null hypothesis of no difference between the distributions of the treated and untreated groups against alternatives that involve a shift of the k-dimensional center of the multivariate distribution. Although Hotelling's T^2 offers protection against a wide variety of alternatives, it is not particularly sensitive to alternatives that entail a shift in just one of the dependent variables.

Boyett and Shuster [1977] show that a more powerful test against such alternatives is based on the permutation distribution of the test statistic

$$\max_{1 \leq j \leq K} \frac{(\bar{X}_{1.j} - \bar{X}_{2.j})}{SE_j},$$

a statistic first proposed in a permutation context by Chung and Fraser [1958], where SE_j is a pooled estimate of the standard error of the mean of the jth variable.

The *maximum t-test*, defined above, is one-to-one related to the Tippett nonparametric combination of several univariate t tests (defined in Section 5.2.1), when all tests have the same degrees of freedom. If the tests do not have the same degrees of freedom, the maximum t-test becomes difficult to apply (see Exercise 9.10); whereas Tippett's combination of permutation p-values has no such limitation.

9.4.2 Block Effects

When we have more than two treatments to compare, an alternative statistic studied by Gerig [1969, 1975] is the multivariate extension of Friedman's chi-square test in which ranks take the place of the original observations, creating an unconditional permutation test.

The experimental units are divided into B blocks each of size I with the elements of each block as closely matched as possible with respect to extraneous variables. During the design phase, one individual from each block is assigned to each of the I treatments. We assume that K (possibly) dependent observations are made simultaneously on each subject. To test the hypothesis of identical treatment effects against translation-type alternatives, we first rank each individual variable separately within each block, ranking them from 1 to I (smallest to largest). The rank totals $T_{i.k}$ are computed for each treatment

i and each variable k. The use of ranks automatically rescales each variable so that the variances (but not the covariances) are the same.

Let T denote the $I \times K$ matrix whose ikth component is $T_{i.k}$. Noting that the expected value of $T_{i.k}$ is $(K+1)/2$, let V denote the matrix whose components are the sample covariances

$$V_{st} = \frac{\sum_{b=1}^{B}\sum_{i=1}^{I} T_{ibs}T_{ibt} - K(K+1)^2/4}{n(K-1)}$$

By analogy with Hotelling's T^2, the test statistic is $T^{\mathrm{T}}V^{-1}T$ (Gerig [1969]). Gerig [1975] extends these results to include and correct for random covariates.

9.4.3 Runs Test

Friedman and Rafesky [1979] provide a multivariate generalization of the distribution-free, two-sample tests of Wald–Wolfowitz and Smirnov, used for testing $F_X = F_Y$ against the highly nonspecific alternative $F_X \neq F_Y$. In both the univariate and the multivariate versions of these two-sample tests, one measures the degree to which the two samples are segregated within the combined sample. In the univariate version, one forms a single combined sample, sorts and orders it, and then

a) counts the number of runs in the combined sample, or
b) computes the maximum difference in cumulative frequency of the two types within the combined sample.

For example, if $x = (1,3,6)$ and $y = (2,4,5)$, the ordered combined sample is 1, 2, 3, 4, 5, 6, that is, an x followed by $y\,x\,y\,y\,x$, and has five runs.

Highly segregated samples will give rise to a small number of runs (and a large maximum difference in cumulative frequency), while highly interlaced distributions will give rise to a large number of runs (and a very small difference in cumulative frequency). Statistical significance, that is, whether the number of runs is significantly small, can be determined from the permutation distribution of the test statistic.

To create a multivariate version of these tests, we must find a way to order observations that have multiple coordinates. The key to this ordering is the *minimal spanning tree* described by Friedman and Rafesky [1979] and given here:

Each point in Figure 9.1a corresponds to a pair of observations, height and weight, say, which were made on a single subject. We build a spanning tree between these data points as in Figure 9.1b, by connecting the points so that there is exactly one path between each pair of points, and so that no path closes back on itself in a loop. Obviously, we could construct a large number of such trees. A minimal spanning tree is one for which the sum of the lengths of all the paths is a minimum. This tree is unique if there are no ties among the $N(N-1)/2$ interpoint distances.

9.4 Alternative Statistics   179

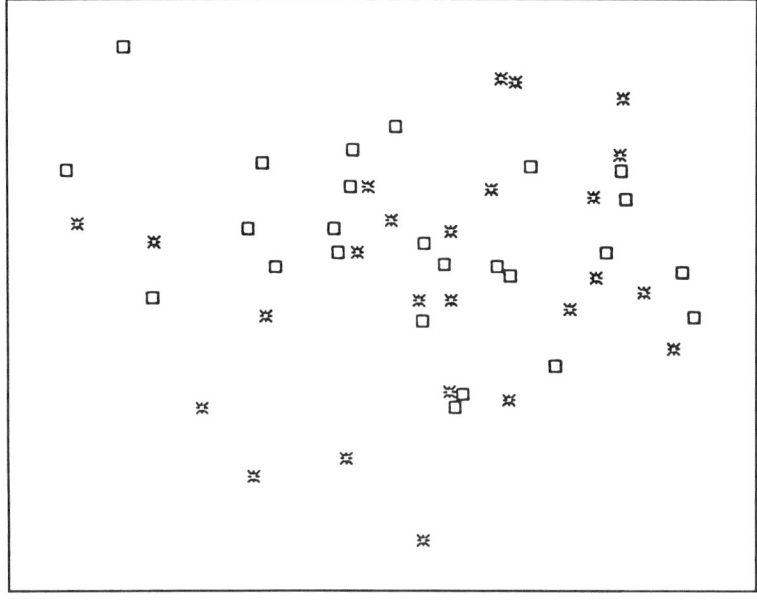

a

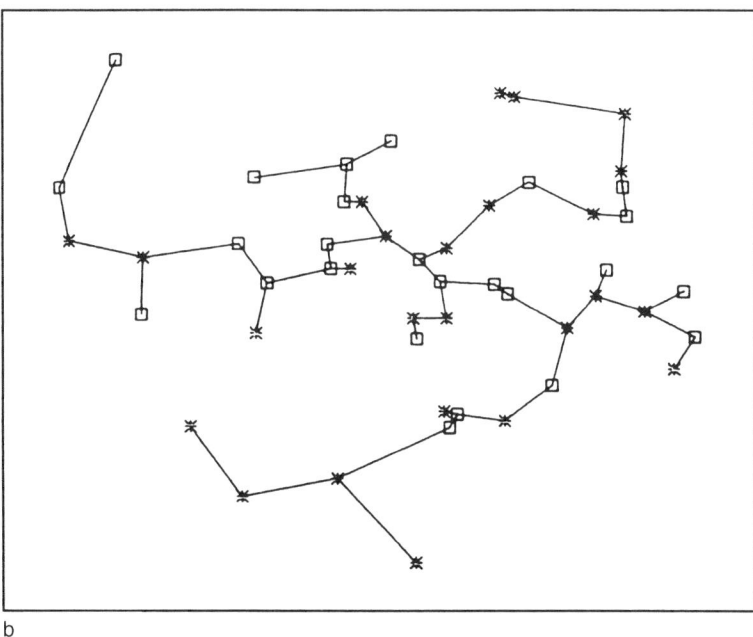

b

Fig. 9.1. Building a minimal spanning tree. From "Multivariate generalizations of the Wald–Wolfowitz and Smirnov two-sample narrative has test" by J.H. Friedman and L.C. Rafsky, *Annals of Statistics*; 1979; **7**: 697–717. Reprinted with permission from the Institute of Mathematical Statistics.

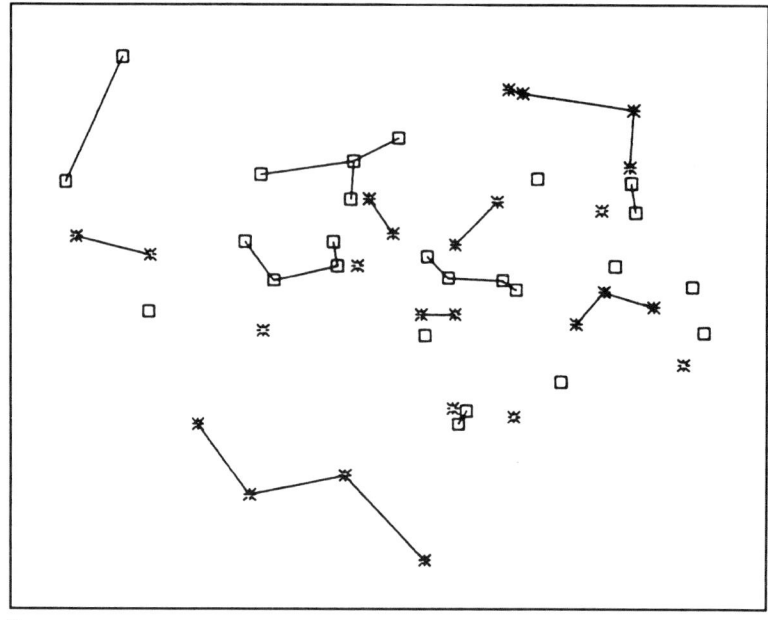

Fig. 9.1. (*continued*).

Before computing the test statistic(s) in the multivariate case, we first construct the minimal spanning tree for the combined sample. Once the tree is complete, we can generate the permutation distribution of the runs statistic through a series of random relabelings of the individual data points. After each relabeling, we remove all edges for which the defining nodes originate from different samples. Figure 9.1c illustrates one such result.

Although it can take a multiple of $N \times N$ calculations to construct the minimal spanning tree for a sample of size N, each determination of the multivariate runs statistic takes only a multiple of N calculations. For large samples a normal approximation to the permutation distribution may be used; the expected value and variance of the runs statistic are the same as in the univariate case.

9.4.4 Which Statistic?

We've now considered three multivariate test statistics for testing hypothesis based on one or two samples. Which one should we use? To detect a simultaneous shift in the means of several variables, use Hotelling's T^2; to detect a shift in any of several variables, use the maximum t; and to detect an arbitrary change in a distribution (not necessarily a shift) use either Pesarin's method of nonparametric combination or Friedman and Rafesky's multivariate runs test.[3] See Exercise 9.9.

9.5 Repeated Measures

In many experiments, we want to study the development of a process over a period of time, such as the growth of a tumor or the gradual progress of a cure. If our observations are made by sacrificing different groups of animals at different periods of time, then time is simply another variable in the analysis that we may treat as a covariate. But if all our observations are made on the same subjects, then the multiple observations on a single individual will be interdependent. And all the observations on a single subject must be treated as a single multivariate vector.

9.5.1 An Example

Higgins and Noble [1993] analyze an experiment whose goal was to compare two methods of treating beef carcasses in terms of the treatments' effect on pH measurements of the carcasses taken over time. Treatment level B is suspected to induce a faster decay of pH values. Formally, we wish to test a hypothesis of no difference between the treatments against the alternative that $XB[t]$ is stochastically smaller than $XA[t]$ for some time t.

Observed data are:

		\multicolumn{6}{c}{t}
		0	1	2	3	4	5
Treatment A	A1	6.81	6.16	5.92	5.86	5.80	5.39
	A2	6.68	6.30	6.12	5.71	6.09	5.28
	A3	6.34	6.22	5.90	5.38	5.20	5.46
	A4	6.68	6.24	5.83	5.49	5.37	5.43
	A5	6.79	6.28	6.23	5.85	5.56	5.38
	A6	6.85	5.51	5.95	6.06	6.31	5.39
Treatment B	B1	6.64	5.91	5.59	5.41	5.24	5.23
	B2	6.57	5.89	5.32	5.41	5.32	5.30
	B3	6.84	6.01	5.34	5.31	5.38	5.45
	B4	6.71	5.60	5.29	5.37	5.26	5.41
	B5	6.58	5.63	5.38	5.44	5.17	5.62
	B6	6.68	6.04	5.62	5.31	5.41	5.44

---

[3] Tests proposed by van Putten [1987] and Henze [1988] offer advantages over Friedman–Rafesky in some cases.

Although normality of these observations may be assumed, the variances and covariances surely vary with time so that the two-way ANOVA model is not appropriate. Instead, we may proceed as follows: First, we standardize the observations, subtracting the baseline value at $t = 0$ from each one. At each time point, the resulting differences are exchangeable. Treating each time point separately, the resulting p-values are as follows:

$T = 1$	$T = 2$	$T = 3$	$T = 4$	$T = 5$
0.01056	0.000127	0.000309	0.000395	0.06803

Using Fisher's nonparametric combination rule, the combined p-value for the global hypothesis is 0.000127. We can conclude that decay of treatment B is faster than that of A, even though at the last time point, $T = 5$, substantially the same distribution of pH values is observed ($p = 0.068$).

9.5.2 Matched Pairs

Puri and Shane [1970] study the multivariate generalization of paired comparisons in an incomplete blocks design (see Section 6.4). Their procedure is a straightforward generalization of the multivariate one-sample test developed by Sen and Puri [1967]; see also Sen [1967, 1969].

For simplicity, suppose we have only a single block. We consider all possible permutations of the signs of the individual multivariate observations. If $\{\boldsymbol{X}_i, \boldsymbol{Y}_i\}$ is the K-dimensional vector of multivariate observations on the ith matched pair, and $\boldsymbol{Z}_i$ is the vector of differences $(Z_1, \ldots, Z_K)$, then our permutation set consists of matrices of differences of the form $((-1)^{j1}\boldsymbol{Z}_1, \ldots, (-1)^{jn}\boldsymbol{Z}_n)$ where $-\boldsymbol{Z}_i = (-Z_{i1}, \ldots, -Z_{iK})$.

Depending on the hypothesis and alternatives of interest, one may want to apply an initial set of linear transformations to each separate coordinate, that is, to replace Z_{ij} by $Z'_{ij} = a_j + b_j Z_{ij}$. Puri and Shane studied the case in which the individual variables were replaced by their ranks, with each variable being ranked separately.

Of course, if the sample size n is large, or we can assume that $\{\boldsymbol{X}_i, \boldsymbol{Y}_i\}$ is drawn from a multivariate normal distribution, then we can use the principle of invariance to find an optimal test. Our observations on the ith subject may be written in the form $\{X_{i,1}, \ldots, X_{i,q}, X_{i,q+1}, \ldots, X_{i,2q}\}$. We are given that $E(X_{i,j}) = \mu_{i,j} = \mu_j$ and our null hypothesis is that $\mu_j = \mu_{j+q}$. The distinction from the approach adopted in Section 9.2 is that, while in both cases each of the K column vectors $\boldsymbol{\mu}_j = (\mu_{1j}, \ldots, \mu_{nj})$ is known to lie in a given s-dimensional subspace Ω where $s < n$, in the matched pairs case $s = 1$, and the hypothesis to be tested is that each of the n row vectors $\boldsymbol{\mu}_i = (\mu_{i,1}, \ldots, \mu_{i,2q})$ lies in a q-dimensional subspace of Ω. The derivation provided by Lehmann [1986, pp 466–469] yields a UMP-invariant test based on Hotelling's T^2. The cut-off value for our test may be determined parametrically or via the more robust permutation distribution.

Pesarin [2001; Section 9.7] describes the analysis of multivariate paired observations when there are randomly missing data.

9.5.3 Response Profiles

We may ask at least three questions about response profiles involving more than two observations on each experimental unit: (1) Are the response profiles the same for the various treatments? (2) Are the response profiles parallel? (3) Are the response profiles at the same level?

A "yes" answer to question 1 implies "yes" answers to questions (2) and (3), but we may get a "yes" answer to 2 even when the answer to (3) is "no."

One simple test of parallelism entails computing the successive differences $z_{j,i} = x_{j,i+1} - x_{j,i}$ for $j = 1, 2; i = 1, \ldots, I-1$ and then applying the methods of 9.2 or 9.3 to these differences. Of course, this approach is applicable only if the observations on both treatments were made at identical times.

To circumvent this limitation and to obtain a test of the narrower hypothesis (1), we follow Koziol et al. [1981] and suppose there are N_i subjects in group i. Let X_{tj}^i, $t = 1, 2, \ldots, T; j = 1, 2, \ldots, N_i$ denote the observation on the jth subject in group i at time t. Not all the X_{tj}^i may be observed in practice; we will only have observations for N_{it} of the N_i in the ith group at time t. If X_{tj}^i is observed, let R_{tj}^i be its rank among the $N_{.t}$ available values at time t. Set $S_{it} = \sum_j R_{tj}^i / N_{it}$.

If luck is with us so that all subjects remain to the end of the experiment, then $N_{it} = N_i$ for all t and each i, and we may adopt as our test statistic

$$L_N = \sum_i N_i \boldsymbol{S}_i^\mathrm{T} \boldsymbol{V}^{-1} \boldsymbol{S}_i,$$

where $\boldsymbol{S}_i$ is a $T \times 1$ vector with components $(S_{i1}, \ldots, S_{iT})$ and $\boldsymbol{V}$ is a $T \times T$ covariance matrix whose stth component is

$$vst = N^{-1} \sum_{i=1}^{I} \sum_{j=1}^{N_i} R_{sj}^i R_{tj}^i$$

This test statistic was proposed and investigated by Puri and Sen [1966, 1969, 1971].

9.5.4 Missing Data

If we are missing data, and missing data is almost inevitable in any large clinical study since individuals commonly postpone or even skip follow-up appointments, then no such simplified statistic presents itself. Zerbe and Walker [1977] suggest that each subject's measurements first be reduced to a vector of polynomial regression coefficients with time as the independent variable. The subjects needn't have been measured at identical times or over identical periods, nor does each subject need to have the same number of observations. Only the number of coefficients (the rank of the polynomial), needs to be the same for each subject. Thus, we may apply the equations of Koziol et al. to these vectors of coefficients though we cannot apply the equations to the original data.

We replace the m_k observations on the kth subject $\{X_{ki}, i = 1, \ldots, m_k\}$, with a set of $J + 1$ coefficients $\{b_{kj}, j = 0, \ldots, J\}$. While the m_k may vary, the number J is the same for every subject; of course, $J < m_k$ for all k. The $\{b_{kj}\}$ are chosen so that for all k and i,

$$X_{ki} = b_{kj} + t_{ki}b_{kj} + \cdots + (t_{ki})^J b_{kj},$$

where the $\{t_{ki}, i = 0, \ldots, mk\}$ are the observation times for the kth subject.

This approach has been adopted by a number of practitioners including Albert et al. [1982], Chapelle et al. [1982], Goldberg et al. [1980], and Hiatt et al. [1983]. Multiple comparison procedures based on it include Foutz et al. [1985] and Zerbe and Murphy [1986]. An SAS/IML program to do the calculations is available (Nelson and Zerbe [1988]).

9.5.5 Bioequivalence

Zerbe and Walker's solution to the problem of missing data suggests a multivariate approach we may use with any time course data. For example, when we do a bioequivalence study, we replace a set of discrete values with a "smooth" curve. This curve is derived in one of two ways: 1) by numerical analysis, 2) by modeling. The first way yields a set of coefficients, the second a set of parameter estimates. Either the coefficients or the estimates may be treated as if they were the components of a multivariate vector and the methods of this chapter applied to them.

Here is an elementary example: Suppose you observe the time course of a drug in the urine over a period for which a linear model would be appropriate. Suppose further that the chief virtue of your measuring system is its low cost so that measurement errors are significant. Consequently, you take a series of measurements on each patient about half an hour apart and then use least squares methods to derive a best-fitting line for each patient. That is, you replace the set of measurements $\{X_{ijk}\}$ where $i = 0$ or 1 denotes the drug, $i = 1$ for generic drug, $j = 1, \ldots, J$ denotes the subject, and $k = 1, \ldots, K_j$ denotes the observation on a subject with the set of vectors $\{\boldsymbol{Y}_{ij} = (a_{ij}, b_{ij})\}$

where a_{ij} and b_{ij} are the intercept and slope of the regression line for the jth subject in the ith treatment group.

Using the computer code in Section 9.3, you calculate the mean vector and the covariance matrix for the $\{Y_{ij}\}$, and compute Hotelling's T^2 for the original observations and for a set of random arrangements. You use the resultant permutation distribution to determine whether the time courses of the two drugs are similar.

9.6 Exercises

1. One can increase the power of a statistical test in three ways: a) make additional observations; b) make more precise observations; c) add covariates. Discuss this remark in the light of your own or someone else's experimental efforts.
2. Suppose that $U_{ij}(j = 1, \ldots, n_i;\ i = 1, \ldots, I)$ are independent multivariate normally distributed observations $N(\mu_i, A^{-1})$ and we wish to test the hypothesis that $\mu_1 = \cdots = \mu_I$. What are the values of n, s, and r?
3. If X denotes a matrix whose rows are independent multivariate normal vectors, C is an orthogonal matrix, and $Y = CX$, then the rows of Y are independent multivariate normal vectors. [Hint. Find the covariance of Y_{ij} and Y_{km}.]
4. Prove that if X denotes a matrix whose rows are independent multivariate normal vectors whose components have expectation zero, C is an orthogonal matrix, and $X^* = CX$, then $X^\mathrm{T} X$ is a maximal invariant with respect to the group of such orthogonal transformations.
5. a) If $n - s < J$, the matrix S is singular.
   b) If $r + n - s \leq J$, the only test that is invariant with respect to the groups of transformations in Section 9.2. is $\varphi(y, u, z) \equiv \alpha$.
6. Let Y and Z denote matrices whose rows are independent multivariate normal vectors, each component of which has expectation zero. Let $V = Y^\mathrm{T} Y$ and $S = Z^\mathrm{T} Z$. Consider the group G of nonsingular $p \times p$ matrices M, such that $Y^* = YM$ and $Z^* = ZM$. Show that the roots of the determinant equation $|V - \lambda S| = 0$ are maximal invariants with respect to G. [Hint: To prove invariance, consider $|M^\mathrm{T}(V - \lambda S)M|$. To prove the roots are maximal invariants, you will need to show that S is positive definite and utilize known properties of quadratic forms.]
7. Show that in the one-sample case the statistic $(n-1)W$ minimizes mean-square losses.
8. The following blood chemistry data are taken from Werner et al. [1970]. The full data set is included with the BMDP^{TM} statistical package. (An asterisk ($*$) denotes missing data.)

1	2	3	4	5	6	7	8	9
2381	22	67	144	N	200	43	98	54
1946	22	64	160	Y	600	35	*	72
1610	25	62	128	N	243	41	104	33
1797	25	68	150	Y	50	38	96	30
1149	53	*	178	N	227	39	*	50
575	53	65	140	Y	220	40	107	46
2271	54	66	158	N	305	42	103	48
39	54	60	170	Y	220	35	88	63

The variables are

1. identification number
2. age in years
3. height in inches
4. weight in pounds
5. uses birth control pills?
6. cholesterol level
7. albumin level
8. calcium level
9. uric acid level.

A potential hypothesis of interest is whether birth control usage has any effect on blood chemistries. As the nature of such hypothetical effects very likely depends upon age and years of use, before testing this hypothesis using a permutation method, you might want to divide the data into two blocks corresponding to young and old patients.

You could test several univariate hypotheses using the methods of Chapter 3, for example, the hypothesis that using birth control pills lowers the albumin level in blood. You might want to do this now to see if you can obtain significant results. As the sample sizes are small, the univariate observations may not be statistically significant. But by combining the observations that Werner and his colleagues made on several different variables to form a single multivariate statistic, you may obtain a statistically significant result, that is, if, indeed, taking birth control pills does alter blood chemistries.

You might also want to compare results using Hotelling's T^2 and the maximum-t statistic described in Section 9.4.

9. To help understand the sources of insolvency, Trieschman and Pinches [1973] compared the financial ratios of solvent and financially distressed insurance firms. A partial listing of their findings is included in the following table. Are the differences statistically significant? Be sure to state the specific hypotheses and alternative hypotheses you are testing.

	Solvent Companies					Insolvent Companies			
	V1	V2	V3	V4		V1	V2	V3	V4
1	0.056	0.398	1.138	0.109	9	0.059	1.168	1.145	0.732
2	0.064	0.757	1.005	0.085	10	0.054	0.699	1.052	0.052
3	0.033	0.851	1.002	0.118	11	0.168	0.845	0.997	0.093
4	0.025	0.895	0.999	0.057	12	0.057	0.592	0	0.057
5	0.050	0.928	1.206	0.191	13	0.337	0.898	1.033	0.088
6	0.060	1.581	1.008	0.146	14	0.230	1	1.157	0.088
7	0.015	0.382	1.002	0.141	15	0.107	0.925	0.984	0.247
8	0.079	0.979	0.996	0.192	16	0.193	1.120	1.058	0.502

V1 = agents balances/total assets
V2 = (stocks−cost)/(stocks−market value)
V3 = (bonds−cost)/(bonds−market value)
V4 = expenses paid/net premiums written

10. You wish to test whether a new fuel additive improves gas mileage and ride quality in both stop-and-go and highway situations. Taking 12 vehicles, you run them first on a highway-style track and record the gas mileage and driver's comments. You then repeat on a stop-and-go track. You empty the fuel tanks and refill, this time including the additive, and again run the vehicles on the two tracks.

The following data were supplied in part by the Stata Corporation. Use Hotelling's T^2 to test whether the additive affects gas mileage on the two tracks. Then use Pesarin's combination method to test whether the additive affects either gas mileage or ride quality.

Id	bmpg1	ampg1	rqi1	bmpg2	ampg2	rqi2
1	20	24	0	19	23.5	1
2	23	25	0	22	24.5	1
3	21	21	1	20	20.5	0
4	25	22	0	24	20.5	−1
5	18	23	1	17	22.5	1
6	17	18	−1	16	16.5	−1
7	18	17	0	17	16.5	0
8	24	28	1	23	27.5	0
9	20	24	0	19	23.5	1
10	24	27	0	22	25.5	0
11	23	21	0	22	20.5	0
12	19	23	1	18	22.5	1

bmpg1   track 1 before additive
ampg1   track 1 after additive
  rqi1 ride quality improvement track 1
bmpg2   track 2 before additive
ampg2   track 2 after additive
  rqi2 ride quality improvement track 2

11. You are studying a new tranquilizer that you hope will minimize the effects of stress. The peak effects of stress manifest themselves between 5 and 10 minutes after the stressful incident, depending on the individual. To be on the safe side, you've made observations at both the 5- and 10-minute marks.

Subject	Pre-stress	5-minute	10-minute	Treatment
A	9.3	11.7	10.5	Brand A
B	8.4	10.0	10.5	Brand A
C	7.8	10.4	9.0	Brand A
D	7.5	9.2	9.0	New drug
E	8.9	9.5	10.2	New drug
F	8.3	9.5	9.5	New drug

How would you correct for the pre-stress readings? Is this a univariate or a multivariate problem? List possible univariate and multivariate test statistics. Perform the permutation tests and compare the results.

12. Mueller [1962] measured free fatty acid levels in the blood of 10 schizophrenics and 10 "normal" subjects at various intervals after injection with insulin. Is the time course of change in fatty acid levels the same in the two groups?

	Normals					Schizophrenics			
	Time in Minutes					Time in Minutes			
	0	15	30	45		0	15	30	45
1	41	30	38	30	1	51	29	19	19
2	30	45	27	30	2	22	31	28	34
3	22	21	26	17	3	38	38	38	41
4	37	34	27	27	4	38	47	38	47
5	30	22	19	29	5	30	18	23	31
6	34	23	33	24	6	50	41	33	50
7	27	21	29	17	7	37	31	26	26
8	32	29	17	9	8	30	18	23	23
9	32	26	20	20	9	26	21	23	34
10	33	29	24	33	10	29	29	46	36

10
Clustering in Time and Space

In this chapter, you learn how to detect clustering in time and space and to validate clustering models. You use the generalized quadratic form in its several guises including Mantel's U and Mielke's multiresponse permutation procedure to work through a series of applications in atmospheric science, epidemiology, ecology, and archeology.

10.1 The Generalized Quadratic Form

10.1.1 Mantel's U

Mantel's U [Mantel, 1967] $\sum\sum a_{ij}b_{ij}$ is perhaps the most widely used of all multivariate statistics. In Mantel's original formulation, a_{ij} is a measure of the time or temporal distance between items i and j, while b_{ij} is a measure of the spatial distance. As an example, suppose the pair (t_i, l_i) represents the day t_i on which the ith individual in a study came down with cholera and $l_i = (l_{i1}, l_{i2})$ denotes her position in space. For all i, j, set $a_{ij} = 1/(t_i - t_j)$ and
$$b_{ij} = 1/\sqrt{(l_{i1} - l_{j1})^2 + (l_{i2} - l_2)^2}$$

A large value for U would support the view that cholera spreads by contagion from one household to the next. How large is large? As always, we compare the value of U for the original data with the values obtained when we fix the i's but permute the j's as in $U' = \sum\sum a_{ij}b_{i\pi(j)}$.

10.1.2 An Example

An ongoing fear among many parents is that something in their environment—asbestos or radon in the walls of their house, or toxic chemicals in their air and ground water—will affect their offspring. Table 10.1 is extracted from data collected by Siemiatycki and McDonald [1972] on congenital neural tube defects.

Table 10.1. Incidents of pairs of anencephalic infants by distance and time months apart.

km apart	<1	<2	<4
<1	39	101	235
<5	53	156	364
<25	211	652	1516

Eyeballing the gradient along the diagonal of this table, one might infer that births of anencephalic infants occur in clusters. One could test this hypothesis statistically using the methods of Chapter 8 for ordered categories, but a better approach, since the exact time and location of each event is known, is to use Mantel's U. The question arises as to which measures of distance and time we should employ. Mantel [1967] reports striking differences between one analysis of epidemiologic data in which the coefficients are proportional to the differences in position and a second approach (which he recommends) to the same data in which the coefficients are proportional to the reciprocals of these differences.[1] Using Mantel's approach, a pair of infants born 5 km and three months apart contribute $(1/3)(1/5) = 1/15$ to the correlation. Summing the contribution from all pairs, then repeating the summing process for a series of random rearrangements, Siemiatycki and McDonald conclude that the clustering of anencephalic infants is not statistically significant.

10.2 Applications

By appropriately restricting the values of a_{ij} and b_{ij}, the definition of Mantel's U can be seen to include several of the standard measures of correlation including those usually attributed to Pearson, Pitman, Kendall, and Spearman [Hubert, 1985]. Mantel's U has been rediscovered frequently, often without proper attribution (see Whaley [1983]). In this section we consider three diverse approaches to the problem of assessing the presence of clustering in space and time. In each case, the permutation distribution of the quadratic form is used to provide a baseline against which the behavior of the observations may be assessed.

10.2.1 The MRPP Statistic

One such variant is the MRPP or *multiresponse permutation procedure* [Mielke, 1979], which is used in applications as diverse as weather and the spatial distribution of archaeological artifacts. The MRPP uses the

---

[1] One further caveat: Mantel's U fails completely if the spatial distribution of the underlying population is also changing with time [Roberson and Fisher, 1986].

permutation distribution of between-object distances to determine whether a classification structure has a nonrandom distribution in space or time. With large samples, a Pearson Type III curve based on the first three (or four) exact moments may be used in place of the permutation distribution [Mielke, Berry, and Brier, 1981].

An example of the application of the MRPP arises in the assignment of antiquities (artifacts) to specific classes based on their spatial locations in an archaeological dig. Presumably, the kitchen tools of "primitives"—woks and Cuisinarts—should be found together, just as a future archaeologist can expect to find TV, DVD, and stereo side by side in a neolithic living room.

Following Berry et al. [1980, 1983], let $\Omega = \{\omega_1, \ldots, \omega_N\}$ designate a collection of N artifacts within a site; let $X_{1i}, \ldots, x_{ri}$ denote the r coordinates for the site space for artifact ω_i; let $S_1, \ldots, S_{g+1}$ represent an exhaustive partitioning of the N artifacts into $g+1$ disjoint classes (the $(g+1)$st being reserved for not yet classified items); and let n_j be the number of artifacts in the jth class.

Define the Euclidean distance between two artifacts,

$$\delta_{ij} = \left[\sum_{k=1}^{r}(X_{ki} - X_{kj})^2\right]^{1/2}.$$

Define the average between-artifact distance for all artifacts within the ith class,

$$\zeta_i = \frac{2}{n_i(n_i - 1)} \sum_{i<j} \delta_{ij} \phi_i(\omega_i) \phi_i(\omega_j),$$

where $\phi_i(\omega)$ is an indicator function that has value 1 if $\omega \in S_i$ and 0, otherwise.

The test statistic is the weighted within-class average of these distances,

$$\Delta = \sum_{i=1}^{g} n_i \zeta_i / K,$$

where $K = \sum_{i=1}^{g} n_i$.

The permutation distribution associated with Δ is taken over all $(N!/\prod_{i=1}^{g+1} n_i!)$ allocations of the N artifacts to the $g+1$ classes with the same numbers $\{n_i\}$ assigned to each class.

Empirical power comparisons between MRPP rank tests and with other rank tests are made by Tracy and Tajuddin [1985] and Tracy and Khan [1990].

10.2.2 The BW Statistic of Cliff and Ord

As a second application of generalized correlation, suppose we want to measure the degree to which the presence of some factor in an area (or time period) increases the chances that this factor will be found in a nearby area.

The BW statistic of Cliff and Ord [1973] is defined as $\sum\sum \delta_{ij}(x_i - x_j)^2$, where

$$x_i = \begin{cases} 1, & \text{if the } i\text{th area has the characteristic,} \\ 0, & \text{otherwise.} \end{cases}$$

$$\delta_{ij} = \begin{cases} 1, & \text{if the } i\text{th and } j\text{th areas are adjacent} \\ 0, & \text{otherwise.} \end{cases}$$

10.2.3 Equivalances

The generalized quadratic form has been rediscovered and redefined in many different guises. Whaley [1983] shows that Mantel's U and the BW statistic are equivalent to the MRPP for testing purposes. A third equivalent example is the k-dimensional runs test of Friedman and Rafsky [1979] studied in Section 9.4.3.

10.2.4 Extensions

Mantel's U is quite general in its application. The sets of coefficients $\{a_{ij}\}$ and $\{b_{ij}\}$ need not represent positions or changes in time and space.

In a completely disparate application in sociology, Hubert and Schultz [1976] observers studied k distinct variables in each of a large number of subjects. Their object was to test a specific sociological model for the relationships among the variables. This time, the $\{a_{ij}\}$ in Mantel's U are elements of the $k \times k$ sample correlation matrix while the $\{b_{ij}\}$ are elements of an idealized or theoretical correlation matrix derived from the model. A large value of U supports the model, a small value rules against it.

10.2.5 Another Dimension

Vecchia and Iyer [1989] generalized the MRPP for use in the comparison of several linear models. In the words of these authors.

> Regarding algebraic quantities useful to detect concentrations of points within distinct groups, one might have asked: *When are two points concurrent?* The answer, that they coincide whenever the distance *between them is zero* motivates the definition of the MRPP statistic in terms of interpoint distance.
>
> Extending this approach, for example, to the *comparison of straight line relations*, the analogous geometric argument is that three points are colinear only if their triangular *are a is zero*.

The statistic used in Vecchia and Iyer's new test is a symmetric volume: A real-valued function, symmetric in its $n+1$ arguments, that is zero if and only if the Euclidean volume of the simple formed by the arguments is zero. An immediate application for this statistic is in assessing the consistency of multiclinic designs. Some of this statistic's asymptotic properties are considered in Vecchia and Iyer [1991].

10.3 Alternate Approaches

10.3.1 Quadrant Density

Following Mead [1974], we overlay an area (possibly irregular) with a grid and divide it into squares. We then group the squares into K regions so there is an equal number of squares in each region. Finally, we count the number n_i of events (nests, animals) observed within each region and form the test statistic $S = \sum n_i^2$.

As we are working with the squares of counts, S takes its largest value if the counts are clustered by region.

We permute the squares among regions and compute S each time, and accept the alternative that there is clustering if only a small percentage of the permutations yields values of S that are as large as S_o, the sum for the original arrangement of squares.

Suppose we have only eight squares, which we group into 2 regions corresponding to the counts 0, 1, 2, 0 and 3, 4, 5, 2. Clusering is evident. $S_o = 9 + 196 = 205$. We rearrange the squares so that the counts within each region are 0, 1, 4, 5 and 0, 2, 2, 3. $S = 100 + 49 = 149$. Continuing in this fashion, we see that S_o is an extreme value and we reject the null hypothesis that the counts are distributed uniformly.

10.3.2 Nearest-Neighbor Analysis

Following Ripley [1981], let $\{p_j\}$ be a set of points in a region where specific events have been observed (cases of leukemia, birds nests, and so forth). Let $q_i\{p_j\}$ denote the distance from the point p_j to its ith nearest-neighbor, and let q_i denote the mean of these distances. Now, overlay the area with a grid so as to divide it into squares. Permute the squares; determine q_i for the permutation, and compare it with q_{io} for the original observations. The question remains as to which of the q_i to use for testing purposes.

10.3.3 Comparing Two Spatial Distributions

Upton [1984] objects to Mead's procedure observing that the result depends strongly on how the regions are defined, particularly for irregular areas and when there are missing data. Syrjala's [1993] use of a cumulative distribution based on the work of Zimmerman [1993] overcomes this objection and, moreover, allows us to extend the procedure to compare two distinct distributions.

Again, we overlay the region with a grid and divide it into squares whose centers are at the points (x_k, y_k). Define the density $d_i(x_k, y_k) = n_i/N$, where $N = \sum n_i$; in order to compare two populations, we normalize the density so that $\gamma_i(x_k, y_k) = \frac{d_i(x_k,y_k)}{\sum d_j(x_k,y_k)}$ and define

$$F_i[x_k, y_k] = \sum_{S_k} d_i(x, y),$$

where the sum $\sum_{S_k}$ is taken over the region $S_k = \{x \leq x_k; y \leq y_k\}$. Our test statistic $\Gamma = \sum_k (F_1 - F_2)^2$, and to obtain its permutation distribution, we evaluate all 2^K permutations of the two species at the K points of the grid $\{x_k, y_k\}$.[2]

10.4 Exercises

1. Show that Pitman's correlation is a special case of Mantel's U.
2. List at least two applications for Vecchia and Iyer's test.

---

[2] Since the value of this statistic also depends on the location of the origin, we may define F_{j1}, F_{j2}, Γ_j for $j = 1, \ldots, 4$ corresponding to the placing of the origin at each of the four corners of a (nearly) rectangular region. Our test statistic then would be the average of the four values, $\Gamma = \sum \Gamma_i / 4$.

11
Coping with Disaster

In this chapter you receive practical guidelines for coping with the many catastrophes that confront the applied statistician:

- subjects who miss an appointment;
- subjects who disappear completely and mysteriously in the middle of an experiment;
- incomplete questionnaires;
- covariates after the fact;
- outlying observations whose extreme and questionable values suggest they may have been recorded incorrectly;
- off-scale and other censored values that cannot be determined with precision.

11.1 Missing Data

The effects of missing data depend upon the nature of the study and the type of analysis. In some instances, for example, in the analysis of the k-sample comparison by permutation means, missing data may have no effect upon the analysis other than to reduce the power of the test. In other, more complex designs, missing data may result in an unbalanced design in which several factors are confounded with one another. In most, though not all, of these latter cases, no special statistical procedures are required, *providing* we are careful in how we interpret the results. We must identify which effects are confounded with one another, a main effect with an interaction, say. In other studies, and one such example was examined in Section 7.7.1, we may have to abandon permutation and parametric procedures altogether and consider using the bootstrap.

The majority of experimental designs belong to the correctable category. We proceed with a permutation rather than a parametric analysis using a revised set of marginal constraints that reflect the actual rather than the hoped-for sample sizes. And in analyzing the results, we acknowledge that one or more higher-order interactions may have contaminated the observed effects.

Consider an example we studied in Chapter 7, the effect of sunlight and fertilizer on crop yield. Suppose that one of the observations in the low sunlight, medium fertilizer group, the 22 noted in parentheses in the table below, is missing from the study.

Effect of sunlight and fertilizer on crop yield

		Fertilizer	
Sunlight	LO	MED	HI
	5	15	21
LO	10	(22)	29
	8	18	25
	6	25	55
HI	9	32	60
	12	40	48

The test statistic for the main effect of sunlight is the sum of the observations at the low level, $S = 23 + (15 + 18) + 75 = 131$. Such an extremely low value is found in only a small handful of the rearrangements in which we swap observations at random between the low and high groups. The number of rearrangements after correcting for the missing data item is $\binom{17}{8}$. The reduction from the hoped for $\binom{18}{8}$ rearrangements reduces the power of the test. But the reduction is irrelevant in this instance as we are rejecting the hypothesis. (Had we accepted the null hypothesis, we would have been forced to consider whether a larger sample size might have enabled us to detect an effect.)

A missing data item in only one of the groups means that the main effect of sunlight is partially confounded with the interaction between sunlight and fertilizer. But our common sense strengthened by a glance at the table tells us that the confounding also is irrelevant in this instance.

One other word of caution: A variety of software is available today to help you determine optimal sample size. But such software does not take into account the possibility of missing data, of failures in recruitment, and long-term retention. Always use the numbers such software provides as starting points, not as final estimates.

The preceding discussion was based on the implicit assumption that dropouts occur at random. If the dropout rate is directly related to the treatment, we must either abandon the study or modify our scoring system explicitly to account for the dropouts (see, for example, Entsuah [1990]).

A further example of using the permutation distribution to cope with missing data is given in Section 12.2.6. Section 12.5 details the use of the bootstrap when all other methods fail.

11.2 Covariates After the Fact

After World War II, public policy makers in the United States did a slow about face on the dangers of tobacco smoke. The changes in policy accelerated during the 1970s. One moment it seemed the cigarette was the ultimate symbol of masculinity and the next it was the primary cause of emphysema, hypertension, lung cancer, and fetal defects. One month you could design a 400-patient, 6-week, 50-variable clinical study with the full support of a Food and Drug Administration panel, and the next the panel would be asking if you'd corrected for the smokers in the control group. Of course you had not—not in those days.

Today, we know that smoking is harmful, but "number of cigarettes smoked per week" is only one of hundreds of possible covariates. Regardless of how many covariates you have controlled or matched in putting together a clinical study, there are sure to be one or two more covariates that you didn't think of—that no one thought of, that no one could have envisioned, that is, until the day after your 300-page report on the study was sent to the printers.

All is not lost. It is still possible to make a comparison among treatment groups using the method of permutations by restricting the rerandomizations to those with specific after-the-fact design matrices.

Using the method due to Rosenbaum [1984], described at length in Section 6.4, we block the data into smokers and nonsmokers (or lemon eaters and non-lemon eaters), and then randomize separately within each block.

Restricting the number of randomizations may reduce the power of the test. (It may also increase it by eliminating a source of variability; see Section 6.4.2.) As a result, we may need to add more subjects and an additional clinical center to the study to justify and confirm any negative results.

11.2.1 Observational Studies

An extreme example of the use of an after-the-fact covariate comes when we attempt to create matched pairs from two groups that were part of an observational study. In an observational study, the groupings themselves are after the fact. The subjects are not randomly assigned to treatment or control but are merely "observed" to belong to one group or the other. Through the use of after-the-fact covariates, we hope to reduce or eliminate any built-in biases.

An example provided by Rosenbaum [1998] is that of a study in humans of the effect of vasectomy on the risk of myocardial infarction. Obviously, we do not have the luxury (nor the authority, thankfully) to select a random sample of patients for a mandatory vasectomy, but must analyze the data as they lie. We can take advantage of concurrent data on obesity and smoking history (both of which are known to affect the risk of myocardial infarction) to help find matched case-controls so as to reduce the between-sample variance. See Rosenbaum [1998] for methods for dealing with imperfect matching.

While no justification for the use of restricted randomization is required when the covariates are built in to the experimental design, formal justification for the use of Rosenbaum's method after the fact requires us to make three assumptions.

First, for all observations, the observed treatment assignment $z(z = j$ if the unit is assigned to treatment j) and the vector $\boldsymbol{r} = (r_1, \ldots, r_j)$ of potential responses to treatment of that unit are conditionally independent given the vector of observed covariates. Second, regardless of the values taken by the covariates, all treatment assignments are possible. And third, the conditional probability $e[X]$ of receiving a particular treatment given a vector of observed covariates X, follows a logistic model [Cox and Shell, 1989], that is,

$$\log\left\{\frac{e[X]}{(1-e[X])}\right\} = \beta^T f(X),$$

where $f(X)$ is a known but arbitrary vector-valued function of X. Since $f(X)$ is arbitrary, this latter condition is not particularly restrictive.

All three of these assumptions are satisfied if the covariates did not affect the treatment assignment. For example, obesity and smoking history would satisfy these conditions if they were not factors in the patient/physician decision to have or perform a specific treatment.

11.3 Outliers

Consider the set of observations 0, 1, 2, 3, 19. Does the 19 represent a genuine response to treatment, the response we have been looking for, or is it an outlier—a typographical error or a bad reading that will only lead us astray? In the first case, we will want to utilize the data just as they are; in the second, we will want to modify or perhaps even to discard the questionable reading.

Shall we deal with such outliers on a one-by-one basis? Or should we establish a policy that will automatically adjust for and diminish the effect of outliers? Ad hoc rejection of suspect data could lead to charges of bias. A systematic policy can be adjusted for sample size and power determinations.

We consider seven policies here:

1) preserving the original data;
2) using ranks in place of the original observations, thus diminishing the effects of outliers;
3) replacing the observations/ranks by scores derived from some standard distribution, e.g., the order statistics of a standardized normal distribution;
4) applying a robust tail-compression transformation to all the data;
5) using an L_1 test;

6) censoring extreme observations;
7) deleting extreme observations.

Whichever policy we elect, the permutation method will be more robust to outliers than a test based on a parametric distribution. The influence functions of a two-sample permutation test are always bounded above, even if the influence functions of the corresponding parametric test are unbounded from above and below [Lambert, 1981]. Our only concern need be the selection of a test statistic that is both practical and optimal.

11.3.1 Original Data

"The Method of Randomization applied to the original observation produced stunningly efficient tests which were dismally impractical."

[Bradley, 1968]

Despite these discouraging words from James V. Bradley, I almost always make use of the original observations rather than their transform.

The exception that proves the rule is in my analysis of the Renis data considered in Exercise 5 of Chapter 3 and in Good [1979]. In that study, I used a preliminary logarithmic transformation, but it was to equalize the variances in the two samples, not to eliminate large values.

The computational difficulties to which Bradley alluded have largely been resolved through advances in computer technology between 1968 and today; the efficiency of the permutation test remains. The power and high relative efficiency of the permutation test comes from its use of exact values. Throw away one of the observations or replace it with its rank or a trimmed value and you reduce the power of the corresponding test. The gain in power is particularly evident when there is a mixture of responders and nonresponders [Good, 1979]; but see Boos and Browne [1986].

On the other hand, a single extreme observation often can have a disproportionate effect. Given the observations 0, 1, 2, 3, 19, would you rather guesstimate the population mean as 2 or 2.5 than estimate it using the sample mean of 5? By taking ranks or applying some other tail-compressing transformation to all the observations, we can "democratize" the data so that each data item has a relatively equal influence upon the final calculation. (See also Hampel et al. [1986].)

11.3.2 Ranks

Suppose we have two samples: The first control sample takes values 0, 1, 2, 3, 15. The second treatment sample takes values 3.1, 3.5, 4, 5, and 6. Does the second sample include larger values than the first?

When we rank the data giving the smallest observation a rank of 1, the next smallest the rank of 2, and so forth, the first sample includes the ranks

1, 2, 3, 4, 10, and the second sample includes the ranks 5, 6, 7, 8, 9. Does the second sample include larger values than the first?

Applying the two-sample comparison described in Section. 3.2 to the ranked data, we conclude at the 10% level that the second sample is significantly larger. The sums of the ranks in the original first sample, 20, is as large or larger in just 19 of the $\binom{10}{5} = 252$ rearrangements.

Obviously, taking ranks diminishes the effects of outliers. Taking ranks has a second advantage from the computational point of view: When we take ranks, the results are unconditionally distribution-free. As we are working with the same values—the ranks, over and over regardless of the actual values of the observations—we can tabulate the significance levels of our test statistics (at least for small samples) and avoid lengthy computations. And we may determine analytically when a sample of ranks is large enough that its permutation distribution may be replaced by an asymptotic approximation. It's not surprising that much of the literature on distribution-free tests is devoted to an analysis of the permutation distributions of ranked data.

The cost of using ranks is a loss of power, that is, a diminished probability of detecting a real difference between the distributions under test. But it is not a great loss. To achieve the same power as the permutation or parametric t-test with very large samples, the Mann–Whitney test—a two-sample comparison that uses ranks in place of the original observations—requires only 3% or 4% more observations. Cheap, if the units are widgets; expensive, if the units are patients or rare Rhesus monkeys.

11.3.3 Scores

If we are testing against almost normal alternatives, we can improve on the power of the Mann–Whitney test by using normal scores in place of ranks.

In the general case, we replace the rank of the ith observation, r_i, say, by the expected value of the r_ith largest value in a sample of n values drawn from the distribution F, $F^{-1}[r_i/(n+1)]$, where F is our best guess of how the observations are really distributed (see also David [1970, p. 65]).

A good guess will produce an optimal test, and, sometimes, even a "bad" guess can be close to optimum. For example, Chernoff and Savage [1958] show that the normal-scores test, in which ϕ is the Gaussian distribution, has a minimum asymptotic efficiency of 1 relative to the usual t-test regardless of the true underlying distribution.

Bell and Doksum [1965] provide detailed comparisons of the rank and normal scores tests in a variety of settings. In Bell and Doksum [1967] they provide conditions under which the normal-scores test is minimax.

Hajek and Sidak [1967] show that, in general, optimal scores for tests of location are based on the scores

$$a(j) = -\frac{f'(F^{-1}[u])}{f(F^{-1}[u])},$$

where $u = j(N+1)$, and f and F are the density and cumulative distribution functions, respectively, of the underlying distribution. For optimal rank tests of scale, the scores are

$$a(j) = 1 - \frac{F^{-1}[u]f'(F^{-1}[u])}{f(F^{-1}[u])}.$$

11.3.4 Robust Transformations

A robust transformation preserves sample values at the center of a distribution while shrinking those in the tails. As one example [Maritz, 1981], consider

$$\phi(u) = u/(1+u^2).$$

For u small $\phi(u)$ is approximately u. For $u < 1, \phi(u)$ is a slowly increasing function of u. If we replace x_i, by $\phi(x_i)$ in computing the mean, then large values will make virtually no contribution to the total.

As a second example (Huber [1972]), take

$$\phi(u) = (1 - \exp[-u])/(1 + \exp[-u]).$$

Again $\phi(u)$ is approximately u for u small, and is bounded between 0 and 1.

In a complex experimental design, the transformation may be applied to the residual rather than the original observation. For example, to test whether $Y = bX$, one would apply ϕ to $y' = y - bx$, rather than to y.

If you are uncertain which transformation to use, you can reduce the effect of extreme values in some cases simply by switching to a statistic based on the absolute differences $|x_i - y_i|$ in place of the squared differences $(x_i - y_i)^2$. The final choice should be dictated by your loss function (see Section 10.4).

If extreme values are unlikely, as is the case with normal alternatives, then a robust transformation will have little or no effect on the power of a test. See Maritz [1981] and Lambert [1985] for further discussion.

11.3.5 Use an L_1 Test

A test based on the absolute values of the deviations about the median rather than the squares of the deviations about the mean is less likely to be affected by extreme values. Such a test is also the appropriate one to use with a first-order loss function. See Wilson [1978], Mielke [1986], Dodge [1987], Wang and Scott [1994], Cade and Richards [1996], and Mielke and Berry [1997].

11.3.6 Censoring

Lambert [1985] offers a two-sample test that is both robust and powerful. First, we order the data so that

$$X_{(1)} < \cdots < X_{(n)} \quad \text{and} \quad Y_{(1)} < \cdots < Y_{(m)}.$$

To test against the alternative that the Y's are larger on the average than the X's, we replace each X_i and Y_j that is less than $k_1 = X_{(n\beta_1)}$ by k_1 and each X_i and Y_j that is greater than $k_2 = Y_{(n\beta_2)}$ by k_2 and then carry out the usual permutation test based on the sum of the observations in the first sample. Note that the censoring values are determined by the data themselves. Unfortunately, there can be more than one "right" choice for β_1 and β_2, and the computations are far from straightforward. One possible compromise is to let $k_1 = X_{(2)}$ and $k_2 = Y_{(m-1)}$ for samples of 15 or less.

11.3.7 Discarding

The most extreme method of dealing with outliers is to discard them. Although Welch and Guiterrez [1988] obtain narrower confidence intervals in matched-pairs designs through the use of permutation applied to trimmed means, there are two objections to this method. First, the resultant test is unlikely to be exact ([Romano, 1990] Theorem 3.3). Second, discarding data reduces the power of the test. In Good [1991], we improve on the power of the Welch–Guiterrez test by treating the outliers as if they were censored. Our approach is described in more detail in the next section.

11.4 Censored Data

We may not be able to make all our measurements with the same precision.

In a radioimmune assay, for example, the typical concentration curve has a sigmoidal shape with flat regions at the two extremes. In the lower, flat region, of the curve, estimation is difficult, if not impossible. While binding values elsewhere may be determined to one part in a billion, in this region they merely are recorded as "below minimum:"

Here is a second example: In many clinical studies, it is neither possible nor desirable to follow all patients to the end of their lifespans. Limiting the duration of the study cuts the costs of observation and puts promising new materials and processes into immediate service. But while some lifespans will be known with precision, others can be noted only as "exceeded treatment period."

In each of these examples some of the data have been censored.

11.4.1 GAMP Tests

When observations are censored, the most powerful test typically depends on the alternative, so that it is not possible to obtain a uniformly most powerful test.

Good [1989, 1991, 1992] found that by establishing a *region of indifference*, it may be possible to obtain a permutation test that is close to the most powerful test, "almost most powerful," regardless of the underlying parameter values.

Suppose we wish to perform a test of a hypothesis F against a series of alternatives $F_1, F_2, \ldots$. To obtain a test that is globally almost most powerful (GAMP), we proceed in three stages.

First, we use the likelihood ratio to obtain a locally most powerful unbiased α-level test of the hypothesis F against the alternative F_1. We repeat this procedure for each alternative F_i to obtain a family of rejection regions $\{R_i\}$.

Next, we form two regions: (i) A *rejection region* $R \subseteq \bigcap_i R_i$ that contains only events common to all the rejection regions of the preceding family; and (ii) *an acceptance region* A that contains only events common to all the acceptance regions.

Last, we construct a permutation test whose p-value is determined by assigning each rearrangement of the data to one of *three* regions: rejection (R), acceptance (A), or *indifference* (I). While we cannot determine the p-value of their new-test exactly, we can bound it:

$$\Pr\{R|X\} \leq p \leq 1 - \Pr\{A|X\}.$$

In Good [1992], I showed that GAMPs exist when the joint log-likelihood of the observations takes the particularly simple form $S_U * f(\theta) + N_C * g(\theta)$ where S_U and N_C are the sum of the uncensored observations and the number of censored observations in the treatment sample, respectively, and f and g are monotone functions of θ. Examples include normally distributed, exponentially distributed, and gamma distributed random variables subject to Type I censoring.

A permutation (or rerandomization) approach is utilized.

There are two distinct cases, which I term left- and right-censoring, though the actual directions—left or right—will depend upon the alternative. To fix ideas, suppose we have samples from two populations and are testing a null hypothesis H: $F_2 = F_1$ against stochastically larger alternatives K: $F_2(x) = F_1(x - \delta)$. With left-censoring we can assign x a precise value only if $x \geq c$; for example, radioimmune assay involves left-censoring. With right-censoring, we can assign x a precise value only if $x \leq c$; for example, reliablility studies usually involve right-censoring.

To eliminate any dependence on the zero point of the underlying scale, we transform the data before we derive the permutation distribution; from each of the orginal observations we subtract $\bar{X}_U$, the mean of the uncensored observations in the sample taken from G where $X'_{ij} = X_{ij} - \bar{X}_U$ for $i = 1, 2, j = 1, \ldots, n_i$, and $S'_{U_0} = 0$; the transformed observations are censored at $c' = c - \bar{X}_U$. Next, we compute S_{U_0} and N_{C_0} for the original treatment sample and permute repeatedly, computing S_U and N_C for each permuted sample.

With left-censoring, we assign a permutation to the rejection region R if $S_U \geq S_{U_0}$ and $N_C \geq N_{C_0}$. We assign it to the acceptance region A if $S_U < S_{U_0}$ and $N_C \leq N_{C_0}$. We assign it to the indifference region I, otherwise.

With right-censoring, we impute the value c to the censored observations. Let $k = N_C - N_{C_0}$. We assign a permutation to the rejection region R if $S_U + kc \geq S_{U_0}$. We assign it to the acceptance region A if $S_U + kc < S_{U_0}$. We assign it to I, the indifference region, otherwise.

The indifference region is small enough in most instances to permit effective decision-making [Good, 1989]. As the sample size increases, the GAMP test converges in probability to a UMP-unbiased test [Good, 1992]. In the rare case where the result does lie in the indifference region, we recommend taking additional observations.

The application of permutation methods to censored data was first suggested by Kalbfleisch and Prentice [1980], who sampled from the permutation distribution of censored data to obtain estimates in a process akin to bootstrapping.

For a survey of other permutation tests that have been applied to censored data, see Schemper [1984]. Conditional rank tests for randomly censored survival data are described by Andersen et al. [1982] and Janssen [1991].

11.4.2 Fishery and Animal Counts

GAMP tests may also be applied in the analysis of fish and animal counts when a large proportion of readings are zero.

11.5 Censored Matched Pairs

As we showed in Section 6.4.2, the sensitivity of an experiment can be increased through the use of matched pairs. But it may happen that an exact observation cannot be made for one or more subjects, the only available information being that the required measurement is greater or less than some known value. Often this censoring process is accidental, but in many toxicology studies and reliability trials, it is a matter of deliberate design: The experimenter trades the cost of enrolling a larger number of subjects at the onset of the experiment for a shortened study period.

Suppose $z = y - x$ is the difference between the (transformed) observations on the two members of a pair, and that observations are not recorded if they exceed C on the (transformed) scale. As noted by Sampford and Taylor [1959], any pair provides information on the distribution of z in one of the following four forms:

(i) both y and x are observed, so that z is determined exactly;
(ii) x is observed, but we only know that y exceeds C; that is, $z > C - x$, so we say z is *upper-censored*;
(iii) y is observed, but we only know that x exceeds C; that is $z < y - C$, so we say z is *lower-censored*;
(iv) both x and y exceed C, so that no information is available on z for this pair; the sample size is effectively reduced.

While cases (ii) and (iii) provide less information than case (i), they are not uninformative, and a variety of hypothesis testing methods have been proposed for capitalizing on the information they provide. Good [1991] developed an "almost" most powerful distribution-free method based strictly on the data at hand. To see how this method is applied, assume that the first observation in each pair has the distribution F and the second has the distribution G. The hypothesis, unless stated to the contrary, is that $F \geq G$. The alternative is that $F < G$.

11.5.1 GAMP Test for Matched Pairs

The globally almost most powerful (GAMP) test for matched pairs represents a simple extension of the GAMP test for two independent samples derived in Good [1989, 1992]. Record U, the number of upper-censored pairs in the original sample, and Z, the sum of the uncensored z's in the original sample. Randomize the observations, permuting the treatment labels within each pair, and let U' and Z' be the corresponding statistics for the permuted sample.

If $U' \geq U$ and $Z' \geq Z$, then assign the permuted sample to the rejection region R.

If $U' \leq U$ and $Z' < Z$, then assign the permuted sample to the acceptance region A.

Otherwise, assign the permuted sample to a region of indifference.

Repeat the randomization process for all possible permutations (or for a suitably large number N of randomly selected permutations) and let f_R, f_A, and f_I be the frequency with which permutations are assigned to the rejection, acceptance, and indifference regions, respectively.

This method of construction ensures that the acceptance region A of the GAMP test is contained in the acceptance regions of each of the most powerful α-level permutation tests of a simple hypothesis $G = F = F^*$ against the simple alternative $G^* = G > F = F^*$. Similarly, the rejection region R of the GAMP test is contained in the rejection regions of each of the most powerful α-level permutation tests.

$f_R \leq p \leq N - f_A$, where p is the significance level of any member of the family of most powerful permutation tests of a simple hypothesis against a simple alternative. Thus, a test of the composite hypothesis $F \leq G$ against the composite alternative $F > G$ based on the bounds defined by A and R is globally almost most powerful.

In practice, an investigator using a GAMP will elect one of three courses of action: 1) accept the null hypothesis, noting the bounds on the p level; 2) reject the hypothesis in favor of a stochastically larger alternative; 3) in order that p might be known with greater certainty, elect to take additional observations. If you require exact significance levels to make power comparisons with other tests, you must randomize on the indifference region as follows.

If f_R is greater than the desired α-level, accept the null hypothesis. If $N-f_A$ is less than the desired α-level, reject. If neither condition holds, choose a random number $Z = U(0,1)$ and reject the hypothesis if $Z \leq (N_\alpha - f_R)/(N - f_R - f_A)$, accepting it otherwise.

11.5.2 Ranks

When data are heavily censored, you can improve on this method by replacing the original observations with ranks. Two approaches suggest themselves: In the first, which we term "post-ranking," compute the differences z or each pair, then rank these differences in absolute value dividing the highest ranks among the censored observations. Denote by Z the rank sum that correspond to those pairs in which y is known to be larger than x. As in the GAMP test, now randomize the observations, permuting the treatment labels within each pair, and denote by Z' the new rank sum. Assign this randomization to R, I, or A according to whether $Z' >, =,$ or $< Z$. As with the GAMP test, reject H in favor of K if only a small proportion of rerandomizations are assigned to R; randomize on the indifference region I to obtain a test at a specific significance level p.

Postranking has the drawback that if, say, 2 is the censoring point, the difference "censored – 1.99" is automatically assigned a higher rank than the difference "1.99 – 0." To avoid this difficulty, in a second approach, which we term *preranking*, first rank the individual observations, again dividing the highest ranks among the censored observation. Next, compute the differences of the ranks within each pair, and, as a third and final step, rank the absolute values of the differences. The drawbacks of this second, preranked approach are computational: you must rank the data twice and you must correct for ties during the second ranking.

When the underlying distribution is normal and censoring is heavy, the preranked permutation test provides the greatest sensitivity [Good, 1991].

When the underlying distribution is normal and censoring is light, or when the underlying distribution is exponential, the GAMP test is preferable.

The strength of the GAMP lies in its use of exact values rather than ranks—thus its effectiveness with heavy-tailed distributions, like the exponential, which have many extreme values. The GAMP is also the most readily computed. Its weakness lies in its dependence on a region of indifference whose size varies from sample to sample.

11.5.3 One-Sample: Bootstrap Estimates

If you are willing to assume the underlying distribution(s) are symmetric, then these methods for paired comparisons may also be applied to hypotheses based on a single sample. If censoring is one-sided, we are forced to censor on the opposite side in order to obtain an exact test. If you (1) are unwilling to assume symmetry, and/or to throw away data through censoring, (2) have 15 or more observations (30 would be better) and (3) are willing to assume that all observations are drawn from the same distribution, then you may apply Efron's [1981] bootstrap method of extending the Kaplan-Meir estimates.

11.6 Adaptive Tests

In an adaptive test [Hogg and Lenth; 1984], we compute several different test statistics, but make use only of the one we estimate to be the most powerful. For example, we could compute both a t-test and a robust test based on an M-estimate and, after the fact, use the one that seems best suited to the data. With some adaptive methods, the frequency of Type I error may increase as a result of this selection procedure. But with Donegani's method [1991] applied to two permutation tests, we can obtain a single test that is both exact and equal in power asymptotically to the most powerful of the two tests.

Let T_1, and T_2 be the two tests and let c_1, and c_2, the "criteria," be two positive real functions defined on the vector of observations X such that if $c_1(X) < c_2(X)$, then T_1 is preferable to T_2. Suppose that large values of either test statistic indicate a departure from the null hypothesis. Proceed in four steps as follows.

1. Evaluate $c_1(X)$, $c_2(X)$ and let 'opt' refer to the index of the criterion having the smaller value.
2. partition the set Pr of all possible rearrangements of the data into two sets,

$$P_1 = \{\pi : c_1(\pi X) < c_2(\pi X)\},$$
$$P_2 = \{\pi : c_1(\pi X) < c_2(\pi X)\}.$$

3. Let H_{opt} be the randomization distribution obtained by evaluating the optimal test statistic T_{opt} on each element of the set that contains the original rearrangement.
4. Reject the null hypothesis at the level α if T_{opt} exceeds the 100-αth percentile of H_{opt}. In other words, if $c_1(x) < c_2(X)$ restrict attention to those rearrangements that are in P_1.

Let N_i denote the number of rearrangements in P_i. Let C_i denote the choice of the statistic T_i. Then

$$\Pr\{R|H\} = \Pr\{R|H, C_1\}\Pr\{C_1|H\} + \Pr\{R|H, C_2\}\Pr\{C_2|H\}$$
$$= α(N_1/(N_1 + N_2)) + α(N_2/(N_1 + N_2))$$
$$= α$$

Donegani [1991] shows that in this case the adaptive procedure is asymptotically optimal and, in the case of matched pairs, that it is optimal with as few as nine pairs of observations.

11.7 Exercises

1. Prove that ranking the data will eliminate any distortions brought about by a nonlinear measuring device. That is, prove that the ranks of the observations are invariant under any continuous, strictly increasing transformation. (We take advantage of this result in a multivariate analysis in which we use ranks to bring several disparate variables together on a single common scale; see Section 9.1.)
2. Show that an exact one-sample permutation test for singly censored data can exist only if you deliberately censor the data from the other side.
3. Let $x_1, \ldots, x_n$ be a sample from the exponential distribution with density be $\frac{1}{b}e^{-x/b}$, $b > 0$. If you have a scintillation counter at hand, you can generate just such a sample by recording the time elapsed between counts. Alternately, you may stand on a street corner or at night club entrance and record the number of seconds before the next redhead or the next BMW goes by. If you have access to a computer, use its random number generator and take the logarithms of the random numbers you generate. Guesstimate the mean waiting time b before you start. Test your guesstimate using (a) the original observations, (b) ranks, (c) normal scores, and (d) the data remaining after you've thrown out all observations that are three times the guesstimated value. Compare your results with the different statistical procedures for sample of size 5, 6, and 7.

12
Solving the Unsolved and the Insolvable

In this chapter, we consider the problem of developing optimal solutions for yet-to-be-encountered problems, problems for which test statistics are not immediately to be found in the pages of this or any other text. First, we review the criteria for test statistics that we first encountered in Chapters 2 and 3. Then, we consider some permutation test statistics that have been developed in some highly specialized situations. Last, to be used when all else fails, we consider bootstrap confidence intervals.

12.1 Key Criteria

In virtually all the instances we have studied to this point in our text, the "obvious" test statistic is one that tends to be very large under the alternative (or very small), while under a null hypothesis no value is more likely than any other. The formal justification for this approach comes from the fundamental lemma of Neyman and Pearson, and if our statistic is sufficient for the parameters we are testing, then we can be almost certain we've made the correct choice.

12.1.1 Sufficient Statistics

Recall that a statistic $T(X)$ is *sufficient* for a parameter θ if the conditional distribution of X given T is independent of θ. Once we have calculated the value of a sufficient statistic or statistics, we may be able to throw away the original observations, for frequently, a sufficient statistic(s) can provide us with all the information a sample has to offer.

An example we have encountered many time in the derivation of permutation tests is that of the order statistics $x_{(1)} \leq x_{(2)} \leq \cdots \leq x_{(n)}$. If we know these order statistics, we know as much about the unknown distribution as we would if we had the original observations in hand.

Another commonly encountered example is that of the mean of a sample of independent, identically Poisson-distributed random variables; this statistic is sufficient to represent the mean of the underlying Poisson distribution. Likewise the mean of a sample of normally distributed random variables is sufficient to represent the mean of the underlying normally distributed population. But there is a distinction: In the first example, the Poisson, the sample mean possesses all the information the sample has to offer with regard to the underlying single-parameter distribution. In the second case, a normal distribution depends on two parameters, the population mean and the population variance. We need to compute both the sample mean and the sample variance to obtain all the information a sample from a normal distribution has to offer.

Even in the case of a normal distribution, as it is a member of an exponential family, we were able to derive in Chapter 3 a test of the population mean, conditional on the value of a sufficient statistic.

In selecting a statistic to test a hypothesis about a population parameter θ, we look first at those statistics that are sufficient for θ.

12.1.2 Three Stratagems

Occasionally—the k-sample comparison of means when $k > 2$ is an excellent example—the use of sufficient statistics alone will not reduce consideration to a single statistic. Three stratagems may help us. We may

- restrict the alternatives;
- consider the loss function;
- invoke impartiality.

12.1.3 Restrict the Alternatives

In the k-sample case, by restricting attention to ordered alternatives, we were able to obtain a UMP-unbiased test (Theorem 6.2). In the next example, that of an $r \times 1$ contingency table, we cannot derive a most powerful test that will protect us against all alternatives, but we can use the likelihood ratio to derive a most powerful test against those alternatives that are of immediate interest. The approach lends itself to any set of data for which we have knowledge of an underlying model.

Suppose the hypothesis to be tested is that certain events (births, deaths, accidents) occur randomly over a given time interval. If we divide this time interval into m equal parts and p_i denotes the probability of an event in the ith subinterval, the null hypothesis becomes $H: p_i = 1/m$ for $i = 1, \ldots, m$. Our test statistic is

$$\chi^2 = mn \sum_{i=1}^{m} \left(v_i - \frac{1}{m}\right)^2,$$

where v_i is the relative frequency of occurrence in the ith interval.

0	1	2	3	$m-1$
v_0	v_1	v_2	v_3	v_{m-1}

To determine whether this test statistic is large, small, or merely average, we examine the distribution of χ^2 for all sets of frequencies $\{v_i\}$ that satisfy the two conditions

1) $v_i \geq 0, i = 1, \ldots, m$;
2) $\sum v_i = 1$.

We reject the hypothesis if the fraction of tables for which $\chi^2 \leq \chi_0^2$ is less than α.

We can obtain a still more powerful test when we know more about the underlying model and, thus, are able to focus on a narrower class of alternatives.

Suppose, in contrast to the previous example, that we use the m categories to record the results of n repetitions of a series of $m-1$ trials, that is, we let the ith category correspond to the number of repetitions which result in exactly $i-1$ successes. If our hypothesis is that the probability of success is .5 in each individual trial, then the expected number of repetitions resulting in exactly k successes is $\pi_k[.5] = n \binom{m}{k} (.5)^m$

If we proceed as we did in the preceding example, then our test statistic would be

$$S_1 = \chi^2 = n \sum_{k=1}^{m} \frac{(v_k - \pi_k[.5])^2}{\pi_k[.5]}.$$

Such a test provides us with protection against a wide variety of alternatives. But from the description of the problem we see that we can restrict ourselves to alternatives for which

$$\pi_k[p] = n \binom{m}{k} (p)^k (1-p)^{m-k}.$$

Fix, Hodges, and Lehmann [1959] show that a more powerful test statistic against such alternatives is

$$S = S_1 - S_2,$$

where

$$S_2 = \min_p \sum_{i=1}^{m} \frac{(v_i - p_i[p])^2}{\pi_i[p]}.$$

The parametric form of the distribution of S is difficult if not impossible to obtain analytically, except for very large sample sizes; as always, we can approximate the permutation distribution by Monte Carlo means, assigning the $\sum v_i$ items to the m categories at random and computing S_2 for each such rerandomization.

12.1.4 Consider the Loss Function

The loss function should be a key factor in the selection of a statistical test. As we saw in Chapter 2, a statistical problem is defined by three elements:

1) The class $\mathcal{P} = \{P_\theta, \theta \in \Omega\}$ to which the probability distribution of the observations is assumed to belong;
2) The set D of possible decisions $\{d(\boldsymbol{X})\}$ one can make in light of the observations $\boldsymbol{X}$,
3) The loss $L(d(\boldsymbol{X}), \theta)$, expressed in dollars, human lives, or some other quantifiable measure, that results when we make the decision d whith θ being true.

When you and I differ in our assessment of the loss function, we are likely to differ in our assessment of the significance of Type I and Type II error and, hence, in our choice of test statistic.

Even when we don't know the exact values taken by a loss function, we have some idea about its form. In many testing situations, for example, in the analysis of variance and in some matched pair applications, the traditional test statistic (or discrepancy measure in Mehta and Patel's terminology) is a function of the square of the distance between the observed or estimated values and the hypothesis. Yet the natural measure is the distance itself. A statistical procedure that minimizes the expected value of the one may not minimize the expected value of the other [Mielke and Berry, 1982, 1983].

The principal reason for using the square of the distance is that it yields a maximum likelihood solution when the underlying distribution is normal. An assumption of normality may or may not be justified while maximum likelihood itself can only be justified on the grounds of convenience.

A second and more compelling reason for using the square of the distance in the data space would be that the loss function, a discrepancy measure in the parameter space, is also proportional to the square. But if we are uncertain about the form of the loss function, would it not be more natural to utilize a test statistic that is linear in both the data and parameter spaces? A first-order statistic will be more robust than a second-order statistic in the face of questionably large deviations [Dodge, 1987].

With parametric tests, we are too often restricted by the availability of tables from which we can obtain critical values. The permutation approach frees us to choose the test statistic that is best suited to the problem at hand. If a second-order statistic is called for, we may use it, and if a first-order statistic is more appropriate, we may take advantage of it, instead. Through the use of resampling methods, we are free to choose the statistic best suited to the problem.

Recall from Chapter 6 that if we have more than two levels of a factor; we have a choice of at least three test statistics. Select the optimal statistic in accordance with both the alternatives of interest and the underlying loss function.

12.1.5 Impartiality

If your measurements are made in feet, would you expect to reach the same conclusions as you would if your measurements were made in inches? What if you discover *after* you report your results that you forgot to rezero the measurement device so that each of your readings is off by exactly 0.0123 gram. Would you still believe that your decision to accept the hypothesis is correct? If your answer to both these questions is an unconditional "yes," then you are already applying the *principle of invariance*, implicitly if not explicitly.

Many statistical problems involve symmetries. In the examples we've considered so far, the observations are exchangeable, so that the order in which we made these observations is irrelevant. Our test statistic(s) should and do reflect this same symmetry. The sample mean and sample variance are good examples of statistics that are symmetric in the underlying variables. Symmetry and invariance are related, as we saw in Chapter 7. The mathematical expression of symmetry is invariance under a suitable group of transformations. In generating an optimal test, look for test statistics that preserve the structure and symmetry of a problem.

12.2 The Permutation Distribution

Many common statistical problems defy conventional parametric analysis simply because the distributions of the resultant test statistics are not well tabulated. Or, worse, we settle for a less-than-optimal statistic simply because a table for the less-than-optimal statistic is readily available—the chi-square statistic (Chapter 8.3.1) and its misapplication to sparse contingency tables is one obvious example.

We need not settle for less than the best. Given a sufficiently powerful computer and the time needed to perform the necessary calculations, we can always obtain the permutation distribution of the statistic that best separates the hypothesis from the alternative.

The purpose of this section is to describe a number of practical applications in animal behavior, atmospheric science, education, epidemiology, molecular genetics and sociology, where permutation distributions have provided new and more powerful solutions.

12.2.1 Ensuring Exchangeability

For the permutation method to provide meaningful results, the stochastic portion of the observations or their transforms must be exchangeable. If the model for our observations is $Y_i = A + B_i + \epsilon_i$, where A and the $\{B_i\}$ are known or unknown constants and the $\{\epsilon_i\}$ are random variables, then to test hypotheses concerning Y or A or the $\{B_i\}$, the $\{\epsilon_i\}$ need be exchangeable.

If we have a set of observations $\{X[t], t = 1,\ldots, n\}$ where $X[t] = a + bX[t-1] + z[t]$ and the $\{z[t]\}$ are i.i.d., then the variables $\{Y[t], t = 2,\ldots, n\}$ where $Y[t] = X[t] - bX[t-1]$ are exchangeable.

Dependent noncolinear normally distributed variables with the same mean are transformably exchangeable, for as the covariance matrix is non-singular, we may use the inverse of this matrix to transform the original variables to independent (and, hence, exchangeable) normal ones. By applying two successive transformations, we can obtain an exact permutation test of the non-null, two-sample, univariate hypothesis for dependent normally distributed variables providing the covariance matrix is known. Unfortunately, as Commenges [2001] shows, whether we accept or reject in a specific case may depend on the transformation we have chosen.

Michael Chernick notes the preceding result applies even if the variables are collinear. Let R denote the rank of the covariance matrix in the singular case. Then there exists a projection onto an R-dimensional subspace where R normal random variables are independent. So if we have an N-dimensional ($N > R$) correlated and singular multivariate normal distribution, there exists a set of R linear combinations of the original N variables so that the R linear combinations are each univariate normal and independent of one other.

12.2.1.1 Test for Parallelism

Suppose we know that the behavior of the expected value of a variable over time has a specific functional form f, and we wish to test whether two such time curves are parallel even though we do not know the value of the intercepts. That is, we are given that

$$y_{ik} = a_i + b_i f(t_{ik}) + \epsilon_{ik} \quad \text{for } i = 1, 2; \ k = 1, \ldots, n_i$$

where the errors $\{\epsilon_{ij}\}$ are exchangeable. To obtain an exact permutation test for H: $b_1 = b_2$, we need to eliminate the $\{a_i\}$ while preserving the exchangeability of the residuals. Writing $x_{ij} = f(t_{ij})$ for simplicity, we know that under the null hypothesis

$$\bar{y}_{i.} = a_i + b\bar{x}_{i.} + \bar{\epsilon}_{i..}$$

Define

$$y' = \frac{1}{2}(\bar{y}_1 - \bar{y}_2); \ x' = \frac{1}{2}(\bar{x}_1 - \bar{x}_2); \ \epsilon' = \frac{1}{2}(\bar{\epsilon}_1 - \bar{\epsilon}_2); \ a' = \frac{1}{2}(a_1 + a_2).$$

Define

$$y'_{1k} = y_{1k} - y' \text{ for } k = 1 \text{ to } n_1 \quad \text{and} \quad y'_{2k} = y_{2k} + y' \text{ for } k = 1 \text{ to } n_2.$$

Define

$$x'_{1k} = x_{1k} - x' \text{ for } k = 1 \text{ to } n_1 \quad \text{and} \quad x'_{2k} = x_{2k} + x' \text{ for } k = 1 \text{ to } n_2.$$

Then
$$y'_{ik} = a' + bx'_{ik} + \epsilon'_{ik} \quad \text{for } i = 1, 2; \ k = 1, \ldots, n_i.$$

Two cases arise. If the original predictors were the same for both sets of observations, that is, if $x_{1k} = x_{2k}$ for all k, then the errors $\{\epsilon'_{ik}\}$ are exchangeable and we can apply the method of matched pairs. Otherwise, we need to proceed as follows: First, estimate the two parameters a' and b by least squares means. Use these estimates to derive the transformed observations $\{y'_{ik}\}$. Then test the hypothesis that $b_1 = b_2$ using a two-sample comparison. If the original errors were exchangeable, then the errors $\{\epsilon'_{ik}\}$, though not independent, are exchangeable, also, and this test is exact.

Now suppose
$$y_{ik} = A_i Z_k + b_i x_{ik} + \epsilon_{ik} \quad \text{for } i = 1, 2; \ k = 1, \ldots, n_i$$

where Z_k is a column vector of covariates with A_i a row vector of the corresponding coefficients. Defining A'_i as the mean of A_1 and A_2, then

$$y'_{ik} = A' Z_k + b x'_{ik} + \epsilon'_{ik} \quad \text{for } i = 1, 2; \ k = 1, \ldots, n_i$$

and we have analogous results for the general case.

12.2.1.2 Linear Transforms That Preserve Exchangeability

Recall that a permutation π is isomorphic to a one-to-one function of the set of natural numbers $\{1, \ldots, n\}$ onto itself. We denote by P_n the set of all such permutations and by $R(\pi)$ the $n \times n$ matrix whose ijth entry is equal to 1 if $\pi(i) = j$ and equal to zero, otherwise. In this notation, an n-component random vector x is said to have exchangeable components if $R(\pi)x$ has the same distribution as x for every permutation π in P_n.

Dean and Verducci [1990] show the following.

Theorem 12.1. *A linear transformation $B: \mathcal{R}^n \to \mathcal{R}^m$ preserves exchangeability if and only if for every permutation τ in P_m, there exists a permutation π in P_n such that $BR(\pi) = R(\tau)B$.*

Theorem 12.2. *A linear transformation $B: \mathcal{R}^n \to \mathcal{R}^m$ satisfies the conditions of Theorem 12.1 if and only if B can be represented in the form $[B_1, \ldots, B_t]R(\sigma)$ where σ is in P_n, and for each $I = 1, \ldots, t$, B_i is an $m \times n_i$ matrix ($\sum n_i = n$) satisfying the following conditions: If the first column of B_i contains the distinct elements $d_1, \ldots, d_k$ with multiplicities $m_1, \ldots, m_k$ (so that $\sum m_i = m$), then the n_i columns of B_i consist of the $n_i = m!/\prod m_j!$ distinct permutations of the first column.*

12.3 New Statistics

In this section, we consider several novel statistics that arise in specific applications. Though their theoretical distributions cannot be readily calculated, p-values can be determined by reference to a permutation or bootstrap distribution.

12.3.1 Nonresponders

An elementary example is a statistic I proposed for use when there is a response threshold, a common occurrence in pharmacological studies [Good, 1979].

We assume that $X_1, \ldots, X_n$, the controls, are independent and identically distributed with distribution F, while responders in the treatment group are independent and identically distributed as $G[x] = F[x - \delta]$. Unfortunately, not every member of the treatment group is capable of responding to the treatment, with the result that we are forced to test the hypothesis $G = F$ against contaminated alternatives of the form

$$G = \pi F[x - \delta] + (1 - \pi) F[x], \quad \text{with } 0 < \pi \leq 1.$$

The conventional statistics for the two-sample comparison—Student's t and the Wilcoxon test—are subject to a loss of power in the presence of nonresponders. This reduction in power of the t-test is due to two factors: (1) A decrease in the absolute difference between the means of the two testing groups, and (2) an increase in the variance of the treatment sample. This last change is the key to the selection of a new test statistic:

$$v(p) = p' \frac{nm}{n+m}(X - Y)^2 + (1-p)S_y^2.$$

This new statistic has two components: The first is proportional to the difference $(X - Y)$ in the means of the two samples, and the second to S_y, the variance of the treatment sample.

Barring the availability of an independent test for response, the p used in the equation for v is at best only a guess as to the true value of π. Good [1979] found that using a value of $p = 0.67$ appears to offer relatively good protection against a broad range of values of π. Boos and Browne [1986] question whether the gain in power is really worth all the extra computation. An increase in power can mean a decrease in sample size with fewer experimental subjects placed at risk and a shortened study time with more rapid dissemination of important results. An increase in computation time puts the strain where it belongs—on the computer.

12.3.1.1 Extension to K-samples

Mielke and Berry [1994] have extended Good's result to k samples, choosing as their test statistic

$$S = \sum_{k=1}^{K} \frac{n_k}{N} \binom{n_k}{2}^{-1} \sum_{i<j} \Delta_{ij} \phi_k(i) \phi_k(j),$$

where n_k is the number of observations in the kth sample, $\phi_k(j)$ is 1 if j belongs to the kth sample and zero otherwise, and $\Delta_{ij} = |x_i - x_j|^m, m > 0$; typically, $m = 1$ or $m = 2$.

12.3.2 Animal Movement

Let $\{(w_i, x_i); i = 1, \ldots, n\}$ denotes a series of paired observations on the successive positions of two organisms in space. We would like to know if the movements of the two organisms are independent or coordinated. The ecological literature favors a test of independence based on the ratio of the actual distance traveled to the distance from the starting point:

$$R_1 = \frac{\sum\{(w_{i+1} - w_i)^2 + (x_{i+1} - x_i)^2\}}{\sum(w_i^2 + x_i^2)},$$

Our own intuition suggests a more powerful test of the hypothesis of independence would result from using either

$$R_2 = \frac{\sum(w_i - x_i)^2}{\sum(w_i^2 + x_i^2)},$$

the ratio of the successive distances of the two organisms from each other and from the starting point, or

$$R_3 = \frac{\sum(w_{i+1} - w_i)(x_{i+1} - x_i)}{\sum(w_i^2 + x_i^2)},$$

the traditional measure of correlation.

We also favor R_2 and R_3 on the grounds of simplicity. To compute the permutation distribution of R_1, we need to rearrange both sets of movements $\{w_i\}$ and $\{x_i\}$. To compute the permutation distribution of R_2 or R_3, we only need to rearrange one set of movements. Whatever statistic we choose, we may use its permutation distribution to obtain a test of statistical significance.

12.3.3 The Building Blocks of Life

In a fascinating state-of-the-art biological application, DNA sequencing, Karlin et al. [1983] use permutation methods to assess the significance of certain repeated patterns of nucleic acids in several viruses.

DNA, the self-replicating molecule that is the basis of life on Earth, is assembled from four specific nitrogenous bases—adenine, guanine, thymine, and cytosine. The sequence in which these bases occur in the DNA molecule determines the structure of the organism. The triplet of deoxyribonucleotides guanine-adenine-cytosine leads to the production of the amino acid aspargine, for example. At issue is whether certain repeated patterns involving multiple copies of lengthy nucleotide sequences is also significant or merely the result of chance. Studying the distribution of repeated patterns that result when one randomly reassigns the labels on the nucleotides while preserving the total numbers of each label, Karlin et al. conclude that the observed patterns are statistically significant. Hasegawa, Krishino, and Yano [1988] approach an analogous problem in DNA sequencing using bootstrap methods. The unraveling of the biological significance of the patterns continues to be an important research problem.

12.3.4 Structured Exploratory Data Analysis

A further illustration of this principle is given by Karlin and Williams [1984] in their use of permutation methods in a structured exploratory data analysis (SEDA) of familial traits. A SEDA has four principal steps:

1) The data are examined for heterogeneity, discreteness, outliers, and so forth, after which they may be adjusted for covariates (as in Section 6.4.3) and the appropriate transform applied (as in Section 6.4.3).
2) A collection of summary SEDA statistics are formed from ratios of functionals.
3) The SEDA statistics are computed for the original family trait values and for reconstructed family sets formed by permuting the trait values within or across families.
4) The values of the SEDA statistics for the original data are compared with the resulting permutation distributions.

As one example of a SEDA statistic, consider the OBP, the offspring-between-parent SEDA statistic:

$$\frac{\sum_i^N \sum_j^{K_i} |O_{ij} - (M_i + F_i)/2|}{\sum_i^N |F_i - M_i|}.$$

In family $i = 1, \ldots, I$, F_i and M_i are the trait values of the father and mother (the cholesterol levels in the blood of the father and mother, for example), while O_{ij} is the trait value of the jth child, $j = 1, \ldots, K_i$.

To evaluate the permutation distribution of the OBP, we consider all permutations in which the children are kept together in their respective family units, while we either:

a) randomly assign to them a father and (separately) a mother; or
b) randomly assign to them an existing pair of spouses. The second of these methods preserves the spousal interaction. Which method we choose will depend upon the alternative(s) of interest.

12.3.5 Comparing Multiple Methods of Assessment

We are often forced to combine several methods of assessment; one obvious example is in quality control; another is in grading students: Is an "A" in statistics equivalent to an "A" in Spanish? Direct comparisons are difficult, if not impossible, when students are free to choose their own courses. Table 12.1, reproduced with permission from Manly [1988], illustrates some of the problems associated with free choice: Missing data are one obvious problem. A second, hidden problem is that there is no guarantee that a student who is good in statistics will do equally well in Spanish.

The solution to both problems is to develop some kind of aggregate measure, compute this measure separately for each course, and then check to see how the distribution of this measure is affected by random relabelings of the students.

Table 12.2, also taken from Manly, illustrates the computation of just such a measure for the course in F. (The names of the actual courses have been changed to letters to protect the identities of overly-generous and overly-stingy graders.) The students are arranged in Table 12.2 in order of increasing mean grade. Each student's mark in course F is subtracted from that student's mean grade and the differences are cumulated.

If the *marks* in the various subjects are comparable, then each random rearrangement of an individual student's marks is equally likely. For example, under the null hypothesis, student 6, who we see from Table 12.1 received marks of 75, 46, 45, and 64 in subjects A, C, E, and F might just as easily have received marks of 64, 45, 75, and 46 in those same subjects. Had this been the case, the CUMSUM score for subject F would have been 67.2 rather than 85.2. By looking at all possible arrangements of each student's marks, we obtain a permutation distribution against which the CUMSUM score for the original arrangement can be assessed.

If the original score does not represent an extreme value, we conclude that the marking for subject F is consistent with the marking for the other subjects.

If, on the other hand the original CUMSUM score does represent an extreme value, our next step is to rescale the marks for subject F, subtracting and/or dividing by a constant. We repeat the test procedure using the rescaled values. And, in a manner akin to the way in which we derive a confidence interval (see Section 3.2), we continue testing and rescaling until all the marks in all the courses have been brought into alignment. Then, we may safely combine the assessments.

Table 12.1. Examination Results from Seven Examination (Subjects A–G) for 64 Students[†]

	Subject									Subject							
Std	A	B	C	D	E	F	G	M	Std	A	B	C	D	E	F	G	M
1	70	—	—	—	—	—	60	65.0	33	75	—	63	67	66	51	—	58.5
2	61	—	38	—	—	—	—	49.5	34	70	—	50	—	65	—	—	67.5
3	94	—	92	42	—	—	—	93.0	35	—	—	38	—	—	—	60	60.6
4	73	—	62	—	—	—	—	59.0	36	61	59	88	—	83	—	—	53.0
5	63	—	62	—	45	—	66	63.7	37	84	—	23	—	—	42	—	78.5
6	75	—	46	—	—	64	—	57.5	38	—	—	29	28	—	—	51	32.5
7	38	—	9	—	—	—	—	23.5	39	52	—	14	—	—	—	—	40.0
8	70	—	41	—	—	—	—	55.5	40	43	—	64	—	—	—	—	28.5
9	59	—	38	—	—	—	—	48.5	41	70	—	—	—	—	—	59	67.0
10	79	—	73	—	—	—	—	76.0	42	64	—	26	—	—	—	—	61.5
11	56	50	—	—	44	—	65	55.0	43	58	—	86	—	—	—	—	42.0
12	78	—	—	—	—	—	—	64.0	44	77	—	—	—	92	—	85	81.5
13	68	—	65	—	—	—	—	66.5	45	—	—	76	43	—	—	—	88.5
14	—	—	29	—	—	—	36	32.5	46	90	—	56	—	—	—	—	83.0
15	—	—	48	61	—	—	—	54.5	47	—	—	—	—	—	—	72	49.5
16	75	—	80	—	—	—	—	77.5	48	—	94	98	94	—	—	—	72.0
17	58	—	—	—	57	—	—	58.0	49	98	85	—	95	—	—	—	96.0
18	—	—	—	—	—	92	—	57.0	50	—	—	—	—	—	—	—	90.0
19	—	—	—	—	—	—	—	92.0	51	90	—	—	—	—	—	—	90.0
20	70	—	40	—	42	80	—	55.0	52	55	—	29	—	—	—	—	42.0
21	62	—	40	—	—	—	—	56.0	53	65	—	39	—	—	—	—	52.0
22	78	—	48	—	—	—	66	63.0	54	60	—	—	—	—	—	—	60.0
23	—	—	40	—	—	80	—	53.0	55	—	—	—	—	58	90	—	74.0
24	—	—	72	—	65	—	56	72.3	56	90	—	52	—	—	—	—	71.0
25	45	—	—	—	54	—	—	50.5	57	91	63	84	83	—	—	—	80.3
26	—	—	60	—	70	—	—	57.0	58	90	—	92	—	—	—	—	91.0
27	—	—	78	—	35	—	—	74.0	59	64	—	41	—	—	—	—	52.5
28	—	—	—	—	74	67	79	51.0	60	20	—	1	—	—	—	—	10.5
29	—	—	81	—	—	—	—	78.0	61	45	—	26	—	—	—	—	35.5
30	64	—	32	—	—	—	—	48.0	62	91	75	79	82	—	—	—	81.8
31	96	—	91	—	—	—	—	93.5	63	60	—	56	66	—	92	—	58.0
32	70	—	65	—	—	—	—	67.5	64	—	—	—	—	—	—	—	79.0

[†] A dash indicates that the student concerned did not sit the examination: Std, student number; M, student mean mark.
Note 1. From Manly [1988]. Reprinted with permission from the Royal Statistical Society.

Table 12.2. CUMSUM calculations for the subject F marks of Table 12.1*

Student	F mark	Mean	Difference	CUMSUM
38	42	32.5	9.5	9.5
28	67	51.0	16.0	25.5
21	80	56.0	24.0	49.5
6	64	57.5	6.5	56.0
33	51	58.5	−7.5	48.5
24	80	72.3	7.7	56.2
55	90	74.0	16.0	72.2
64	92	79.0	13.0	85.2
19	92	—	—	—

*Student 19 only took subject F. There is therefore no comparison possible with other subjects and no contribution to the CUMSUM. From "The comparison and scaling of student assessment marks in several subjects," B.F.J. Manly, *Applied Statistics*: 1988; 37: 385–95. Note: Reprinted with permission from the Royal Statistical Society.

12.4 Model Validation

12.4.1 Regression Models

We consider two methods of validation, via the bootstrap and via permutation tests.

12.4.1.1 Via the Bootstrap

Gail Gong [1986] was the first to use the bootstrap for this purpose. She took a data set for which the results of a multiple logistic regression had already been published. She applied the same stepwise regression technique the original authors had to a series of bootstrap samples taken from the original sample. A few of the independent variables that the original stepwise regression procedure had selected were incorporated in each of the resulting regression models. But just as often, a variable that appeared in an analysis of one bootstrap sample proved insignificant and was dropped in the analysis of a second bootstrap sample.

This bootstrap method can be used in modeling to determine which, if any, variables are essential to a modeling effort and which are dispensable. It can also be used to determine whether the original sample size is adequate given the number of independent variables.

12.4.1.2 Via Permutation Tests

Before beginning to develop a regression model, divide the data at random into two parts, one of which will be used for model development and estimation, the other for validation. Our *goodness-of-fit metric* G is

$$G = \frac{\sum_{k \in \{\text{validation}\}} (Y_{\text{observed}} - Y_{\text{predicted}})^2}{\sum_{k \in \{\text{estimation}\}} (Y_{\text{observed}} - Y_{\text{predicted}})^2},$$

where the summation in the numerator is taken over all the observations in the validation data set and the summation in the denominator is taken over all the observations in the estimation data set.

This ratio will almost always be larger than unity, as the estimation data set was used to choose the variables that went into the model. Consequently, we cannot refer to tables of the F-distribution. We *can* derive the permutation distribution of G as follows:

Divide the original data set into two parts at random a second time, but use the estimation set only to calculate the values of the coefficients. Use the same model you used before; that is, if $\log[X]$ was used in the original model, use $\log[X]$ in this new one. Compute G a second time.

Repeat this resampling process several hundred times. If the original model is appropriate for prediction purposes, it will provide a relatively good fit to most of the data sets, if not, the goodness-of-fit statistic for our original estimation set will be among the largest of the values, since its denominator will be among the smallest.

12.4.2 Models With a Metric

The general circulation models of the Earth's atmosphere and oceans used in weather and current prediction are of mind-boggling complexity, while the available data are all too finite. Priesendorfer and Barnett [1983] confront the problem of model-reality comparison studies of general circulation models head on by developing their own triple of metrics. In Figure 12.1a and b which illustrates some of their concepts, the set D represents actual on-site data while M corresponds to a computer-generated model.

Rerandomization is accomplished in two steps. First, the data from D and M are combined into a single data set. Then, this combined set is repeatedly subdivided at random into sets of the same size as the original D and M. The resultant reference distributions for each of the three metrics are used to assess the agreement of the model with reality.

How good is the Priesendorfer–Barnett test? The answer to this question illustrates the value of the permutation approach to the scientist and engineer whose primary training is not in statistics. For the answer does not depend on the abilities of Priesendorfer and Barnett as statisticians—the calculations in their test are straightforward—but on their abilities as meteorologists and oceanographers. Their test of statistical significance will be a good one, *if* they have selected the appropriate metric and the appropriate variables.

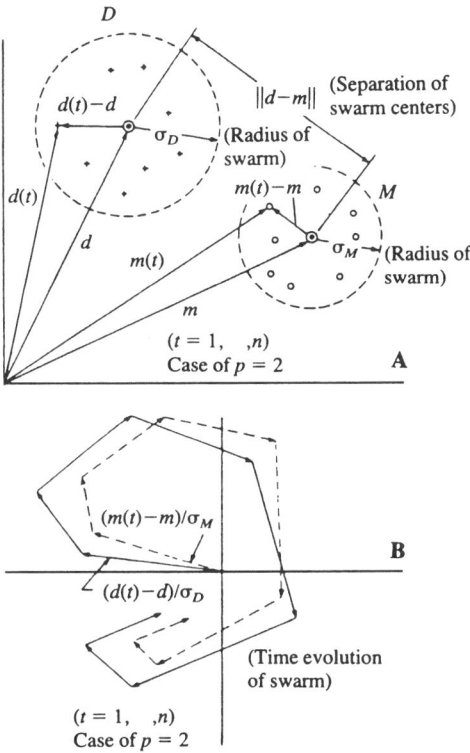

Fig. 12.1. The geometric meaning of the trinity statistics SITES, SPRED, and SHAPE. The statistic SITES is essentially a dimensionless measure of the separation of data swarm centroids, while SPRED is a dimensionless measure of the differences in the root-mean-square radii of the swarms. The statistic SHAPE is a combined measure of the time evolution of the data swarms (and their associated maps). Note: From "The numerical model/reality intercomposition tests using small-sample statistics," by R.W. Priesendorfer and T.P. Barnett, which appeared in *Journal of the Atmospheric Sciences*; 1983; 40: 1884–96. Reprinted with permission from the American Meteorological Society.

12.5 Bootstrap Confidence Intervals

When all else fails, the bootstrap may provide the confidence intervals we need to make a decision. Parametric and permutation methods typically restrict us to sufficient statistics related to location parameters and dispersions. The bootstrap can provide tests for means, medians, interquartile deviations, and percentiles of a distribution. It can provide a confidence interval for the ARE Hodges–Lehmann estimator of the location parameter, the median of the pairwise averages

$$\hat{\Delta} = \text{median}_{i \leq j} \frac{(X_j + X_i)}{2}.$$

As we learned in Chapter 3, a hypothesis test can always be derived from a confidence interval and vice versa.

Permutation and parametric tests require us to make assumptions about the underlying distributions. With the bootstrap we can focus entirely on the parameters. When testing a hypothesis regarding the mean of a population from which observations are drawn, a parametric test requires that all observations come from the same distribution; a permutation test requires that all observations come from a symmetric distribution. The bootstrap requires only that all observations have the same mathematical expectation—their distributions are not otherwise restricted.

We avoid using the bootstrap for the most part because its desirable properties are all asymptotic. With small samples, the bootstrap distribution may be quite unrealistic. Increasing the number of bootstrap samples will not help, but there are several techniques that can result in much improved bootstrap confidence intervals. We consider in turn the Hall–Wilson criteria (or bootstrap-t) and the bias-corrected percentile.

12.5.1 Hall–Wilson Criteria

Recall that a desirable point estimate is both accurate and precise. An accurate point estimate is one that is close to the true value of the parameter being estimated. A precise estimate is one that varies little as we go from sample to sample. For an interval estimate to be desirable, it must have a high probability of covering the true value of the parameter and a relatively low probability of covering any false values. The narrower the interval, the more likely we are to achieve the latter objective. The boundaries of desirable interval estimates also should vary as little as possible from sample to sample.

Alas, the confidence interval for the sample median that we derived by bootstrap means in Chapter 3 is not the best possible. Not only is this interval wider than it should be (with the result that the probability of making a Type II error by including a false value is high), but the probability of making a Type I error by failing to include the correct value may be much larger than the stated value. In other words, the probability that the interval (P_{05}^*, P_{95}^*) covers the true value may be much less than 90 percent.

Suppose θ is the parameter we wish to estimate and $\hat{\theta}$ is our estimator. The first of Hall and Wilson's proposals to increase the accuracy of our bootstrap confidence intervals is to use the distribution of the differences $\hat{\theta}^* - \hat{\theta}$ rather than the distribution of the estimate $\hat{\theta}^*$ based on the bootstrap sample alone.

Now suppose $\hat{\sigma}$ is an estimate of the scale of $\hat{\theta}$, and $\hat{\sigma}^*$ the value of $\hat{\sigma}$ computed for the bootstrap sample. To reduce the width of our confidence intervals (thus decreasing the probability of a Type II error), Hall and Wilson [1991] propose we scale each of these differences by $\hat{\sigma}^*$. Instead of looking at the differences $\hat{\theta}^* - \hat{\theta}$, we are to look at the distribution of the Studentized differences $(\hat{\theta}^* - \hat{\theta})/\hat{\sigma}^*$, where $\hat{\sigma}^*$ is to be estimated by bootstrapping from the bootstrap sample. These guidelines can sometimes lead to rather bizarre results—see Exercise 12.8.

12.5.2 Bias-Corrected Percentile

The bias-corrected interval represents a substantial improvement for one-sided confidence intervals, though it is still suspect. The idea behind this interval comes from the observation that percentile bootstrap intervals are most accurate when the estimate is symmetrically distributed about the true value of the parameter and the tails of the estimate's distribution drop off rapidly to zero. The symmetric mono-modal normal distribution represents this ideal.

Suppose θ is the parameter we are trying to estimate, $\hat{\theta}$ is the estimate based on the original sample, and $\hat{\theta}^*$ is the estimate obtained from the bootstrap sample.

Let $K_B[x] = P\{\hat{\theta}^* \leq x\}$ where P is the probability conditioned on the observed sample. The uncorrected $1 - \alpha$ lower confidence based on the bootstrap would be $\theta_L = K_B^{-1}[\alpha]$.

Suppose, now we are able to come up with a monotone increasing transformation m such that $m(\hat{\theta})$ is normally distributed about $m(\theta)$ regardless of F, the distribution of the observations. We could use this normal distribution to obtain an unbiased confidence interval, then apply a back-transformation to obtain an almost-unbiased confidence interval. We shall see in what follows that we don't actually have to perform these operations to obtain the desired result.

Given such an m, the $1 - \alpha$ lower confidence bound for θ would be $m^{-1}(m(\hat{\theta}) + z_\alpha)$ where z_α is the αth percentile of an $N(0,1)$ distribution. In practice, a more accurate lower confidence bound is given by

$$\theta_{mL} = m^{-1}(m(\hat{\theta}) + z_\alpha + z_0) \tag{12.1}$$

where z_0 is termed the "bias" of m.

Let $K_B[x] = P\{\hat{\theta}^* \leq x\}$ where $\hat{\theta}^*$ is the estimate obtained from the bootstrap sample, and P is the probability conditioned on the observed sample. $K_B[\hat{\theta}] = P\{(m(\hat{\theta}^*) - m(\hat{\theta}) + z_0) \leq z_0\} = \Psi(z_0)$ where Ψ is the $N(0,1)$ distribution function. This implies

$$z_0 = \Psi^{-1}[K_B[\hat{\theta}]] \tag{12.2}$$

Moreover,

$$1 - \alpha = \Psi(-z_\alpha) = P\{(m(\hat{\theta}^*) - m(\hat{\theta}) + z_0) \leq z_\alpha\}$$
$$= P\{\hat{\theta}^* \leq m^{-1}(m(\hat{\theta}) - z_0 - z_\alpha)\}.$$

Or, equivalently, for $0 < \alpha < 1$,

$$K_B^{-1}[\alpha] = m^{-1}(m(\hat{\theta}) - z_0 - z_\alpha). \tag{12.3}$$

From equations (12.1) and (12.3) we see that $\theta_{mL} = K_B^{-1}[\Psi[2z_0 + z_\alpha]]$.

From Equation (12.2), we derive the bootstrap bias-corrected lower confidence bound $\theta_{BCL} = K_B^{-1}\{\Psi[2\Psi^{-1}[K_B[\hat{\theta}]] + z_\alpha]\}$ (Efron [1981]). To improve the rate of convergence, Efron [1987] proposed a bootstrap-accelerated bias-corrected method.

12.6 Exercises

1. Suppose you wish to compare two groups of observations. Would it be better to compare them using the two-sample comparison of Section 3.3 or the matched pairs technique of Section 3.6? Is your decision rule an "always..." or does it depend on how the observations are dispersed and the relative importance of the covariates used to do the matching?
2. Suppose you have discarded the n original observations in the sample, keeping only the n order statistics when you obtain independent evidence that the data are normally distributed. Can you still compute the sample mean and variance?
3. Suppose you have multiple observations on each subject, some in feet, some in inches, some in pounds. Should they all be transformed to a common unit of reference before you begin your multivariate analysis? What transformation(s) should you use?
4. What statistic(s) remain invariant under an arbitrary monotone-increasing transformation of the observations? Is this result relevant to the preceding question?
5. Ninety-nine percent of all scientists ignore the loss function and make do with a significance level and (hopefully) a minimum power level against one or two selected alternatives. Reconsider the statistical analyses you performed recently. What was the loss function in each instance? Were the test statistics you selected appropriate for this loss function?
6. a) Can the four k-sample statistics $F1$, $F2$, $F3$, and R introduced in Section 4.2.2 be made equivalent to one another if we eliminate terms that are invariant under permutations?
   b) If your answer to the previous question is "no," will there be data sets for which tests based on $F1$, $F2$, and R lead to different conclusions?
   c) How would you decide which of these three statistics to use?
   d) Are you free to compute the permutation distributions of $F1$, $F2$, and R for a specific data set and then choose the statistic that does the best job of proving your point?
   e) Suppose you were an examiner at the Food and Drug Administration. How would you react to a submission the authors of which had done as in 6d?
   f) If you were one of those authors, how would you justify your choice of test statistic to an examiner at the FDA?
   g) Throughout this text, we have tried to justify our choice of statistic on the grounds that the resultant test was (i) unbiased, (ii) most powerful,

(iii) minimized losses, or was (iv) invariant under transformations of location and scale. Do these criteria satisfy your own instincts? What other criteria can you suggest?

7. Unbiasedness and invariance represent two complementary but distinct approaches to testing. Which principle would you apply in the following situations:
   a) Comparing two Poissons.
   b) Comparing two binomials.
   c) Testing the hypothesis that the variance $\sigma = \sigma_0$ against the alternative $\sigma \neq \sigma_0$ when the observations come from a normal distribution.
   d) Given the X_i are $N(\mu, \sigma)$, testing the hypothesis that $\mu/\sigma \leq 3$.

8. Although the Hall–Wilson corrections are widely accepted, sometimes they can produce idiotic results. Obtain a Hall–Wilson corrected bootstrap interval estimate for the population mean using the following sample:
   0 0 0 0 7.53 0 0 0 15.77 0 0 0 0 7.53 6.16 0 0 0 0 18 0 5.71 5.71 0 7.78 0 7.03 0 10.22 0 12 19.07 15.50 0 0 0 0 0 3.81 6.10 3 10.78 0 10.44 0 0 0 0 0 0 4 0 0 0 103.05 0 0 0 0 12 0 0 0.

9. Suppose we know that $y_k = a + bt_k + \epsilon_k$ for $k = 1, \ldots, n$, where the residual errors $\{\epsilon_k\}$ are exchangeable. Now suppose we replace a and b by their least squares estimates, that is, by the values of a and b that will minimize the sum $\sum(y_k - a - t_k)^2$. Show that if we rewrite the equation for y_k in terms of these least squares estimates and t_k, that the resulting residuals $\{\epsilon'_k\}$ will still be exchangeable.

10. Prove Theorem 12.2.

13
Publishing Your Results

McKinney et al. [1989] report that more than half the published articles that apply Fisher's exact test do so improperly. Our own survey of some 50 biological and medical journals supports their findings. This chapter provides you with a positive prescription for the successful publication of the results of testing procedures. First, we consider the rules you must follow to ensure that your data can be analyzed by statistical methods. Then, we provide you with a number of simple rules to prepare your report for publication.

13.1 Design Methodology

It's never too late to recheck your design methodology. Recheck it now in the privacy of your office rather than before a large and critical audience. All our testing methods rely on the independence and/or the exchangeability of the observations. Were your observations. independent of one another? What was the experimental unit? Were your subjects/plots assigned at random to treatment? If not, how was randomization restricted? With complex multi-factor experiments, you need to list the blocking variables and describe your randomization scheme.

13.1.1 Randomization in Assignment

Are we ever really justified in exchanging labels among observations? Consider an experiment in which we give six different animals exactly the same treatment. Because of inherent differences among the animals, we end up with six different measurements, some large, some small, some in between. Suppose we arbitrarily label the first three measurements as "controls" and the last three as "treatment." These arbitrary labels are exchangeable and thus the probability is 1 in 20, that the three "control" observations will all be smaller than the three "treatment." Now suppose we repeat the experiment, only this time we give three of the animals an experimental drug and three a saline solution. To be sure of getting a positive result, we give the experimental drug to those animals who got the three highest scores in the first experiment. Not fair, you say. Illegal! Illegitimate! No one would ever do this in practice.

In the very first set of clinical data I received for statistical analysis was brought by a young surgeon. He described the problems he was having with his chief of surgery. "I've developed a new method for giving arteriograms, which I feel can cut down on the necessity for repeated amputations. But my chief will only let me try out the technique on patients that he feels are hopeless. Will this affect my results?" It would and it did. Patients examined by the new method had a very poor recovery rate. But, of course, the only patients who'd been examined by the new method were those with a poor prognosis. The young surgeon realized that he would not be able to test his theory until he was able to assign patients to treatment at random.

Not incidentally, it took us three more tries until we got this particular experiment right. In our next attempt, the chief of surgery—Mark Craig of St. Eligius in Boston—announced that he would do the "random" assignments. He finally was persuaded to let me make the assignment using a table of random numbers. But then he announced that he, and not the younger surgeon, would perform the operations on the patients examined by the traditional method to make sure "they were done right." Of course, this turned a comparison of methods into a comparison of surgeons and intent.

In the end, we were able to create the ideal "double blind" study: The young surgeon performed all the operations, but the incision points were determined by his chief after examining one or the other of the two types of arteriogram.

13.1.2 Choosing the Experimental Unit

The exchangeability of the observations is a sufficient condition for a permutation test to be exact. It is also a necessary condition for the application of any statistical test. Suppose you were to study several pregnant animals that had been inadvertently exposed to radiation (or acid rain or some other undesirable pollutant) and examine their offspring for birth defects. Let $X_{ij}, i = 1, \ldots, I; j = 1, \ldots, J$, denote the number of defects in the jth offspring of the ith parent; let $Y_i = \sum_j X_{ij}, i = 1, \ldots, I$ denote the number of defects in the ith litter. The $\{Y_i\}$ *may* be exchangeable (we would have to know more about how the data were collected to determine this). The $\{X_{ij}\}$ are not: The observations within a litter are interdependent; what affects a parent affects all her offspring. In this experiment, the litter is the correct experimental unit.

The viewpoints of the observer and the statistician can be quite different. If we wear two hats—serving both as observer and statistician, recognition of this distinction can be painful. For example, in a typical toxicology study a pathologist may have to examine three to five slides at each of 15 to 20 body sites in each of three to five animals—several hours of labor—just to get an effective sample size of $n = 1$.

13.1.3 Determining Sample Size

As noted in Chapter 2, the number of observations must be large enough that the resultant hypothesis test will have sufficiently high probability (power) of detecting effects that are of scientific and/or practical interest. Before you start, specify the significance level, the minimum effect of interest, and the desired power for that effect, then use one of the methods described in Section 14.10 to determine the appropriate sample size.

You may need to conduct your experiment in several stages, using your initial efforts as a basis for estimating the population parameters needed in the power calculations.

13.1.4 Power Comparisons

When making power comparisons between permutation methods, which yield exact values for significance levels and parametric and bootstrap methods, which yield only approximations, it is essential that the critical values used in power comparisons be chosen so as match the actual significance levels. For example, suppose that the tabulated critical value for a parametric test is c^*, that is, we are to reject the hypothesis if our test statistic $S > c^*$, but the probability that $S > c^*$ under the conditions of our power comparison is $\alpha^* > \alpha$. This would be the case, for example, if the parametric test relied on the assumption of normality, but the test distribution was a mixture of exponentials. Let c denote the true critical value such that the probability that $S > c$ is α. Clearly, $c^* < c$. The power we report for our bootstrap or parametric test will be in inflated to the extent it is based on values of S such that $c^* < S < c$. To forestall such an error, the critical values used in power comparisons should be determined by a preliminary Monte Carlo under the assumptions of the null hypothesis (Zhang and Boos [1994]). See Xu and Lee [2003] for an application of this approach in the analysis of microarray data.

13.2 Preparing Manuscripts for Publication

You've laid the groundwork. You've done the experiment. You've completed the analysis. A few simple rules can help you prepare your article for publication.

13.2.1 Reportable Elements

Reportable elements include descriptions and details of all of the following:

- objectives of your study
- experimental design
    - endpoints
    - surrogate variables
    - control variables
    - covariates and (potential) confounding variables
- hypotheses and principal alternatives
- power and sample size calculations
- data collection methods (describe the experimental unit, any use of clusters when sampling)
- sources of missing data
- exceptions
- validation methods
- statistical analysis.

See Chapter 7 of Good and Hardin [2003] for additional material on each of the above points.

13.2.2 Details of the Analysis

1. State the test statistic explicitly. Reproduce the formulae. If you cite a text, for example, Good [1993], include the page number(s) on which the statistic you are using is defined.
2. State your assumptions. Are your observations independent? Exchangeable? Is the underlying distribution symmetric? Contrary to statements that have appeared in several recent journal articles—we withhold the names to protect the guilty—permutation tests cannot be employed without one or both of these essential assumptions. See Draper et al. [1993], Gastwirht and Rubin [1971], and Hettmansperger [1984] for discussions of this point.
3. If using a permutation test, state which labels you are rearranging. Provide enough detail that any interested reader can readily reproduce your results. In other words, report your statistical procedures in the same detail you report your other experimental and survey methodologies.
4. State whether you are using a one-tailed or a two-tailed test. See Section 8.1.1 for help in making a decision.
5. a) If you detect a statistically significant effect, then provide a confidence interval (see Section 3.3). Remember, an effect can be statistically significant without being of practical or biological significance.
   b) If you do not detect a statistically significant effect, could a larger sample or a more sensitive experiment have detected one? Consider reporting the power of your test. See Sections 2.1.4 and 14.10.

14
Increasing Computational Efficiency

14.1 Seven Techniques

With today's high-speed computers, drawing large numbers of subsamples with replacement (the bootstrap) or without (the permutation test) is no longer a problem, unless or until the entire world begins computing resampling tests at one time! To prepare for this eventuality, and because computational efficiency is essential in the search for more powerful tests, a primary focus of research in resampling today is the development of algorithms for rapid computation.

There are seven main computational approaches, several of which may be and usually are employed in tandem, as follows:

1. The *Monte Carlo*, in which a sample of the possible rearrangements is drawn at random and these samples are used in place of the complete permutation distribution.

2. *Rapid enumeration and selection algorithms*, whose object is to provide a rapid transition from one rearrangement to the next.

3. *Recursive relationships*, which reduce the number of computations.

4. *Branch and bound algorithms* that eliminate the need to evaluate each individual rearrangement.

5. *Gibbs sampling*.

6. Solution through *characteristic functions and fast Fourier transforms*.

7. *Asymptotic approximations*, for use with sufficiently large samples.

In the following sections, we consider each of these approaches in turn.

14.2 Monte Carlo

Instead of examining all possible rearrangements, we can substantially reduce the computations required by examining only a small but representative random sample [Dwass, 1957; Barnard, 1963]. In this process, termed a Monte

Carlo, we proceed in stages: 1) We rearrange the data at random; 2) we compute the test statistic for the rearranged data and compare its value with that of the statistic for the original sample; and 3) we apply a stopping rule to determine whether we should continue sampling, or whether we are already in a position to accept or reject.

The program fragments reproduced in Chapters 3 through 5 of this text use the Monte Carlo approach. In the not necessarily optimal computer algorithm introduced in those chapters, all observations in all subsamples are loaded into a single linear vector $\boldsymbol{X} = \{\boldsymbol{X}[0], \boldsymbol{X}[1], \ldots, \boldsymbol{X}[N-1]\}$. Then, a random number is chosen from the set of integers $0, 1, \ldots, I$ with $I = N-1$ initially. If the number we choose is i, $\boldsymbol{X}[i]$ is swapped with $\boldsymbol{X}[I]$ in a three-step process:

    temp := X[i];
    X[i] := X[i];
    X[N-1] := temp;

and I is decremented. This process is repeated until we have rearranged the desired number of observations and are ready to compute the test statistic for the new rearrangement.

We dont't always need to reselect all N observations. For example, in a two-sample comparison of means, with $N = n + m$, our test statistic only makes use of the last m observations. Consequently, we only need to choose m random numbers each time.

After we obtain the new value of the test statistic, we compare it with the value obtained for the original data. We continue until we have examined N random rearrangements and N values of the test statistic. Typically, N is assigned a value between 100 and 1600, depending on the precision that is desired (see Section 14.2.2 and Marriott [1979]). Through the use of a Monte Carlo, even the most complicated multivariate experimental design can be analyzed in les than a minute on a desktop computer.

14.2.1 Stopping Rules

If a simple accept/reject decision is required, we needn't perform all N calculations, but can stop as soon as it is obvious that we must accept or reject the hypothesis at a specific level. In practice, we use a one-sided stopping rule based on the 10% level. Suppose in the first n rearrangements we observe a fraction H_n with a value of the test statistic that is as, or more, extreme than the value for the original observations. If $H_n > 0.1N$, then we accept the hypothesis at the 10% level. Otherwise, we continue until $n = N$ and report the exact percentage of rejections. Besag and Clifford [1991] and Lock [1991] describe two-sided sequential procedures in which the decision to accept, reject, or continue is made after each rearrangement is examined.

14.2.2 Variance of the Result

The resultant estimated significance level $\hat{p}$ is actually a binominal random variable $B(N,p)$, where N is the number of random rearrangements and p is the true but still unknown value of the significance level. The variance of $\hat{p}$ is $p(1-p)/N$. If p is 10%, then using a sample of 81 randomly selected rearrangements provides a standard deviation for $\hat{p}$ of 1%. A sample of 364 reduces the standard deviation to 0.25%.

The use of a variable in place of a fixed significance level results in a minor reduction in the power of the test, particularly with near alternatives [Dwass, 1957]. In most cases, this reduction does not appear to be of any practical significance; see Vadiveloo [1983], Jockel [1986], Bailer [1989] Edgington [1987], and Noreen [1989].

In a Monte Carlo variant called importance sampling, the rearrangements are drawn with weights chosen so as to minimize the variance. In some instances, when combined with branch and bound techniques, as in Mehta Patel, and Senchaudhuri [1988], importance sampling can markedly reduce the number of samples that are required. See, also, Besag and Clifford [1989].

14.2.3 Cutting the Computation Time

The generation of random rearrangements creates its own set of computational problems.

Each time a data element is selected for use in the test statistic, two computations are required: (1) A random number is selected, and (2) two elements in the combined sample are swappepd.

The ideal futuristic computer will have a built-in random number generator—for example, it might contain a small quantity of a radioactive isotope, with the random intervals between decays producing a steady stream of random numbers. This futuristic computer might also have a butterfly network that would randomly swap 10 or 100 elements of an array in a single pass.

Today, in the absence of such technology, any improvements in computation speed must be brought about through software. Little direct research has been done in the area, although recently Baglivo et al. [1992] reported on techniques for doing many of the repetitive computations in parallel. I did some preliminary work in which I considered a sort of drunkard's walk through the set of rearrangements: The first rearrangement was chosen at random; thereafter, the program stumbled from rearrangement to rearrangement swapping exactly two data elements at random each time. The results were disappointing. Any savings in computation time per rearrangement were more than offset by the need to sample four or five times as many rearrangements to achieve the same precision in the result. I did achieve a substantial increase in efficiency by selecting several separated random bits from each random number.

14.3 Rapid Enumeration and Selection Algorithms

If we are systematic and proceed in an orderly fashion from one rearrangement to the next, we can substantially reduce the time required to examine a series of rearrangements. The literature on this topic is extensive. See, for example, the review by Wright [1984]. We have posted a bibliography on this and related computational topics at http://mysite.verizon.net/res7sf1o/bibcomp.htm.

14.3.1 Matched Pairs

Sometimes we can reduce the number of computations that are required by taking advantage of the way we label or identify individual permutations. In the case of paired comparisons, we readily enumerate each possible combination by running through the binary numbers from 0 to $2^n - 1$, letting the zeroes and ones in each number (obtained via successive right shifts, a single machine-language instruction in most computers) correspond to positive and negative paired differences, respectively.

The shift algorithm, introduced in this context by Baker and Tilbury [1993] for use with discrete data, avoids the need for assembler level programming. The test statistic $T_k \sum_{i=1}^{k} |X_i| - \sum_{i=1}^{k} X_i$ is calculated one variable at time, and an array of counters or bar chart $N[\,]$ is incremented appropriately. At step 0, $T_0 = 0$; we initialize the array of counters, so that $N[0] = 1$ and all other elements are zero. At step 1, we add X_1 to T_0; as X_1 could be either positive or negative, we increment $N[|X_1|]$ by $N[0]$ so that both $N[0]$ and $N[|X_1|]$ are now equal to one. At step 2, we add X_2 to T_1 and increment $N[|X_1| + |X_2|]$ by $N[|X_2|]$ by $N[0]$. Note that if $X_1 = X_2, N[|X_1|] = 2$. We continue in this fashion, so that at the kth step, we increment $N[j]$ by $N[j - |X_k|]$ for $j = \sum_{i=1}^{k} |X_i|, \ldots, |X_k|$.

Censoring actually reduces the time required for enumeration. For if there are n_c censored pairs, then enumeration need only extend over the 2^{n-n_c} values that might be assumed by the uncensored pairs. In computing the GAMP test for paired comparisons, it is easy to see that

$$\Pr\{U' \geq U \text{ and } S' \geq S\} = \Pr\{U' \geq U\}^* \Pr\{S' \geq S\},$$

$$\Pr\{U' \geq U\} = \frac{1}{2^{U+L}} \sum_{k=u}^{U+L} \binom{U+L}{k}$$

The remaining probability $\Pr\{S' \geq S\}$, may be obtained by enumeration and inspection.

14.4 Recursive Relationships

Although tables for determining the significance level of Fisher's exact test are available, in Finney [1948] and Latscha [1953], for example, these are restricted to a few discrete *p*-values. Today, it is usually much faster to compute a significance level than it is to look it up in tables. Beginning with Leslie [1955], much of the subsequent research on Fisher's exact test has been devoted to developing algorithms that would speed or reduce the number of computations required to obtain a significance level.

As one rapid alternative to Equation (6.1), we may use the recursive relationship provided by Feldman and Kluger [1963]: With table entries (a_0, b_0, c_0, d_0), define

$$p = \frac{(a_0+b_0)!(a_0+c_0)!(d_0+b_0)!(d_0+c_0)!}{N!a_0!b_0!c_0!d_0!}.$$

It is easy to see that

$$p_{i+1} = \frac{a_i d_i}{b_{i+1} c_{i+1}} p_i,$$

where $a_i = a_0 - i$.

We may speed the computations of the statistics for unordered $r \times c$ contingency tables considered in Section 6.4 by noting that Q is invariant under permutations that leave the marginals intact. Thus, we may neglect the numerator Q in calculating the permutation distribution and focus on the denominator R (March [1972]).

We may use a recursive algorithm developed by Gail and Mantel [1977] to speed the computations for $r \times 2$ contingency tables. If $N_i(f_{.1}; f_1, f_2, \ldots, f_n)$ denotes the number of tables with the indicated marginals, then

$$N_{i+1}(f_{.1}; f_1, f_2, \ldots, f_n) = \sum_j N_i(f_{.1} - j; f_1, f_2, \ldots, f_n).$$

The algorithms we developed in Chapters 3 and 4 are much too slow, since they treat each observation as an individual value.

Algorithms for speeding the computations of the Freeman–Halton statistics in the general $r \times c$ case are given in March [1972], Gil and Mantel [1977], Mehta and Patel [1983, 1986a, 1986b], and Pagano and Halvorsen [1981]. Details of the Mehta and Patel approach are given in Section 14.4. An efficient method for generating $r \times c$ tables with given row and column totals is provided by Patefield [1981]. See also Agresti, Wackerly, and Boyett [1979] and Streitberg and Rohmed [1986].

The power of the Freeman–Halton statistic in the $r \times 2$ case is studied by Krewski, Brennan, Bickis [1984].

14.5 Focus on the Tails

We can avoid examining all $N!$ rearrangements if we focus on the tails using the internal logic of the problem to deduce the number of rearrangements that yield values of the test statistic as, or more, extreme than the original.

Consider the shift algorithm introduced in the preceding section. Suppose that T^0 is the test statistic for the original data; as T_k is nondecreasing, we only need to keep track of individual values of T_k that are less than T^0. Our modified procedure at the kth step is as follows:

If $\sum_{i=1}^{k} |X_i| < T^0,$ increment $N[j]$ by $N[j - |X_k|]$ for $j = \sum_{i=1}^{k} |X_i|, \ldots, |X_k|;$

otherwise, set $N[T^0] = 2N[T^0]$, then increment $N[j]$ by $N[j - |X_k|]$ for $j = T^0, \ldots, |X_k|$.

Of course, if $N[T^0] > \alpha n$, we would terminate the procedure and accept the null hypothesis.

Green [1977] was the first to suggest a branch and bound method for use in two-sample tests and correlation. Our description of Green's method is based on De Cani [1979].

In the two-sample comparison described in Section 3.6 suppose our test statistic $T = \sum x_{\pi(i)}$ and that the observed value is T_0. We seek $P(T \geq T_0)$, the probability under the null hypothesis that a random value of T equals or exceeds T_0.

Assume that the combined observations are arranged in descending order $X_{(1)} \geq X_{(2)} \geq \cdots \geq X_{(N)}$. To simplify the notation, let Z_i denote the ith order statistic $X_{(i)}$. If the labels (subscripts) on the X's really are irrelevant (as they would be under the null hypothesis) then T can be regarded as a random sample of m of the observations selected at random without replacement from the $\{Z_i\}$.

Suppose we have selected k such values, $Z_{I_1}, \ldots, Z_{I_k}, k < m$. The maximum attainable value of T is obtained by adding to $Z_{I_1} + \cdots + Z_{I_k}$ the $m - k$ largest of the $N - k$ remaining elements. Call this maximum $T(l_1, \ldots, l_k)$. Similarly, the minimum attainable value of T is obtained by adding to $Z_{I_1} + \cdots + Z_{I_k}$ the $m - k$ smallest of the $N - k$ remaining elements. Call this minimum $t(l_1, \ldots, l_k)$. Given $l_1, \ldots, l_k$, we can bound T:

$$t(l_1, \ldots, l_k) \leq T \leq (l_1, \ldots, l_k).$$

There are $\binom{N-k}{m-k}$ sets of m elements of Z whose totals lie between the given bounds.

If $t(l_1, \ldots, l_k) \geq T_0$, then

$$P(T \geq T_0) \geq \binom{N-k}{m-k} \Big/ \binom{N}{m}.$$

If $T_0 > T(l_1, \ldots, l_k)$, then

$$p(T \geq T_0) \leq 1 - \binom{N-k}{m-k} \Big/ \binom{N}{m}.$$

If T_0 lies between the bounds, or if we require an improved bound on $P(T \leq T_0)$, then we can add a $(k+1)$th element to the index set.

Our results apply equally to any test statistic of the form $\sum_{i=1}^{m} f[x_{\pi(i)}]$, where f is a monotone increasing function. Examples of such monotone functions include the logarithm (when applied to positive values), ranks, and any of the other robust transformations described in Chpter 11.

14.5.1 Contingency Tables

A large number of authors have joined in the search for a more rapid method for enumerating the tail probabilities for Fisher's exact test, including Leslie [1955], Feldman and Kluger [1963], Good [1976], Gail and Mantel [1977], Pagano and Halvorsen [1981], and Patefield [1981]. See for example, the review by Agresti [1993]. A quantum leap toward a more rapid method took place with the publication of the network approach of Mehta and Patel [1980]. Their approach is widely applicable, as we shall see below. It has three principle steps:

1. Representation of each contingency table as a path through a directed acyclic network with nodes and arcs.
2. An algorithm with which to enumerate the paths in the tail of the distribution without tracing more than a small fraction of those paths.
3. Determination of the smallest and largest path lengths at each node.

Only the last of these steps is application-specific. Network algorithms have been developed for all of the following:

- $2 \times c$ contingency tables [Mehta and Patel, 1980]
- $r \times c$ contingency tables [Mehta and Patel, 1983]
- the common odds ratio in several 2×2 contingency tables [Mehta, Patel, and Gray, 1985]
- logistic regression [Hirji, Mehta, and Patel, 1987]
- restricted clinical trials [Mehta, Patel, and Wei, 1988]
- linear rank tests and the Mantel–Haenszel trend test [Mehta, Patel, and Senchaudhuri, 1988].

For simplicity, we focus in what follows on the $2 \times c$ contingency table.

14.5.1.1 Network Representation

Define the reference set Γ to be all possible $2 \times k$ contingency tables (see Chapter 6) with row marginals (m, n) and column marginals $(t_1, t_2, \ldots, t_k)$. Thus, each table $x \in \Gamma$, is of the form

x_1	x_2	$\cdots$	x_k	m
x'_1	x'_2	$\cdots$	x'_k	n
t_1	t_2	$\cdots$	t_k	$N.$

For each table $x \in \Gamma$, we may define a discrepancy measure

$$d(x) = \sum_{i=1}^{k} a_i(m_{i-1}, x_i)$$

and a probability

$$h(x) = C^{-1} \prod_{i=1}^{k} \lambda_i(m_{i-1}, x_i)$$

where the partial sum $m_j = \sum_{i=1}^{j} x_i$ and the normalizing constant $C = \sum_{x \in \Gamma} \prod_{i=1}^{k} \lambda_i(m_{i-1}, x_i)$. Important special cases of $d(x)$ and $h(x)$ are

$$d(x) = \prod_{i=1}^{k} a_i x_i$$

for linear rank tests and

$$h(x) = \prod_{i=1}^{k} \binom{t_i}{x_i} \bigg/ \binom{N}{m}$$

for unordered contingency tables.

As in Section 6.3, our object is to compute the one-sided significance level $p = \sum_R h(x)$, where R is the set on which $d(X) \geq d_0$.

First, we represent Γ as a directed acyclic network of nodes and arcs. Following Mehta and Patel [1983], the network is constructed recursively in $k = 1$ stages labeled $0, 1, 2, \ldots, k$. The nodes at the jth stage are ordered pairs (j, m_j) whose first elements is j and whose second is the partial sum of the frequencies in the first j categories of the first row. If there is a total of two observations in the first category, then there will be three nodes at the first stage—$(1, 0)$, $(1, 1)$, $(1, 2)$—corresponding to the three possible distributions of elements m this category.

Arcs emanate from the node (j, m_j); each arc is connected to exactly one successor node. Each path linking $(0, 0)$ with the terminal node (k, m) corresponds to a unique contingency table. For example, the path

$$(0,0) \to (1,0) \to (2,2) \to (3,4) \to (4,4)$$

corresponds to the table

0	2	2	0	4
2	0	0	2	4
2	2	2	2	

.

The total number of paths in the network corresponds to the total number of tables. We could count the total number of tables by tracing each of the individual paths, but we can do better.

14.5.1.2 The Network Algorithm

Our goal in network terms is to quickly identify and sum all paths whose lengths do not exceed $d \cdot h$: for the original unpermuted table. Let $\Gamma_j = \Gamma(j, m_j)$ denote the set of all paths from any node (j, m_j) to the terminal node (k, m). In other words, Γ_j represents all possible completions of those tables in Γ for which the sum of the first j cells of row 1 is m_j. Define the shortest path length

$$SP(j, m_j) = \min_{x \in \Gamma_j} \sum_{i=j+1}^{k} a_i(m_{i-1}, x_i)$$

and the longest path length

$$LP(j, m_j) = \max_{x \in \Gamma_j} \sum_{i=j+1}^{k} a_i(m_{i-1}, x_i).$$

Let $L(\text{PAST})$ denote the length of a path from $(0, 0)$ to (j, m_j). If this path is such that

$$L(\text{PAST}) + LP(j, m_j) \leq d \cdot h,$$

then all similar subpaths from $(0, 0)$ to (j, m_j) of equal or smaller length contribute to the p value. This number can be determined by induction—the details depend on the actual form of d and h, and thus we need not enumerate the tables explicitly. If this path is such that

$$L(\text{PAST}) + SP(j, m_j) \geq d \cdot h,$$

then we can ignore it and all similar paths of equal or greater length—again, without actually enumerating them.

If the path satisfies neither condition, then we extend it to a node at the $(j+1)$th stage, compute the new shortest and longest path lengths, and repeat the calculation.

The shortest and longest path lengths may be determined by dynamic programming in a single backward pass through the network. Dynamic programming is used by Mehta and Patel [1980] in their seminal paper on the topic. Their original approach can be improved upon in three ways:

1. by taking advantage of the structure of the problem;
2. by a Monte Carlo, randomly selecting the successor node at each stage;
3. by a Monte Carlo utilizing importance sampling, that is, weighting the probabilities with which an available node is selected so as to reduce the variance of the resultant estimate of p.

The three approaches can be combined: A highly efficient two-pass algorithm for importance sampling using backward induction followed by forward induction was developed by Mehta, Patel, and Senchaudhuri [1988]. Their new algorithm guarantees that all rearrangements sampled will lie inside the critical region. A result of Joe [1988] also represents a substantial increase in computational efficiency.

14.5.2 Play the Winner Allocation

A network algorithm also proves of value in play-the-winner allocation. Recall from Chapter 6, that in the randomized play-the-winner allocation rule, we begin with A balls of each of two types in an urn. Balls are drawn from this urn with replacement to determine what treatment the next experimental unit will receive. If the response to treatment k is a success, B more balls of type k are placed in the urn, otherwise B more balls of the opposite type are placed there.

The obvious test statistic of the hypothesis $p_0 = p_1$ is $S_n = \sum_{j=1}^{n} x_j Y_j$ where $x_j = 1$ if the jth trial results in success, and $x_i = 0$ otherwise, and $Y_j = 1$ or 0, depending on which treatment the jth experimental subject is assigned. To determine the distribution of S_n we construct a network of $(n+1)$ stages. At stage j the network consists of a set of nodes of the form (N_j, S_j) where $N_j = \sum Y_j$. Let p_{j+1} denote the conditional probability that $Y_{j+1} = 1$ given $Y_1, \ldots, Y_j$ and $x_1, \ldots, x_j$. For RPW(A, B), as in Wei [1989],

$$p_{j+1} = \{A + B(2S_j + j - N_i - \Sigma x_j\}/(2A + B_j).$$

Let L denote the probability of observing the event (N_j, S_j). Let Ω_j denote the set of all triples (N_j, S_j, L_j) with distinct (N_j, S_j). Each such triple generates records $(N_{j+1}, S_{j+1}, L_{j+1})$ where $L_{j+1} = L_j \{p_{j+1}^{Y_{j+1}}(1-p_{j+1})^{1-Y_{j+1}}$. Starting with $\Omega_0 = \{(0, 0, 1)\}$, we can generate the $\{\Omega_j\}$ recursively and derive the permutation distribution of S_n.

14.5.3 Directed Vertex Peeling

This method, due to Cohen and Sackrowitz [1998, 2000], is applicable to $2 \times C$ tables and focuses on the partial sums $Y_j = \sum_{k=1}^{j} f_{1k}, j = 1, 2, \ldots, C-1$. The rejection region is built a table at a time from the vectors $\boldsymbol{y} = \{y_1, \ldots, y_{C-1}\}$ on the convex hull or boundary of the set of vectors until the desired significance level is attained. At every stage, only vertices are considered, and of these, only the vertex corresponding to the largest (smallest) value of the text statistic is considered. In the event of ties, back-up statistics are used.

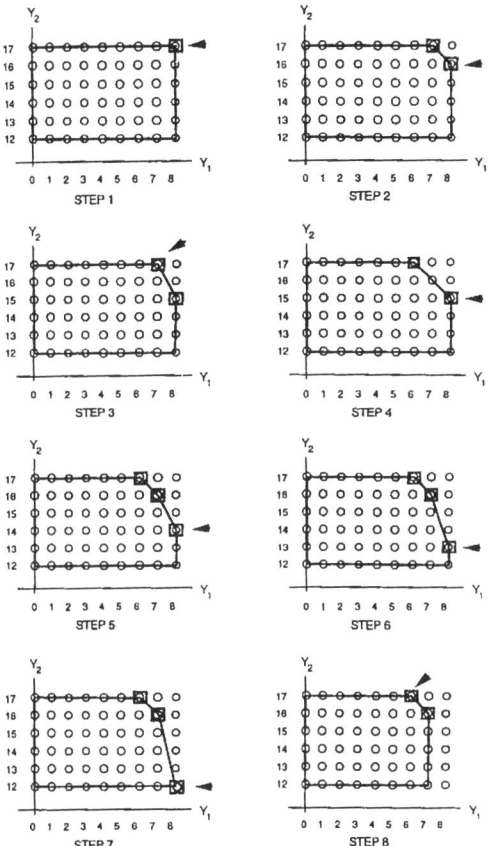

Fig. 14.1. The DVP process for the point (6,17). At each step dark lines indicate the current convex hull and squares indicate the current points eligible for peeling. Arrows indicate the eligible point with the largest chi-square statistic. Reproduced from Cohen and Sackrowitz [2003] with permission from Hodder-Arnold Publishing, London.

Figure 14.1 (reproduced with permission from Cohen and Sackrowitz [2003]) illustrates the application of the method to the 2×3 contingency table with the following marginals:

			17
			15
8	19	5	32

14.6 Gibbs Sampling

Suppose we have a 2×2 table with entries $f_{11}, f_{12}, f_{21}, f_{22}$; assume this table results from a sequence of random variables $X_0, Y_0, X_1, Y_1, \ldots$, each taking the value 0 or 1, where the estimated conditional probabilities of $Y|X$ and $X|Y$ can be expressed in the two matrices

$$A_{y|x} = \begin{pmatrix} \dfrac{f_{11}}{f_{11}+f_{21}} & \dfrac{f_{21}}{f_{11}+f_{21}} \\ \dfrac{f_{12}}{f_{12}+f_{22}} & \dfrac{f_{22}}{f_{12}+f_{22}} \end{pmatrix}, \quad A_{x|y} = \begin{pmatrix} \dfrac{f_{11}}{f_{11}+f_{12}} & \dfrac{f_{12}}{f_{11}+f_{12}} \\ \dfrac{f_{21}}{f_{21}+f_{22}} & \dfrac{f_{22}}{f_{21}+f_{22}} \end{pmatrix}.$$

Using these matrices, generate a single couple y, x. Modify the table (preserving the marginals) to provide for this new entry; if it is not possible to preserve the marginals, do not modify the table. Compute the test statistic, and compare with the original value of the test statistic. Modify the transition matrices to reflect the change and repeat the procedure.

A similar procedure but one guaranteed to converge to the correct result is based on the Gibbs sampler, a technique for generating random variables from a (marginal) distribution indirectly, without having to calculate the density (see Casella and George [1992]). At each step, we draw from the hypergeometric distribution produced by taking a binomial $(p, f_{11} + f_{21})$, and an independent binomial $(p, f_{12} + f_{22})$, conditional on the sum of the two binomials being $f_{11} + f_{12}$. To obtain a new table, we let the computer pick a uniformly distributed random number between 0 and 1 and evaluate the hypergeometric quantile at this number. Methods for its rapid calculation are described in Kolassa and Tanner [1994].

By taking advantage of a second result of Kolassa and Tanner [1994], we can extend the preceding to contingency tables with r rows and c columns.

Let $\{f_{ij}\}$ be an $r \times c$ contingency table with independent entries.

If $i < r$ and $j < c$, then the distribution of element f_{ij}, conditional on all other elements except for those in the last row and column and conditional on all marginals, is the same as the distribution of the first element in the 2×2 table with elements f_{ij}, f_{ic}, f_{rj}, f_{rc}, conditional on all marginals.

Thus, we may proceed from cell to cell, drawing Gibbs samples as described above. Ambiguities arise in how we are to balance the marginals. We may balance cell by cell as we go or we may keep a running tabulation and balance only when sampling is complete. The overage or discrepancy may be assigned in a number of ways: To an adjacent cell, to a following cell, to a cell in the same column or row chosen at random, or to a cell further along in the same column or row chosen at random.

14.6.1 Metropolis–Hastings Sampling Methods

McDonald, Smith, and Forster [1999] proposed the use of Metropolis–Hastings sampling methods for estimating the exact conditional p-value for tests of goodness-of-fit of log-linear models for mortality rates and standardized mortality ratios. Metropolis–Hastings sampling [Hastings, 1970] is a Markov chain Monte Carlo method for generating samples from arbitrary multivariate distributions. The procedure is as follows:

1. Given current table frequencies f, generate a new set f' from some probability distribution $q(f, f')$.
2. Accept f' as the next realization of the chain with probability $a(f, f')$ where

$$a(f, f') = \begin{cases} \min\left\{\dfrac{g(f')q(f',f)}{g(f)q(f,f')}, 1\right\} & \text{if } q(f)q(f,f') > 0, \\ 1 & \text{if } q(f)q(f,f') = 0; \end{cases}$$

otherwise, retain f.

If q is chosen appropriately, then g is the stationary distribution for this Markov chain. See Smith and Roberts [1993] and references therein for details.

The distribution q also should be chosen so as to ensure the conditioning constraints are satisfied by all f'. Otherwise, the acceptance probability $a(f, f')$ for certain proposed moves is zero. Another important consideration is that the resultant Markov chain should be nondegenerate.

Choices of q that are both effective and computationally feasible for specific problem areas have been proposed by Forster, McDonald and Smith [1996, 2003], Diaconis and Sturmfels [1998], and McDonald, Smith, and Forster [1999].

14.7 Characteristic Functions

As the sample size increases, the number of possible rearrangements increases exponentially. For example, in the one-sample test of a location parameter based on n observations, there are 2^n possible rearrangements. When finding the permutation distribution of a statistic that is a linear combination of some function of the original observations, Pagano and Tritchler [1983] show we can reduce the computation time from $C_1 2^n$ to $C_2 n^c$, where c we hope, much less than n.

Their technique requires two steps. In the first, they determine the characteristic function of the permutation distribution through a set of difference equations. This step requires $2Qm(m+n)$ complex multiplications and additions to find the characteristic function at Q points. In the second, they use the basic theorem in Fourier series to invert the characteristic function and determine or approximate the permutation distribution at $U < Q$ different points. This step requires $2Q \log Q$ calculations. Q is normally chosen to be a power of 2 (e.g., 256 or 512) so that one can take advantage of a fast Fourier transform; the exact number will depend on the precision with which one wants to estimate the significance level.

This method is chiefly of historical interest: Branch and bound algorithms offer greater computational efficiency, particularly when coupled with importance sampling. Vollset, Hirji, and Elashoff [1991] found that the fast Fourier transform method can result in considerable loss of numerical accuracy.

14.8 Asymptotic Approximations

14.8.1 A Central Limit Theorem

The fundamental asymptotic result for the permutation distribution of the two-sample test statistic for a location parameter was first stated by Madow [1948] and formalized by Hoeffding [1951, 1952], who demonstrates convergence of the distribution of the Studentized test statistic under the alternative as well as under the null hypothesis.

Let $T_n = T(X_{(1)}, \ldots, X_{(n)})$ be the test statistic, and let μ_n and σ_n^2 be its first and second moments, respectively. Then the permutation distribution F_n of $Z_n = (T_n - \mu_n)/\sigma_n$ obtained by randomly rearranging the subscripts of the arguments of T_n converges to Φ, the Gaussian (normal) distribution function.

This result means that for sufficiently large samples, we can give our computers a rest, at least temporarily, and approximate the desired p-value with the aid of tables of the normal distribution. To use these tables, we need to know the first and second moments of the permutation distribution. Occasionally, with samples of moderate size, we may also need to know and use the third and higher moments in order to obtain an accurate approximation. Moments for the randomized block design are given by Pitman [1937] and Welch [1937]; for the Latin Square by Welch [1937]; for the balanced incomplete block by Mitra [1961]; and for the completely randomized design by Robinson [1983] and Bradbury [1988].

Extensions to, and refinements of, Hoeffding's work are provided by Silvey [1954, 1956], Dwass [1955], Motoo [1957], Erdos and Renyi [1959], Hajek [1960, 1961], and Kolchin and Christyakov [1973]. Asymptotic results for rank tests are given in Jogdeo [1968] and Tardif [1981]. For further details of the practical application of asymptotic approximations to the analysis of complex experimental designs, see Lehmann [1986], Kempthorne et al. [1961], and Ogawa [1963].

14.8.2 Edgeworth Expansions

While the Gaussian distribution may provide a valid approximation to the *center* of the permutation distribution, it is the tails (and the p-values of the tails) with which we are primarily concerned. Edgeworth expansions give good approximations to the tails in many cases. Edgeworth expansions for the distribution function under both the alternative and the null hypotheses have been obtained by Alber, Bickel and Van Zwet [1976], Bickel and Van Zwet [1978], Robinson [1978], and John and Robinson [1983].

Saddle point methods and large deviation results give still better approximations in the tails. Saddle point approximations for the one- and two-sample tests of location as suggested by Daniels [1955, 1958] are derived by Robinson [1982]. Saddle point approximations for use with general linear models for both the permutation distribution and the bootstrap are given by Booth and Butler [1990].

14.8.3 Generalized Correlation

Test statistics for location parameters are almost always linear or first-order functions of the observations. By contrast, test statistics for scale parameters, the chi-square statistic, and the Mantel–Valand statistic for generalized correlation are quadratic (or second-order) functions of the observations. Their limiting distributions are not Gaussian but chi-square or a Pearson Type III distribution [Berry and Mielke, 1984, 1986; Mielke and Berry, 1985]. Other asymptotic approximations for second-order statistics are given by Shapiro and Hubert [1979], O'Reilly and Mielke [1980], and Ascher and Bailar [1982].

14.9 Confidence Intervals

The trial-and-error method of determining confidence intervals described in section 3.2 is time-consuming and confusing and entails a seemingly unending number of calculations. The stepwise approach suggested by Garthwaite [1996] is both systematic and efficient with the need for only a single permutation at each step.

Let $T_0(U)$ be the value of the statistic used to test the hypothesis $\theta = U$, obtained for the actual sample data. Let $T_i(U)$ be the value of the statistic obtained for a random permutation π_i of the data.

Observing $T(U_i)$, we update our estimate of the upper limit at the ith step as follows:

$$\text{if } T_i(U) > T_0(U), \text{ set } U_{i+1} = U_i - c\alpha/i;$$
$$\text{otherwise, set } U_{i+1} = U_i + c\alpha/i,$$

where α is the significance level and c is known as the *step-length constant*.

We continue in this fashion generating exactly one new permutation and evaluating and comparing exactly two values of the test statistic at each step. It is easy to see that the process converges.

$$c = k(U_1 - \theta), \quad \text{where } k = \frac{2}{2\pi z_\alpha - \frac{1}{2}\exp(-z_\alpha^2/2)}.$$

One possible starting guess is $U_1 = \hat{\theta} + z_\alpha s$, where s is the sample standard deviation.

14.10 Sample Size and Power

Suppose we are in the design stages of a study and we intend to use a permutation test for the analysis. How large should our sample sizes be? Our answer will depend on three things:

- the alternative(s) of interest;
- the power desired at these alternatives;
- the significance level.

A not unrelated question arises if we conclude an analysis by accepting the null hypothesis. Does this mean the alternative is false or that we simply did not have a large enough sample to detect the deviation from the null hypothesis? Again, we must compute the power of the test for several alternatives before we are able to reach a decision.

14.10.1 Simulations

One way to estimate the power is by drawing a series of K (simulated) random samples from a distribution similar to that which would hold under the alternative. For each sample, we perform the permutation test at the stated significance level and record whether we accept or reject the null hypothesis. The proportion of rejections becomes our estimate of the power of the test. This proportion is a random variable with the binomial distribution with K trials and a probability β of success in each trial, where β is the (unknown) power of the test to be estimated.

When designing a study, I use $K = 100$ until I am ready to fine tune the sample size, when I switch to $K = 400$ I also study (estimate) the power for at least two distinct alternatives.

For example, when testing the hypothesis that the observations are normal with mean zero against the alternative that they have a mean of at least one, we might sample from alternatives with at least two different variances, say, one with variance equal to unity and one with variance equal to 2, where 1 is our best guess of the unknown variance, and 2 is a worst-case possibility.

When doing an after-the-fact analysis of the power, use estimates of the parameters based on the actual data. If the pooled sample variance is 1.5, then use a best guess of 1.5 and a worst case of 3 or even 4. (The use of a single estimate alone would be misleading; see Zumbo and Hubley [1998]). The final-result may require $8KN$ computations, where N is the average number of resamplings required each time we perform the test.

With such a large number of calculations, it is essential we take advantage of one or more of the computational procedures described in Sections 2 through 6 of this chapter. Oden [1991] offers several recommendations. Gabriel and Hsu [1983] describe an application-specific method for reducing the number of computations required to estimate the power and determine the appropriate sample size.

14.10.2 Network Algorithms

The same network algorithms that we used to determine significance level can also be used to calculate power, provided we can determine the probability of each specific permutation under the alternative; see, for example, Hilton and Mehta [1993], and Mehta, Patel, and Senchaudhuri [1998]. For example, Mehta et al. studied the Cochran–Armitage test for trend, for which the test statistic is $T = \sum_{j=1}^{J} d_i X_j$, where random variable X_j denotes the integer number of responders among the n_j subjects treated at dose d_j and assumes the value x_j. The reference set for permutations is

$$\Gamma_m = \left\{ x : \sum_{j=1}^{J} x_j = m \right\},$$

and its critical region is

$$\Gamma_m(t) = \left\{ x \in \Gamma_m : \sum_{j=1}^{J} d_j x_j \geq t \right\}.$$

For a given significance level α, let $t_\alpha(m)$ be the smallest possible cut-off value such that $\Pr\{T \geq t_\alpha(m) | m, H\} \leq \alpha$. This cut-off is data-dependent through the marginal m and $\beta(m) = \Pr\{T \geq t_\alpha(m) | m, K\}$, so that the unconditional power is $\sum \beta \Pr\{m|K\}$.

Again, we represent the permutation reference set as a network of nodes and arcs constructed in $J + 1$ stages. At any stage j, the network contains a set of nodes of the form (j, m_j), where j represents the jth of the J binomial populations, and m_j is one possible value of the partial sum of responses from the first j populations. Arcs emanate from each node, and each node, and each such arc is connected to a successor node $(j + 1, m_{j+1})$ at stage $j + 1$. When the network is complete, it will terminate with single node (J, m) and each path form $(0, 0)$ to (J, m) represents one and only one response vector (permutation, rearrangement) in Γ_m.

The arc connecting the nodes (j, m_j) with its successor is assigned a rank length based on the Cochran–Armitage statistic, $r_{j+1} = d_{j+1}(m_{j+1} - m_j)$, and two probability lengths $p_{H,j+1}$ and $p_{k,j+1}$ based on their likelihoods under the hypothesis and alternative, respectively. By specifying a path through the network, we automatically know the corresponding response vector x, its test statistic $t(x) = \sum r_j$, and its unnormalized probability under the null hypothesis and alternative, respectively. Any method we use to generate an estimate of the significance level will provide us with an estimate of the power at the same time.

14.11 Some Conclusions

In the Monte Carlo, we compute the test statistic for a sample of the possible rearrangements and use the resultant sampling distribution and its percentiles in place of the actual permutation distribution and its percentiles. The drawback of this approach is that the resultant significance level p' may differ from the significance level p of a test based on the entire permutation distribution. p' is a consistent estimate of p with a standard deviation on the order of $Np(1-p)$ where n is the number of rearrangements considered in the Monte Carlo.

In the original Monte Carlo the rearrangements are drawn with equal probability. In a variant called *importance sampling*, the rearrangements are drawn with weights chosen so as to minimize the variance. In some instances, when combined with branch and bound techniques as in Mehta, Patel, and Senchaudhuri [1988], importance sampling can markedly reduce the number of samples that are required. (See also Besag and Clifford [1989].)

A second drawback of the Monte Carlo is that selecting a random arrangement is itself a time-consuming operation that can take several multiples of the time required to compute the sample statistic. A current research focus is on rapid enumeration and selection algorithms that can provide a fast transition from one rearrangement to the next. To date, all solutions have been highly application-specific.

Branch and bound algorithms eliminate the need to evaluate each rearrangement individually. The network approach advanced by Mehta and Patel [1980, 1983] can cut computation time by several orders of magnitude.

Solutions through characteristic functions are seldom of practical interest. When subsamples are large—and is the size of the subsample or block, not the sample as a whole, that is the determining factor—an asymptotic approximation should be considered. In my experience as an industrial statistician with the pharmaceutical and energy industries, the opportunity to take advantage of an asymptotic approximation seldom arises. In preclinical work, one seldom has enough observations, and in a clinical trial, though the sample size is large initially, one is usually forced to divide the sample again and again to correct for covariates. In practice, contingency tables always have one or two empty cells. The errors in significance level that can result from an inappropriate application of an asymptotic approximation are amply illustrated in Table 6.4.

If you are one of the favored few able to take advantage of an asymptotic approximation, you first will need to compute the mean and variance of the permutation distribution. In some cases you will also need to calculate and use the third and fourth moments to increase the accuracy of the approximation. The calculations are different for each test; for details, consult the references cited in the corresponding sections of this text.

14.12 Software

Software for parametric tests is widely distributed. Less well known are the sources of software for use in bootstrap and permutation testing. We describe some of these in the present section.

14.12.1 Do-It-Yourself

R's sample() function provides for both rearrangements and bootstrap samples. Extensive libraries of bootstrap functions will be found at http://lib.stat.cmu.edu/S/bootstrap.funs and http://statwww.epfl.ch/davison/BMA/library.html. Some code for special purposes is available at http://lib.stat.cmu.edu/S including least absolute deviation and quantile regression, nonlinear regression, nonlinear regression, and a weighted likelihood bootstrap. Download the R interpreter for Windows or Unix from http://cran.r-project.org/.

Resampling Stats. A programming language for beginners and those without (or with minimal) programming experience. Two commands, **Sample** and **Shuffle**, make it easy to bootstrap and rearrange data. Available in several versions including an add-in for use with Excel. For Windows only. Purchase from www.resample.com.

S-PLUS. The next generation of S-PLUS software for bootstrap and other resampling procedures is available for testing. Please contact bootstrap-beta@insightful.com if you are interested in testing this software; indicate your name and what version of S-PLUS you are using (type "version" at the command line).

Statistical Calculator (SC). (DOS, Unix, T800 transputer.) An extensible statistical environment, supplied with over 1200 built-in (compiled C) and external (written in SC's C-like language) routines. Permutation-based methods for contingency tables (chisquare, likelihood, Kendall S, Theil U, kappa, tau, odds ratio), one and two-sample inference for both means and variances, correlation, and multivariate analysis (MV runs test, Boyett/Schuster,

Hoetelling's T. Ready-made bootstrap routines for testing homoscedacity, detecting multimodality, plus general bootstrapping and jack-knifing facilities. http://www.mole-soft.demon.co.uk/.

14.12.2 Complete Packages

14.12.2.1 Freeware

Blossom Statistical Analysis Package. This interactive program for analyzing data utilizing multi-response permutation procedures (MRPP) includes statistical procedures for grouped data, agreement of model predictions, circular-distributions, goodness of fit, least absolute deviation and quantile regression. Programmed by Brian Cade at the U.S. Geological Survey, Midcontinent Ecological Science Center. PC only with online manual in HTML format. http://www.fort.usgs.gov/products/software/blossom/blossom.asp.

SnPM. The Statistical nonParametric Mapping toolbox provides an extensible framework for voxel level non-parametric permutation/randomisation tests of functional Neuroimaging experiments with independent observations. It is most suitable for single subject PET/SPECT analyses, or designs with low degrees of freedom available for variance estimation as it provides the freedom to use weighted locally pooled variance estimates or variance smoothing. http://www.fil.ion.ucl.ac.uk/spm/snpm/.

NPSTAT. Parametric and randomization tests on single factor designs, repeated measures, correlations, and Fisher's exact test. **NPFACT** support 2- and 3-factor designs. DOS only. http://home.rmci.net/rmay/ npstat.html.

14.12.2.2 Shareware

GoodStats. Permutation tests for 2-sample comparison, correlation, k-sample comparison with ordered or unordered populations, plus the only software available using synchronized permutations for analysis of $2 \times K$ designs. DOS only. Windows version in preparation. Download from http://mysite.verizon.net/res7sf1o/GoodStat.htm.

14.12.2.3 $$$$

Stata provides a comprehensive set of parametric statistics routines plus subroutines and pre-programmed macros for bootstrap, density estimation, and permutation tests. Programmable with many flexible graphics routines. (Windows, Unix) Stata Corp, 702 University Drive East, College Station TX 77840. 800/782-8272. www.stata.com.

StatXact is a must for the exact analysis of contingency tables (categorical or ordered data). Includes power and sample size calculations. Versions for Windows or Unix. Also available as an add-on module for both SAS and SPSS. Cytel Software Corporation, 675 Massachusetts Avenue, Cambridge, MA 02139. 617-661-2011. www.cytel.com.

NPC TEST is the only statistics program on the market today that provides for multi-variable analysis by permutation means. Cutting edge, but has yet to be validated. A demonstration version, SAS macro, and S-Plus code may be downloaded from http://www.stat.unipd.it/~pesarin/software.html.

14.13 Exercises

1. Most microcomputer-based random number generators use multiplicative congruence to produce a 16-bit unsigned integer between 0 and 2^{15}. Yet in the two-sample comparison, for example, we only use one of the 15 bits, the least significant bit, in selecting items for rearrangement. Could we use more of the bits? That is, are some or all of the bits independent of one another? Write algorithm(s) that take advantage of multiple bits.
2. Apply the Mehta and Patel approach to the following 3×2 contingency table:

3	1	0
1	2	1

   Compute the marginals for this table. Draw a directed graph in which each node corresponds to a 3×2 table whose marginals are the same as those of the preceding table. Choose a test statistic (see Chapter 8.3). Identify those nodes which give rise to a value of the test statistic less than that of the original table.
3. Suppose you are interested in the theoretical alternative

4/6	1/6	1/6
1/6	4/6	1/6

   How big a sample size would you need to insure that the probability of detecting this alternative was 80% at the 10% significance level? (Hint: Use a six-sided die to simulate the drawing of samples.)
4. There are many algorithms for generating all possible permutations (see, for example, Durstenfeld [1964], Boothroyd [1967], Berry [1979]). Show that these would be grossly inefficient for generating distinct rearrangements.
5. There are many algorithms for generating random combinations, for example, Chase [1970], and Bebbington [1975]. Develop an optimal algorithm for generating all possible combinations of 2 samples. Can your algorithm be extended to k-samples?

6. Can you develop an optimal algorithm for generating symmetric rearrangements of a 2^k factorial design with $k > 2$? (Optimal algorithms for generating symmetric rearrangements of $2 \times C$ designs are incorporated in GoodStats at http://mysite.verizon.net/res7sf1o/GoodStats.htm.)

Appendix

Theory of Testing Hypotheses

This appendix is provided as a service to the reader who desires a mathematically rigorous foundation for the theory of testing hypotheses in the continuous case. In contrast to the balance of the present volume, a basic knowledge of measure theory and complex variables is essential on the reader's part. A brief review of probability theory is provided in the next section, followed by sections containing proofs of the Neyman–Pearson lemma in the continuous case and related theorems for tests of one- and multiparameter exponential families. The concluding sections summarize essential findings for the resampling methods including asymptotic rates of convergence.

A.1 Probability

Imagine that we have before us a spinner such as one often finds in board games, a metal needle on a pivot in the center and an outer circle that is marked from 1 to 360. If the spinner moves freely, each of the numbers 1 to 360 is equally likely and the actual probability of the needles pointing to 271, say, is 1/360.

Of course, the needle is unlikely to point exactly to 271, but to somewhere near enough to 271 that we call it 271, anyway. That is, the event "needle points to 271" is actually an aggregate of a noncountable number of possible outcomes such as "needle points to 271.00003065300...."

Let $\mathcal{L}$ denote the totality of all possible outcomes. In our example $\mathcal{L}$ consists of all real numbers between 0 and 360, and our primary interest is in subsets of $\mathcal{L}$, in aggregates of outcomes such as "the needle points approximately to 271." The *union* of two such aggregates A and B, that is, the set of outcomes that are in either A or B, is denoted by $A \cup B$. The *complement* of A consists of all events in $\mathcal{L}$ that are not in A.

If P is a *probability*, then $P(\mathcal{L}) = 1$ and $P(\emptyset) = 0$ where $\emptyset$ denotes the empty set, devoid of values. Probabilities are *countably additive*, that is, if the events A_i are mutually exclusive, $P(\cup A_i) = \sum P(A_i)$. Unfortunately, when the number of possible outcomes is noncountable, this simple condition actually places a restriction on the events for which a probability can be defined.[1] The sets for which P is defined are said to be *measurable*.

Often, to simplify proofs, we will make use of *finite measures* μ for which $\mu(\mathcal{L}) < \infty$. Clearly a probability measure is a finite measure.

A class of sets $\mathcal{C}$ that contains $\mathcal{L}$, plus the complements and countable unions of all its sets is called a σ-*field*. Such a set is also closed under countable intersections (Exercise A.1).

A probability space consists of the triple $(\mathcal{L}, \mathcal{C}, P)$, where $\mathcal{C}$ is a σ-field of measurable sets P.

In applied work, while one may observe an event ω belonging to $\mathcal{L}$ (for example, the spinner went around three plus times bouncing on its pivot before settling adjacent to 271), what one records is the associated value $X(\omega)$ of a real-valued random variable X. In this instance, $X(\omega) = 271$.

Denote by $\mathcal{X}$ the range of X; $\mathcal{X}$ is a subset of the extended real line $[-\infty, \infty]$. The sets of values of X with which we are most concerned have the form $\{x: a < x \leq b\}$; the resultant σ-*field* β made up of countable unions and complements of such sets is called a *Borel field*. We restrict attention to random variables that are *Borel measurable functions*. We call this set β, and it is the set of events that lead to outcomes in B for which $X^{-1}(B)$ belongs to $\mathcal{C}$. The probability spaces we shall be concerned with have the form $(\mathcal{X}, \mathcal{B}, P^X)$.

Let $Z = (X, Y)$ be a random variable defined over $(\mathcal{X} X \mathcal{Y}, \mathcal{A} X \mathcal{B})$ and suppose that the random variables X and Y have distributions P^X and P^Y defined over $(\mathcal{X}, \mathcal{A})$ and $(\mathcal{Y}, \mathcal{B})$, respectively, Then X and Y are said to be independent if $P^Z = P^X X P^Y$. If $\mathcal{A}' \subseteq \mathcal{A}$ and $\mathcal{B}' \subseteq \mathcal{B}$, show that even if $P^Z(A' \times B') = P^X(A') X P^Y(B')$ for all A' in $\mathcal{A}'$ and B in $\mathcal{B}'$, there may exist a set $A \times B$, with A in $\mathcal{A}$ and B in $\mathcal{B}$ such that $P^Z(A \times B) \neq P^X(A) X P^Y(B)$ (Exercise A.4).

As the intervals $\{x: x \leq b\}$ are in $\mathcal{B}$, the probabilities $F[b] = P^X\{x: x \leq b\}$ are defined for all b. Following Lebesgue, we extend the notion of integration, writing $F[b] = \int_{-\infty}^{b} dF = \int_{-\infty}^{b} f dx$. F has the density f with respect to Lebesgue measure. A real-valued measurable function T is said to be *simple* if it takes only a finite number of values over its range. Suppose T takes on the distinct values $t_1, \ldots, t_n$ on the mutually exclusive sets $B_1, \ldots, B_n$ in $\mathcal{B}$, and that $\sum P(B_i) = 1$. Then $\int T dF = \sum t_i P(B_i)$. Given *any* non-negative measurable function T, there exists a sequence of simple functions T_n converging to it, such that

$$\int T dF = \lim_{n \to \infty} \int T_n dF.$$

---

[1] For examples of nonmeasurable sets, a topic well beyond the scope of this book, see any standard text on measure theory such as Halmos [1997, Section 16] and Gelbaum and Olmsted [2003].

The positive and negative parts of any measurable function are also measurable (Exercise A.6). If the integrals of both the positive and negative parts of a measurable function T are finite, then T is said to be *integrable* and its expectation to exist and be finite.

In most cases, we shall be taking n successive observations and be concerned with probability distributions over n-dimensional Euclidean spaces, $F[b_1,\ldots,b_n] = P\{X_1 \leq b_1,\ldots,X_n \leq b_n\}$, the associated sample space $(\mathcal{X}, \mathcal{B}, P^X)$ where the observations in $\mathcal{X}$ are vector-valued, and measurable transformations $\mathcal{T}$ from $(\mathcal{X}, \mathcal{B})$ into some $(\mathcal{T}, \mathcal{B}')$. $\mathcal{T}$ is a *measurable transformation* if for all B' in $\mathcal{B}'$, the inverse $C = T^{-1}(B') = \{x \colon x \text{ in } \mathcal{X}, T[x] \text{ in } B'\}$ is in $\mathcal{B}$. T is a *statistic* if it is a measurable transformation that does not depend on the specific measure P^X.

If the measurable transformation $T(x)$ is integrable, then its expectation is defined by $E(T) = \int T(x)dF$ and both $\int_{T>0} T(x)dF$ and $\int_{T<0} T(x)dF$ are finite.

A.2 The Fundamental Lemma

We are now in a position to extend the fundamental lemma of Neyman and Pearson outlined in Section 3.1 to the class of all distributions. Recall that a hypothesis test φ is simply a decision rule that takes values between 0 and 1. When $\varphi(x) = 1$, we reject the hypothesis and accept the alternative; when $\varphi(x) = 0$, we accept the hypothesis and reject the alternative; and when $\varphi(x) = p$, with $0 < p < 1$, we flip a coin that has been weighted so that the probability is p that it will come up heads—whence we reject the hypothesis—and $1 - p$ that it will come up tails, whence we accept the hypothesis.

Theorem A.1. *Let P_0 and P_1 be probability distributions possessing densities p_0 and p_1 with respect to a measure μ.*

a) *For testing the hypothesis P_0 against the alternative P_1, there exists a test φ and a constant k such that*

$$E_0\varphi(x) = \alpha$$

and

$$\varphi(x) = \begin{cases} 1 & \text{when } p_1(x) > kp_0(x) \\ 0 & \text{when } p_1(x) < kp_0(x). \end{cases}$$

b) *A test that satisfies these conditions for some k is most powerful for testing P_0 against the alternative P_1 at level a.*

c) *If φ is most powerful for testing P_0 against P_1 at level a, then for some k, the test φ satisfies these conditions except on a set that is assigned probability zero by both distributions and unless there exists a test at a smaller significance level whose power is 1.*

Proof. Such a test exists as can be seen by letting

$$\varphi(x) = \begin{cases} 1 & \text{when } p_1(x) > k'p_0(x), \\ \dfrac{\alpha - P\{p_1(X) > k'p_0(X)|P_0\}}{P\{p_1(X) = k'p_0(X)|P_0\}} & \text{when } p_1(x) = k'p_0(x), \\ 0 & \text{when } p_1(x) < k'p_0(x). \end{cases}$$

φ is well-defined except possibly on a set of probability zero and $E[\varphi(x)| P_0] = \alpha$. □

Now suppose φ^* is any other test with $E[\varphi^*(x)|P_0] \leq \alpha$. Denote by $S+$ the set of values for which $\varphi(x) - \varphi^*(x) > 0$ and $S-$ the set for which $\varphi(x) - \varphi^*(x) < 0$. $\int (\varphi - \varphi^*)(p_1 - kp_0)d\mu = \int_{S+}(\varphi - \varphi^*)(p_1 - kp_0)d\mu + \int_{S-}(\varphi - \varphi^*)(p_1 - kp_0)d\mu$. From the definition of φ it is easy to see that both latter integrals must be greater than or equal to zero and therefore the difference in power between the two tests $\int(\varphi - \varphi^*)p_1 d\mu \leq k\int(\varphi - \varphi^*)p_0 d\mu \leq 0$.

Finally, suppose φ^* and not φ is the most powerful at level $-\alpha$ for testing P_0 against P_1. Let $S = S+ \cap S-$, on which φ^* and φ differ with the set $\{x\colon p_1(x) \leq p_0(x)\}$. Unless $\mu(S) = 0$, $\int_S(\varphi - \varphi^*)(p_1 - kp_0)d\mu > 0$, which would mean φ is most powerful, a contradiction. Then $\mu(S) = 0$ and the two tests are the same, establishing the final part of the theorem.

If $E\varphi^*|P_0 < \alpha$ and $E\varphi^*|P_1 < 1$, it would be possible to include in the rejection region additional points and thereby increase the power, a contradiction. Thus, either $E\varphi^*|P_0 = \alpha$ or $E\varphi^*|P_1 = 1$.

A.3 Two-Sided Tests

In Chapter 3 we considered possible tests of the hypothesis $H_1\colon \theta \leq \theta_0$ against the alternative hypothesis $K_1\colon \theta > \theta_0$. But we also may require tests of the following:

$$\begin{aligned} H_2\colon \theta = \theta_0 \quad &\text{vs} \quad K_2\colon \theta \neq \theta_0, \\ H_3\colon \theta \leq \theta_1 \text{ or } \theta \geq \theta_2 \quad &\text{vs} \quad K_3\colon \theta_1 < \theta < \theta_2, \\ H_4\colon \theta_1 \leq \theta \leq \theta_2 \quad &\text{vs} \quad K_4\colon \theta < \theta_1 \text{ or } \theta > \theta_2. \end{aligned}$$

We can formulate unbiased permutation tests of H_1 and H_2, but not of the remaining hypotheses/alternative pairs. The test of H_2 would have the form

$$\varphi(T) = \begin{cases} 0 & \text{if } C_1 < T < C_2, \\ 1 & \text{if } T < C_1 \text{ or } T > C_2. \end{cases}$$

As with tests of H_1, the choice of the test statistic T will depend upon the alternatives of interest. And we saw in Chapter 8, this choice will also depend upon the nature of the loss function.

A.3.1 One-Parameter Exponential Families

We can obtain UMP or UMPU parametric tests in each instance if our observations come from a distribution that is a member of the one-parameter exponential family $dP_\theta^X(x) = C(\theta)\exp[Q[\theta]T(x)]h[x]d\mu$, where $Q[\theta]$ is monotone increasing.

In contrast to the one-sided case, we have not one but two side conditions, $E_{\theta_1}[\varphi(x)] = E_{\theta_2}[\varphi(x)] = \alpha$ and we require the following preliminary results:

Lemma A.1. *Let β denote the power of the most powerful level-α test $(0 < \alpha < 1)$ for testing P_0 against P_1. Then $\alpha < \beta$ unless $P_0 = P_1$.*
The proof is left as an exercise (A.10).

The following lemma is an extension of the Neyman–Pearson lemma (Theorem A.1) to the case of multiple side conditions.

Lemma A.2. *Let $f_1, \ldots, f_{m+1}$ be real-valued functions, integrable μ. The set M of points in M-dimensional space whose coordinates are*

$$\left(\int \varphi f_1 d\mu, \ldots, \int \varphi f_m d\mu\right)$$

for some critical function φ is convex and closed. If $(c_1, \ldots, c_m)$ is an inner point of M, then there exist constants $k_1, \ldots, k_m$ and a critical function φ satisfying

$$\varphi(x) = 0 \quad \text{when } f_{m+1}(x) < \sum_{i=1}^m k_i f_i(x),$$

$$\varphi(x) = 1 \quad \text{when } f_{m+1}(x) > \sum_{i=1}^m k_i f_i(x),$$

$$\int \varphi f_i d\mu = c_i \quad \text{for } i = 1, \ldots, m,$$

for which $\int \varphi f_{m+1} d\mu$ is a maximum.

Proof. That M is closed follows from the weak compactness theorem. M is convex, for if φ_1 and φ_2 are critical functions, so is $\alpha\varphi_1 + (1-\alpha)\varphi_2$ for any $0 \leq \alpha \leq 1$. The balance of the proof is essentially geometric in nature. See Lehmann [1986, pp 96–99]. □

Lemma A.3. *Let $p_\theta(x)$ be a family of densities on the real line with monotone likelihood ratio in x; suppose moreover that $p_\theta(x) > 0$ for all θ and x, and that $p_{\theta'}(x)/p_\theta(x)$ is strictly increasing in x for $\theta < \theta'$. Let ψ be a function with a single change of sign; specifically, suppose there exists a value x' such that $\psi(x) \leq 0$ for $x < x'$ and $\psi(x) \geq 0$ for $x > x'$. Moreover,* suppose $P\{\psi(x) \neq 0\} > 0$. If $E_{\theta_1}[\psi(x)] = 0$, then $E_\theta[\psi(x)] < 0$ for $\theta < \theta_1$ and $E_\theta[\psi(x)] > 0$ for $\theta > \theta_1$.

Proof. Let $\theta' < \theta''$. Since $p_\theta(x) > 0$ for all θ and x, $0 < p_{\theta''}(x')/p_{\theta'}(x') = c < \infty$. Since the density has monotone likelihood ratio in x, this ratio is bounded above by its value at x' on the interval $(-\infty, x')$ when $\psi(x) \leq 0$ and attains its minimum at x' on the interval $[x', 0)$ when $\psi(x) \geq 0$. Thus,

$$E_{\theta''}[\psi(x)] = \int \psi p_{\theta'}(x) p_{\theta''}(x)/p_{\theta'}(x) d\mu$$

$$\geq c \int_{-\infty}^{x'^-} \psi p_{\theta'}(x) d\mu + c \int_{x'}^{\infty} \psi p_{\theta'}(x) d\mu = c E_{\theta''}[\psi(x)].$$

Consequently, $E_{\theta'}[\psi(x)] > 0$ implies $E_{\theta''}[\Psi(x)] > 0$ and vice versa. Letting $\theta_1 = \inf\{\theta: E_\theta[\Psi(x)] > 0\}$, the lemma follows. □

We are now in a position to show that for testing the hypothesis H_3: $\theta \leq \theta_1$ or $\theta \geq \theta_2$ against K_3: $\theta_1 < \theta < \theta_2$ in the one-parameter exponential family, there exists a UMP test given by

$$\varphi(x) = \begin{cases} 0 & \text{when } C_1 < T(x) < C_2 \\ \gamma_j & \text{when } T(x) = C_j, j = 1, 2. \\ 1 & \text{when } T(x) < C_1 \text{ or } T(x) > C_2 \end{cases} \quad (A.1)$$

where the C_j and γ_j are determined by

$$E_{\theta_1}[\varphi(x)] = E_{\theta_2}[\varphi(x)] = \alpha. \quad (A.2)$$

As $dP_\theta^X(x) = C(\theta)\exp[Q[\theta]T(x)]h[x]d\mu$, we can restrict attention to the sufficient statistics $T(X)$ whose distribution is $dP_\theta^T(t) = C(\theta)\exp[Q[\theta]t]d\nu$. Let M denote the set of all points $\{E_{\theta_1}[\psi(T)], E_{\theta_2}[\psi(T)]\}$ as ψ ranges over the totality of critical functions. By Lemma A.1, if $0 < \alpha < 1$, the set M contains points (α, μ) and (α, ν) with $\mu < \alpha < \nu$; in fact, it contains all points (μ, μ) with $0 < \mu < 1$ so that (α, α) is an inner point of M. Hence by Lemma A.2, there exist constants k_1, k_2, and a test $\varphi(x) = \psi_0(T(x))$ satisfying the restrictions A.2 such that $\psi_0(t) = 1$ when $k_1 C(\theta_1)\exp[Q[\theta_1]t] + k_2 C(\theta_2)\exp[Q[\theta_2]t] < C(\theta')\exp[Q[\theta']t]$ or, equivalently, when $a_1\exp(b_1 t) + a_2\exp(b_2 t) < 1$, with $b_1 < 0 < b_2$.

Both a_1 and a_2 must be positive (Exercise A.12), showing that the test satisfies equation (A.1). The C_j and γ_j are uniquely determined by Equations (A.1) and (A.2). For suppose we can find two tests φ and φ^* satisfying these conditions with, say, $C_1^* < C_1$ and $C_2^* < C_2$. Let $\psi = \varphi - \varphi^*$. This function has one change of sign; the conditions of Lemma A.3 are satisfied, and (A.2) implies $\varphi = \varphi^*$ almost everywhere.

Let $\theta' < \theta_1$; apply Lemma A.2 to minimize $E_{\theta'}[\varphi(x)]$ subject to $E_{\theta_1}[\varphi(x)] = \alpha$. Dividing through by $\exp[Q[\theta_1]]$, the desired test is seen to have rejection region of the form $a_1 \exp(b_1 t) + a_2 \exp(b_2 t) < 1$, with $b_1 < 0 < b_2$, thus coinciding with the test we already exhibited. By comparison with the test $\psi(t) \equiv \alpha$, we see that for $\theta' < \theta_1$ the desired test φ is such that $E_{\theta'}[\varphi(x)] \leq \alpha$. Similar arguments reveal that for $\theta' > \theta_2$, the desired test φ is such that $E_{\theta'}[\varphi(x)] \leq \alpha$. Applying arguments similar to those of Theorem A.1(b) (Exercise A.13), we see that the test is UMP among those subject to the restrictions (A.2).

To find tests for the remaining hypotheses/alternative pairs, we need the next lemma.

Lemma A.4. *Let φ be any function on (R, B) for which the integral $\int \varphi[x] \exp[\theta T(x)] d\mu$ considered as a function of the complex variables $\theta = \xi + i\eta$ exists for all ξ and is finite. Then the integral is an analytic function of θ in the region R of parameter points for which ξ is an interior point of the natural parameter space, and the derivative of the integral with respect to θ can be computed under the integral sign.*

Proof. Define $\psi(\theta) = \int \exp[\theta T(x)] d\nu$ where $d\nu = \varphi[x] d\mu$. Let ξ^0 be any fixed interior point; there exists $\delta > 0$ such that $\psi(\theta)$ exists and is finite for all θ with $|\xi - \xi^0| \leq \delta$. Consider the difference quotient

$$(\psi(\theta) - \psi(\theta^0))/(\theta - \theta^0) = \int \exp[\theta^0 T(x)][\exp[(\theta - \theta^0) T(x)] - 1]/(\theta - \theta^0) d\nu$$

$$\leq \delta^{-1} \int |\exp[\theta^0 T(x) + \delta |T(x)|]| d\nu$$

$$\leq \delta^{-1} \int |\exp[(\theta^0 + \delta) T(x)] + \exp[(\theta^0 - \delta) T(x)]| d\nu$$

for $|\xi - \xi^0| \leq \delta$. Since the right hand side is integrable it follows from the Lebesgue dominated-convergence theorem that for any sequence of points θ_n^0 tending to θ^0, the difference quotient tends to the first derivative $\int T(x) \exp[\theta T(x)] d\nu$. □

Recall that an *unbiased test* φ is one for which the power function $\beta_\varphi(\theta) = E_\theta[\varphi(x)]$ satisfies $\beta_\varphi(\theta) \leq \alpha$ for θ in Ω_H and $\beta_\varphi(\theta) \geq \alpha$ for θ in Ω_K.

Lemma A.5. *If the distributions P_θ are such that the power function of every test is continuous and if φ^* is a level-α test of H that is UMP among all tests satisfying $\beta_\varphi(\theta) = \alpha$ for all θ in the boundary set of points that are members or limit points of both Ω_H and Ω_K, then φ^* is UMP-unbiased.*

The proof is left as an Exercise (A.14).

For testing H_4: $\theta_1 \leq \theta \leq \theta_2$ versus K_4: $\theta < \theta_1$ or $\theta > \theta_2$ with observations drawn from a member of a one-parameter exponential family, the most powerful unbiased test takes the form

$$\varphi(x) = \begin{cases} 0 & \text{when } C_1 < T(x) < C_2, \\ \gamma_j & \text{when } T(x) = C_j, j = 1,2, \\ 1 & \text{when } T(x) < C_1 \text{ or } T(x) > C_2, \end{cases}$$

where the C_j and γ_j are determined by

$$E_{\theta_1}[\varphi(x)] = E_{\theta_2}[\varphi(x)] = \alpha.$$

By Lemma A.4, the power function of this test is continuous so that Lemma A.5 is applicable. The boundary set consists of θ_1 and θ_2. Our objective is to minimize $E_\theta[1 - \varphi(x)]$ for some θ outside the interval $[\theta_1, \theta_2]$, a problem we solved in deriving a UMP test of H_3 versus K_3. By Lemma A.4, our test is UMPU for testing H_4 versus K_4.

We can apply exactly the same critical function for testing H_2: $\theta = \theta_0$ versus K_2: $\theta \neq \theta_0$, but the C_j and γ_j now are determined by $E_{\theta_0}[\varphi(x)] = \alpha$ and $E_{\theta_0}[\varphi(x)]T(X) = \alpha E_{\theta_0}[T(X)]$.

A.4 Tests for Multiparameter Families

A.4.1 Basu's Theorem

As we saw in Chapter 3, the Neyman and Pearson lemma is readily applied to the derivation of permutation tests as all x are equally likely under P_0, and to one-parameter parametric families with monotone likelihood ratio in some statistic. To obtain parametric, permutation, or bootstrap tests for multiparameter families we need to demonstrate the independence of certain statistics. For example, to obtain tests for the parameters of a normal distribution, we need to show that the mean and variance of a sample of n independent identically normally distributed observations are independent.

Some preliminary definitions are required. A statistic $V(X)$ is said to be *ancillary* if its distribution does not depend on the probability measure P. A statistic $T(X)$ is said to be *complete* for $P \in \mathcal{P}$ if for any Borel-measurable function f, $E(f[T[X]]) = 0$ for all $P \in \mathcal{P}$ implies $f = 0$ except on a set of measure zero with respect to all $P \in \mathcal{P}$. T is said to be *boundedly complete* if the previous statement holds for any bounded Borel-measurable function. Alternatively, the family $\mathcal{P}^T$ of probability distributions may be referred to as *complete* or *bounded complete*, respectively, when these conditions hold.

Theorem A.2. *Let $V(X)$ and $T(X)$ be two statistics where X is distributed according to $P \in \mathcal{P}$. If V is ancillary and T is boundedly complete and sufficient for $P \in \mathcal{P}$, then V and T are independent with respect to any $P \in \mathcal{P}$.*

Proof. Following Shao [2003, p. 112] let B be an event on the range of V. Since V is ancillary, $P\{V^{-1}[B]\}$ is a constant. Since T is sufficient, $E[I_B(V)|T]$ is independent of P. Since $E\{E[I_B(V)|T]\} - P\{V^{-1}[B]\} = 0$ for all $P \in \mathcal{P}$ and T is boundedly complete, $P\{V^{-1}[B]T\} = E[I_B(V)|T] = P\{V^{-1}[B]\}$ almost everywhere with respect to $\mathcal{P}$. Let A be an arbitrary event in the range of T. Then,

$$\begin{aligned} P\{T^{-1}[A] \bigcap V^{-1}[B]\} &= E\{E(I_A(T)I_B(V)|T)\} \\ &= E\{[I_A(T)EI_B(V)|T\} = E\{[I_A(T)P\{V^{-1}[B]\}\} \\ &= P\{T^{-1}[A]\}P\{V^{-1}[B]\} \end{aligned}$$

as was to be proved. $\square$

A.4.2 Conditional Probability and Expectation

The material in this section is a necessary prerequisite for deriving parametric tests. Recall that the conditional probability of an event B given an event A is defined as $P\{B|A\} = P\{A \text{ and } B\}/P\{A\}$. In this section we extend the concept of conditional probability to cases in which a random variable has a continuous distribution function and $P\{A\}$ may be zero.

Let X be a random variable on $(\mathcal{L}, \mathcal{C}, P)$ such that $E[X] = \int X dP$ exists and is finite. Let $\mathcal{A}$ be a sub σ-field of $\mathcal{C}$. The conditional expectation $E[X|\mathcal{A}]$ is a measurable function from $(\mathcal{L}, \mathcal{A})$ to $(R, \mathcal{B})$; $\int_\mathcal{A} E(X|\mathcal{A})dP = \int_\mathcal{A} X dP$ for any A in $\mathcal{A}$. An immediate application of this result is that if T is a statistic, and if the expected value of a measurable function f of X exists and is finite, then $E[f(X)] = EE[f(X)|t]$.

The *conditional probability* of an event A given the value of a statistic $T = t$ is defined as $P\{A|T = t\} = E[I_A(X)|t]$. It is a Borel measurable function of T.

A.4.3 Multiparameter Exponential Families

Theorem A.3. *Let X be distributed according to the exponential family*

$$dP^{X_{\theta,\lambda}}(x) = C(\theta, \lambda)\exp[\theta U(x) + \sum_j \lambda_j T_j(x)]d\mu.$$

Then there exist measures μ_θ and ν_t such that

1) *the distribution of $T = (T_1, \ldots, T_J)$ is an exponential family of the form*

$$dP^{T_{\theta,\lambda}}(t) = C(\theta, \lambda)\exp[\sum_j \lambda_j T_j]d\mu_\theta;$$

2) *the conditional distribution of U given $T = t$ is an exponential family of the form*

$$dP^{U|t_\theta}(u) = C_t(\theta)\exp[\theta u]d\nu_t,$$

and is independent of λ.

To prove this result, we need the following lemma.

Lemma A.6. *Let $(\mathcal{T},\mathcal{B})$ and $(\mathcal{Y},\mathcal{C})$ be Euclidean spaces and sigma fields, and let $P_0^{T,Y}$ be a distribution over the product space $(\mathcal{X},\mathcal{A}) = (\mathcal{T}X\mathcal{Y},\mathcal{B}X\mathcal{C})$. Suppose that another distribution over $(\mathcal{X},\mathcal{A})$ satisfies $dP_1(t,y) = \alpha(y)b(t) dP_0(t.y)$ with $\alpha(y) > 0$ for all y. Then under P_1, the marginal distribution of T, is given by*

$$dP_1^T(t) = b(t)\left[\int \alpha(y) dP_0^{Y|t}(y)\right] P_0^T(t),$$

and the conditional distribution of Y given t is given by

$$dP_1^{Y|t} = \frac{\alpha(y) dP_0^{Y|t}(y)}{\int_{\mathcal{Y}} \alpha(y') dP_0^{Y|t}(y')}.$$

Proof. The form of the marginal distribution follows from the equation

$$P_1\{T \in B\} = E_1[I_B(T)] = E_0[I_B(T)\alpha(Y)b(T)]$$
$$= \int_B b(t)\left[\int_{\mathcal{Y}} a(y) dP_0^{Y|t}(y)\right] dP_0^T(t).$$

For any integrable f,

$$E_1[f(T,Y)] = \int_{\mathcal{T}}\left[\int_{\mathcal{Y}} f(t,y) dP_1^{Y|t}(y)\right] dP_1^T(t)$$
$$= \int_{\mathcal{T}}\left[\int_{\mathcal{Y}} f(t,y) dP_1^{Y|t}(y)\right] b(t) \left[\int_{\mathcal{Y}} \alpha(y) dP_0^{Y|t}(y)\right] dP_0^T(t).$$

Letting $f = I_B(T)$ and noting that $\int_{\mathcal{Y}} \alpha(y) dP_0^{Y|t}(y) > 0$ as $a(y) > 0$, the form of the conditional distribution provided by the theorem is evident.

Proof of Theorem A.3. Let (θ^0,λ^0) be a point of the natural parameter space of the exponential family and let $\mu' = P_{\theta^0,\lambda^0}^X$. Then

$$P_{\theta^0,\lambda^0}^X = C'(\theta,\lambda,\theta^0,\lambda^0)\exp\left[(\theta - \theta^0)U(x) + \sum(\lambda_j - \lambda_j^0)T(x)\right] d\mu'.$$

Applying the lemma, we have

$$d\mu_\theta(t) = \exp(-\sum \lambda_j^0 t)\left[\int (\theta - \theta^0)u dP_{\theta^0,\lambda^0}^{U|t}\right] dP_{\theta^0,\lambda^0}^T$$

and

$$d\nu_t(u) = \exp(-\theta^0) dP_{\theta^0,\lambda^0}^{U|t}.$$

A.4 Tests for Multiparameter Families

We now can show that the results of Section A.3 can be extended to the multiparameter exponential family

$$dP^X_{\theta,\lambda}(x) = C(\theta,\lambda)\exp[\theta U(x) + \sum_j \lambda_j T_j(x)]d\mu.$$

As in Chapter 3, we may restrict attention to the sufficient statistics (U,T).

When $T = t$ is given, we saw in Section A.3 that the conditional distribution of U given t is a member of an exponential family and the results of Section A.3 apply immediately to tests of

H_1: $\theta = \theta_0$ vs K_1: $\theta > \theta_0$,
H_2: $\theta = \theta_0$ vs K_2: $\theta \neq \theta_0$,
H_3: $\theta \leq \theta_1$ or $\theta \geq \theta_2$ vs K_3: $\theta_1 < \theta < \theta_2$,
H_4: $\theta_1 \leq \theta \leq \theta_2$ vs K_4: $\theta < \theta_1$ or $\theta > \theta_2$.

For example, for testing H_4: $\theta_1 \leq \theta \leq \theta_2$ versus K_4: $\theta < \theta_1$ or $\theta > \theta_2$, the most powerful unbiased test takes the form

$$\varphi(x) \begin{cases} 0 & \text{when } C_1(t) < T(x) < C_2(t), \\ \gamma_j(t) & \text{when } T(x) = C_j(t), j = 1, 2, \\ 1 & \text{when } T(x) < C_1(t) \text{ or } T(x) > C_2(t), \end{cases}$$

where the C_j and γ_j are determined by

$$E_{\theta_1}[\varphi(U,T)|t] = E_{\theta_2}[\varphi(U,T)|t] = \alpha \quad \text{for all } t.$$

To show that these same critical functions are appropriate as unconditional UMPU tests, we need the following lemmas: □

Lemma A.7. *Let X be a random vector with probability distribution*

$$dP_\theta(x) = C(\theta)\exp\left[\sum_{j=1}^k \theta_j T_j(x)\right]d\mu$$

and let $\mathcal{P}^T$ be the family of distributions of T as θ ranges over the set ω. Then $\mathcal{P}^T$ is complete provided contains ω a k-dimensional rectangle.

Proof. Following Lehmann [1986, p. 142] let us assume without loss of generality that ω contains the rectangle $R = \{(\theta_1,\ldots,\theta_k): -a \leq \theta_j \leq a, j = 1,\ldots,k\}$ and suppose that $f(t)$ is such that $E[f(T)|\theta] = 0$ for all θ in ω. Then for all θ in ω,

$$\int f^+(t)\exp\sum_{j=1}^k \theta_j t_j d\nu = \int f^-(t)\exp\sum_{j=1}^k \theta_j t_j d\nu$$

where $f^+(t)$ and $f^-(t)$ denote the positive and negative parts of $f = f^+ - f^-$, and ν denotes the measure induced in T-space by the measure μ. In particular, $\int f^+(t)d\nu = \int f^-(t)d\nu$.

Dividing f by a constant, again without loss of generality, we can assume that the common value of these two integrals is 1, so that $dP^+(t) = f^+(t)d\nu$ and $dP^-(t) = f^-(t)d\nu$ are probability measures.

Consider the integrals $\int \exp \sum_{j=1}^{k} \theta_j t_j dP^+(t) = \int \exp \sum_{j=1}^{k} \theta_j t_j dP^-(t)$ as functions of the complex variables $\theta_j = \xi_j + i\eta_j, j = 1, \ldots, k$. By an obvious extension of Lemma A.4, these integrals are analytic functions of θ_j in the strip R_j: $-a < \xi_j < a$, $-\infty < \eta_j < \infty$ of the complex plane. For $\theta_j, j = 2, \ldots, k$ fixed, real and between $-a$ and a, equality of the integrals holds on the line segment $\{(\xi_1, \eta_1): -a < \xi_1 < a, = 0\}$ and therefore can be extended to the strip R_1 in which the integrals are analytic. By induction, the inequality can be extended to the region $\{(\theta_1, \ldots, \theta_k): (\xi_j, \eta_j) \text{ in } R_j \text{ for } j = 1, \ldots, k\}$. In particular, for all real $(\eta_1, \ldots, \eta_k)$

$$\int \exp i \sum_{j=1}^{k} \eta_j t_j dP^+(t) = \int \exp i \sum_{j=1}^{k} \eta_j t_j dP^-(t).$$

These integrals are the characteristic functions of the distributions $P+$ and $P-$ respectively, and by the uniqueness theorem for characteristic functions, the two distributions must coincide. $f^+(t) = f^-(t)$ except on a set of measure zero and hence $f(t) = 0$ a.e. with respect to the family $\mathcal{P}^T$. □

Lemma A.8. *Suppose the statistics (U, T) have the joint distribution $dP^{U,T}_{\theta,\lambda}(u,t) C(\theta, \lambda) \exp[\theta u + \sum_j \lambda_j T_j] d\mu(u,t)$.*

Let

$$\varphi(u,t) = \begin{cases} 0 & \text{when } u < C_0(t), \\ \gamma_j(t) & \text{when } u = C_0(t), \\ 1 & \text{when } u > C_0(t). \end{cases}$$

Then $\varphi(u,t)$ is jointly measurable in u and t.

Proof. Set $F_t(u) = P_{\theta'}\{U \leq u \mid t\}$, the conditional distribution function of U given $T = t$. Then the restriction $E_{\theta'}[\varphi(U,T) \mid t] = \alpha$ for all t is equivalent to $F_t(C) - \gamma[F_t(C) - F_t(C^-)] = 1 - \alpha$ where $F_t(C^-)$ denotes the supremum of $F_t(u)$ on the open interval $(-\infty, C)$. $C = C(t)$ is chosen so that $F_t(C^-) \leq 1 - \alpha \leq F_t(C)$; hence $C(t) = F_{t^{-1}}(1 - \alpha) = \inf\{u: F_t(u) \geq 1 - \alpha\}$. $F_{t^{-1}}(z) \leq u$ if and only if $F_t(u) \geq z$, so that $F_{t^{-1}}(z)$ is t-measurable for any fixed z. Thus $C(t)$ and $\gamma(t)$ will both be measurable providing both $F_t(u)$ and $F_t(u^-)$ are jointly measurable in u and t.

For each fixed u, $F_t(u)$ is a measurable function of t and for each fixed t, it is nondecreasing and continuous on the right. Then if $F_t(u) \geq c$, there exists a rational number r such that $u \leq r < u + 1/n$ and $F_t(r) \geq c$. Denoting the rationals (a countable set) by $r_1, r_2, \ldots$, we can show that

$$\{(u,t): F_t(u) \geq c\} = \bigcap_n \bigcup_t \{(u,t): 0 \leq r_i - u < 1/n, F_t(r_i) \geq c\}.$$

Thus $F_t(u)$ is jointly measurable in u and t, and a similar argument establishes the desired result for $F_t(u^-)$. $\square$

We are now in a position to prove that the critical function

$$\varphi^*(u,t) = \begin{cases} 0 & \text{when } u < C_0(t), \\ \gamma_0(t) & \text{when } u = C_0(t), \\ 1 & \text{when } u > C_0(t), \end{cases}$$

where C_0 and γ_0 are determined by

$$E_{\theta_0}[\varphi(U,T) \mid t] = \alpha \quad \text{for all } t,$$

provides a UMPU test of $H_1: \theta \leq \theta_0$ against $K_1: \theta > \theta_0$ for the multiparameter exponential family.

We already know that the statistic T is sufficient for λ if θ has any fixed value. By Theorem A.3, the distribution of T belongs to the family, $dP^{T_{\theta,\lambda}}(t) = C(\theta, \lambda)\exp[\sum_j \lambda_j T_j]d\mu_\theta$. The natural parameter space Ω of the exponential family is convex. For Ω let $(\theta_1, \ldots, \theta_j)$ and $(\theta'_1, \ldots, \theta'_j)$ be two parameter points for which the integral of the density is finite. Then for any $0 < \alpha < 1$ by Hölder's inequality, $\int \exp[\sum(a\theta_j + (1-a)\theta'_j)T_j(x)]d\mu \leq [\int \exp[\sum(\theta_j T_j(x)]d\mu]\alpha[\int \exp[\sum(\theta'_j T_j(x)]d\mu]^{1-\alpha} < \infty$.

Assuming Ω has full dimension $k+1$, the set $\omega = \{(\theta_0, \lambda)\}$ is convex and contains a k-dimensional rectangle. By Lemma A.7, $\mathcal{P}^T$ is complete; so as in Section 3.5.3, the test has Neyman structure and $E_{\theta_0}[\varphi(U,T) \mid t] = \alpha$ for all t.

By Lemma A.8, φ^* is jointly measurable in t and u, so that its expectation and power exists and is equal to

$$E_{\theta,\lambda}\varphi(U,T) = \int \left[\int \varphi(U,T)dP^{U|t_\theta}(u)\right]dP^{T_{\theta,\lambda}}(t).$$

Earlier in this section, we showed that a test of form φ^* maximizes the power of the conditional test, given by the inner integral separately for each t. Thus, it maximizes the overall power among all tests that are similar on the boundary and is UMP-unbiased.

To prove that the remaining tests are UMP-unbiased, it is necessary to show that they, too, are jointly measurable in t and u. Outlines of such proofs are provided in Lehmann [1986, pp 173–174].

A.5 Exchangeable Observations

Let $\mathcal{P}$ be a family of distributions of $X = \{X_1, \ldots, X_n\}$ that are symmetric in the sense that if π is any permutation of the subscripts $\{1, \ldots, n\}$, then $P\{(X_{\pi(1)}, \ldots, X_{\pi(n)}) \in B\} = P\{(X_1, \ldots, X_n) \in B\}$ for all Borel sets B. Then the random variables $\{X_1, \ldots, X_n\}$ are said to be *exchangeable*.

Permutation tests rely on the assumption of *exchangeability*, that is, under the hypothesis, the joint distribution of the observations is invariant under permutations of the subscripts. If a set of observations are independent, identically distributed (i.i.d.), or if they are jointly normal with identical covariances,[2] then they are *exchangeable*

A caveat is that a set of units may be exchangeable for some purposes and not for others, depending on what is measured and the questions of interest. A simple example suggested by Draper et al. [1993] is a circadian series in which observations *within* days are not exchangeable because of serial correlation, while observations *between* days (at the same point in time) are exchangeable as are the residuals from a model incorporating serial correlation.

Let $G\{x; y_1, y_2, \ldots, y_{n-1}\}$ be a distribution function in x and symmetric in its remaining arguments—that is, permuting the remaining arguments would not affect the value of G. Let the conditional distribution function of x_i given $x_1, \ldots, x_{i-1}, x_{i+1}, \ldots, x_n$ be G for all i. Then the $\{x_i\}$ are exchangeable.[3]

It is easy to see that a set of i.i.d. variables is exchangeable. Or that the joint distribution of a set of normally distributed random variables whose covariance matrix is such that all diagonal elements have the same value σ^2 and all the off-diagonal elements have the same value ρ^2 is invariant under permutations of the variable subscripts.

The requirement for exchangeability in testing arises in either of two ways:

1. Sufficiency—the order statistics are sufficient for a wide variety of problems.
2. Invariance—the joint distribution of the observations is invariant under permutation of the subscripts.

For many testing problems, it as or more important that the underlying model remain invariant under permutations of the subscripts. This can only be accomplished in many cases if we restrict the set of permutations. Recall that in the classic definition[4] a set of n random variables is said to be *exchangeable* if the joint distribution of the variables is invariant with respect to the group S_n of all possible permutations of the subscripts.

In group theory, a set B is said to contain the *generators* of a group G, if all the elements of G can be obtained by repeated application of the elements

---

[2] For additional examples, see Galambos [1986] or Draper et al. [1993].
[3] Or simply exchangeable in the classic definition.
[4] See de Finetti [1930] and Galambos [1986].

of B. A set of permutations P is said to be *weakly mixing* with respect to the elements of a set X, if for every i and j there is at least one permutation π in P that exchanges the ith and jth elements of X.

Define the *weak exchangeability* of a set of random variables as the invariance of their joint distribution with respect to a non-empty subgroup of the set of permutations. Clearly, a set of variables that is exchangeable is also weakly exchangeable. Equally clearly, the converse may not be true.

A.5.1 Order Statistics

If the observations are exchangeable, and the data are ordinal and uni-dimensional, the set of order statistics is sufficient. If $X_1, \ldots, X_N$ are independent and identically distributed with cumulative distribution function F belonging to the class of all absolutely continuous distributions $\mathcal{F}$, then the set of order statistics is complete with respect to $\mathcal{F}$. The set of summands $S = \{\sum X_i, \sum X_{i^2}, \ldots, \sum X_{i^n}\}$ is equivalent to the order statistics in the sense that they both induce the same subfield of the sample space (Exercise A.18). Consider the family of densities

$$f(x) = C(\theta_1, \ldots, \theta_n) \exp(-x^{2n} + \theta_1 x + \cdots + \theta_n x^n),$$

where C is merely a normalizing constant, chosen so that $\int f(x)dx = 1$.

The density of a sample of size n is

$$C^n \exp(-\sum x_i^{2n} + \theta_1 \sum x_i + \cdots + \theta_n \sum x_i^n);$$

these densities constitute a multiparameter exponential family. By Lemma A.7, S is complete for this family and hence for the larger family $\mathcal{F}$.

The order statistics constitute a set of maximal invariants with respect to the group G of all permutations of the components of X (Exercise 6.5). Thus, under certain conditions,

a. UMPU permutation tests may also be UMP invariant (UMPI);
b. UMPU permutation tests may be the most stringent tests of hypotheses.

We saw an example of the former in Section 6.1 when $X_1, \ldots, X_n$ were independent, normally distributed as $N(\mu, \sigma^2)$, and we developed a UMPI/UMPU test of the hypothesis $\sigma \geq \sigma_0$. The power of this test is a constant on each of the sets $\{(\mu, \sigma): -\infty < \mu < \infty, \sigma = \sigma'\}$. As it is the most powerful test on each such set, it is most stringent.

The following example shows that a permutation test may also be a most stringent test even when no impartiality criterion is applied explicitly. Suppose we conduct an experiment with $2n$ subjects in which we randomly assign one half to receive a treatment whose results follow the distribution $N(\mu + u_i, \sigma^2)$,

$i = 1, \ldots, n$ and the other half to a second treatment whose results follow the distribution $N(\eta + u_i, \sigma^2)$, $i = n, \ldots, 2n$. Note that the effects μ and η depend on the treatment, while the effects $\{u_i\}$ depend on the individual who is treated. The UMPU permutation test of the hypothesis $H: \mu = \eta$ against the alternative $K: \mu \neq \eta$ rejects when the absolute value of the difference in means of the two samples is large. The test is symmetric and in particular, the power of the test is the same against the two alternatives (μ^*, η^*, σ) and (η^*, μ^*, σ) and is the maximum attainable of any test against those alternatives. Denote by K_{μ^*} the set consisting of these two alternatives. $K = K_{\mu^*}$. Because the UMPU permutation test attains the maximum possible power for each member of the union, it is most stringent.

A.5.2 Transformably Exchangeable

Suggesting the concept of transformably exchangeable is the procedure for testing a non-null two-sample hypothesis $H: F[x] = G[x - d]$, for if we have two sets of independent observations $\{Z_i\}$ and $\{Y_i\}$ with Z_i distributed as F and Y_i as G, we can obtain an exact test of H if we first transform the variables subtracting 0 from each of the Z_i and 3 from each of the Y_i.

A set of observations (random variables) $\boldsymbol{X}$ will be said to be *transformably exchangeable* if there exists a transformation (measurable transformation) T, such that $\text{T}\boldsymbol{X}$ is exchangeable.[5]

If we have a set of observations $\{X[t], t = 1, 2, \ldots, n\}$ where $X[t] = a + bX[t-1] + z_t$ and the $\{z_t\}$ are i.i.d., then the variables $\{Y[t], t = 2, \ldots, n\}$ where $Y[t] = X[t] - bX[t-1]$ are exchangeable.

Dependent noncolinear normally distributed variables with the same mean are transformably exchangeable, for as the covariance matrix is nonsingular, we may use the inverse of this matrix to transform the original variables to independent (and hence exchangeable) normal ones. By applying two successive transformations, we can obtain an exact permutation test of the non-null two-sample univariate hypothesis for dependent normally distributed variables providing the covariance matrix is known. Unfortunately, as Commenges [2001] shows, whether we accept or reject in a specific case may depend on the transformation we have chosen.

The preceding result applies even if the variables are colinear: Let R denote the rank of the covariance matrix in the singular case. Then there exists a projection onto an R-dimensional subspace where R normal random variables are independent. So if we have an N-dimensional ($N > R$) correlated and singular multivariate normal distribution, there exists a set of R linear combinations of the original N variables so that the R linear combinations are each univariate normal and independent of one other.

---

[5] See Commenges [2001].

A.5.3 Exchangeability-Preserving Transforms

Suppose we wish to test whether two regression curves are parallel even though we do not know the value of the intercepts. We are given that

$$y_{ik} = a_i + b_i x_{ik} + \epsilon_{ik} \quad \text{for } i = 1, 2; \ k = 1, \ldots, n_i$$

where the errors $\{\epsilon_{ik}\}$ are exchangeable. To obtain an exact permutation test for $H: b_1 = b_2$, we need to eliminate the $\{a_i\}$ while preserving the exchangeability of the residuals. We know that under the null hypothesis

$$\bar{y}_{i.} = a_i + b\bar{x}_{i.} + \bar{\epsilon}_{i.}$$

Define

$$y' = \frac{1}{2}(\bar{y}_1 - \bar{y}_2); \ x' = \frac{1}{2}(\bar{x}_1 - \bar{x}_2); \ \epsilon' = \frac{1}{2}(\bar{\epsilon}_1 - \bar{\epsilon}_2); \ a' = \frac{1}{2}(a_1 + a_2).$$

Define

$$y'_{1k} = y_{1k} - y' \text{ for } k = 1 \text{ to } n_1 \quad \text{and} \quad y'_{2k} = y_{2k} + y' \text{ for } k = 1 \text{ to } n_2.$$

Define

$$x'_{1k} = x_{1k} - x' \text{ for } k = 1 \text{ to } n_1 \quad \text{and} \quad x'_{2k} = x_{2k} + x' \text{ for } k = 1 \text{ to } n_2.$$

Then

$$y'_{ik} = a' + bx'_{ik} + \epsilon'_{ik} \quad \text{for } i = 1, 2; \ k = 1, \ldots, n_i.$$

Two cases arise. If the original predictors were the same for both sets of observations, that is, if $x_{1k} = x_{2k}$ for all k, then the errors $\{\epsilon'_{ik}\}$ are exchangeable and we can apply the method of matched pairs.[6] Otherwise, we need to proceed as follows: First, estimate the two parameters a' and b by least squares means. Use these estimates to derive the transformed observations $\{y'_{ik}\}$. Then test the hypothesis that $b_1 = b_2$ using a two-sample comparison. If the original errors were exchangeable, then the errors $\{\epsilon'_{ik}\}$, though not independent, are exchangeable also and this test is exact.

Now suppose

$$y_{ik} = A_i Z_k + b_i x_{ik} + \epsilon_{ik} \quad \text{for } i = 1, 2; \ k = 1, \ldots, n_i$$

where Z_k is a column vector of covariates with A_i a row vector of the corresponding coefficients. Defining A'_i as the mean of A_1 and A_2, then

$$y'_{ik} = A' Z_k + bx'_{ik} + \epsilon'_{ik} \quad \text{for } i = 1, 2; \ k = 1, \ldots, n_i$$

and we have analogous results for the general case.

---

[6] See, for example, Good [2000, p. 51].

Dean and Verducci [1990] characterize the linear transformations that preserve exchangeability. Commenges [2001] characterizes the linear transformations that also preserve the permutation distribution. Clearly any transformation that preserves the ordering of the order statistics preserves exchangeability.

A.6 Confidence Intervals

Let $x = \{X_1, X_2, \ldots, X_n\}$ be an exchangeable sample from a distribution F_θ, which depends upon a parameter $\theta \in \Omega$. A family of subsets $S(x)$ of the parameter space Ω is said to be a family of confidence sets for θ at level $1 - \alpha$ if

$$P_\theta\{\theta \in S(X)\} \geq 1 - \alpha \quad \text{for all } \theta \in H(\theta').$$

The family is said to be unbiased if

$$P_\theta\{\theta' \in S(X)\} \leq 1 - \alpha \quad \text{for all } \theta \in \Omega - H(\theta').$$

The construction of a confidence set from a family of acceptance regions is described in Chapter 3. The following theorem shows us this construction can proceed in either direction.

Theorem A.4. *For each $\theta' \in \Omega$, let $A(\theta')$ be the acceptance region of the level-α test for $H(\theta')$: $\theta = \theta'$, and for each sample point x, let $S(x)$ denote the set of parameter values $\{\theta: x \in A(\theta), \theta \in \Omega\}$. Then $S(x)$ is a family of confidence sets for θ at confidence level $1 - \alpha$.*

Theorem A.5. *If for all θ', $A(\theta')$ is UMPU for testing $H(\theta')$ at level α against the alternatives $K(\theta')$, then for each θ' in Ω, $S(X)$ minimizes the probability*

$$P_\theta\{\theta' \in S(X)\} \quad \text{for all } \theta \in K(\theta')$$

among all unbiased level $1 - \alpha$ family of confidence sets for θ.

Proof A.4. By definition, $\theta \in S(x)$ if and only if $x \in A(\theta)$; hence, $P_\theta\{\theta \in S(X)\} = P_\theta\{X \in A(\theta)\} \geq 1 - \alpha$. □

Proof A.5. If $S^*(x)$ is any other family of unbiased confidence sets at level $1 - \alpha$ and if $A^*(\theta) = \{x: \theta \in S^*(x)\}$, then

$$P_\theta\{X \in A^*(\theta')\} = P_\theta\{\theta' \in S^*(x)\} \geq 1 - \alpha \quad \text{for all } \theta \in H(\theta'),$$

and

$$P_\theta\{X \in A^*(\theta')\} = P_\theta\{\theta' \in S^*(x)\} \leq 1 - \alpha \quad \text{for all } \theta \in \Omega - H(\theta'),$$

so that $A^*(q')$ is the acceptance region of a level-α unbiased test of $H(\theta')$. Since A is *UMPU*,

$$P_\theta\{X \in A^*(\theta')\} \geq P_\theta\{X \in A(\theta')\} \quad \text{for all } \theta \in \Omega - H(\theta');$$

hence, $P_\theta\{\theta' \in S^*(x)\} \geq P_\theta\{\theta' \in S(x)\}$ for all $\theta \in \Omega - H(\theta')$, as was to be proved. $\square$

A.7 Asymptotic Behavior

A major reason for the popularity of the permutation tests is that with very large samples their power is almost indistinguishable from that of the most powerful parametric tests. To establish this result, we need to know something about the distribution of the permutation statistics as the sample size increases without limit. Two sets of results are available to us. The first, due to Wald and Wolfowitz [1947] and Hoeffding [1953] provides us with conditions under which the limiting distribution is normal under the null hypothesis; the second, due to Albers, Bickel, and Van Zwet [1976] and Bickel and Van Zwet [1978] provides conditions under which this distribution is normal for near alternatives.

A.7.1 A Theorem on Linear Forms

Let $S_N = (s_{N1}, s_{N2}, \ldots, s_{NN})$ and $U_N = (u_{N1}, u_{N2}, \ldots, u_{NN})$ be sequences of real numbers, and let $s_{N\cdot} = \sum s_{Nj}/N$; $u_{N\cdot} = \sum u_{Nj}/N$.

The sequences S_N satisfy the condition W if for all integers $r > 2$,

$$W(S_N, r) = \frac{|\sum (s_{Nj} - s_{N\cdot})^r|}{\sum [(s_{Nj} - s_{N\cdot})^2]^{r/2}} \quad \text{is bounded above for all } n.$$

The sequences S_N, U_N jointly satisfy the condition H_1 if for all integers $r > 2$,

$$\lim_N N^{r/2-1} W(S_N, r) W(U_N, r) = 0.$$

The sequences S_N, U_N jointly satisfy the condition H_2, if for all integers $r > 2$,

$$\lim_N N \frac{\max_j (s_{Nj} - s_{N\cdot})^r}{\sum (s_{Nj} - s_{N\cdot})^r} \frac{\max_j (u_{Nj} - u_{N\cdot})^r}{\sum (u_{Nj} - u_{N\cdot})^r}$$

For any value of N let $X = (x_1, x_2, \ldots, x_N)$ be a chance variable whose possible values correspond to the $N!$ permutations of the sequence $A_N = (a_1, a_2, \ldots, a_N)$. Let each permutation of A_N have the same probability $1/N!$, and let $E(Y)$ and $SD(Y)$ denote the expectation and standard deviation of the variable Y, respectively.

Theorem A.6. *Let the sequences $A_N = (a_1, a_2, \ldots, a_N)$ and $D_N = (d_1, d_2, \ldots, d_N)$ for $N = 1, 2, \ldots$, satisfy any of the three conditions W, H_1, and H_2. Let the chance variable L_N be defined as $L_N = \sum d_i x_i$. Then, as $N \to \infty$, $\Pr\{L_N - E(L_N) < tSD(L_N)\} \to 1/\sqrt{2\pi} \int_{-\infty}^{t} e^{-x^2/2} dx$.*

A proof of this result for condition W is given in Wald and Wolfowitz [1944]. The proof for conditions H_1 and H_2 is given in Hoeffding [1953].

This theorem applies to the majority of the tests we have already considered, including:

1) Pitman's correlation $\sum d_i a_i$;
2) the two-sample test with observations $a_1, \ldots, a_{m+n}$, and d_i equal to one if $i = 1, \ldots, m$ and zero otherwise;
3) Hotelling's T with $\{a_{1j}\}$ and $\{a_{2j}\}$ the observations—both sequences must separately satisfy the conditions of the theorem, and $d_i = 1/m$ for $i = 1, \ldots, m$ and $d_i = -1/n$ for $i = m+1, \ldots, m+n$.

A.7.2 Monte Carlo

In practice, we are more likely to utilize a Monte Carlo in which we examine only a random sample of permutations than to enumerate all possibilities. Providing we are concerned only with detecting alternatives that differ from the null hypothesis by some fixed amount—as we would be in virtually every practical situation—the results of the preceding section are still applicable. On the other hand, Dwass [1957] reports that if the alternative is allowed to approach indefinitely close to the alternative as the sample size increases, the power would be adversely affected by the use of a Monte Carlo.

A.7.3 Asymptotic Efficiency

In this section, we provide asymptotic expansions to order N^{-1} for the power of the one- and two-sample permutation tests and compare them with the asymptotic expansions for the most powerful parametric unbiased tests. The general expansion takes the form

$$b_N = c_0 + c_1 N^{-1/2} + c_{2,N} N^{-1} + o(N^{-1}),$$

where the coefficients depend on the form of the distribution, the significance level, and the alternative—but in both the one- and two-sample cases, the expansions for the permutation test and the t-test coincide for all terms through N^{-1}. The underlying assumptions are: (1) The observations are independent; (2) within each sample they are identically distributed; and (3) the two populations differ at most by a shift, $G(x) = F(x - \delta)$, where $\delta \geq 0$. $\beta(p, F, \delta)$ and $\beta(t, F, \delta)$ are the power functions of the permutation test and the parametric t-test, respectively (see Section 2.3). The theorem's other restrictions are technical in nature and provide few or no limitations in practice; e.g., the significance level must lie between 0 and 1, and the distribution must have absolute moments of at least ninth order. We state the theorem for the one-sample case only.

Theorem A.7. *Suppose the distribution F is continuous and that positive numbers C, D, and $r > 8$ exist such that $\int |x|^r dF[x] \leq C$ and $0 \leq \delta \leq DN^{-1/2}$; then if α is neither 0 nor 1, there exists a $B > 0$ depending on C and D, and $ab > 0$ depending only on r such that $|\beta(p, F, \delta) - \beta(t, F, \delta)| \leq BN^{-1/b}$.*

Proof of this result and details of the expansion are given in Bickel and Van Zwet [1976]. The practical implication is that for large samples the permutation test and the parametric t-test make equally efficient use of the data.

Robinson [1989] finds approximately the same coverage probabilities for three sets of confidence intervals for the slope of a simple linear regression based, respectively, on (1) the standardized bootstrap, (2) parametric theory, and (3) a permutation procedure. Under the standard parametric assumptions, the coverage probabilities differ by $o(n^{-1})$, and the intervals themselves differ by $O(n^{-1})$ on a set of probability $1 - O(n^{-1})$.

A.7.4 Exchangeability

The requirement that the observations be exchangeable can be relaxed at least asymptotically for some one-sample and two-sample tests. Let $X_1, \ldots, X_n$ be a sample from a distribution F that may or may not be symmetric. Let $R_n(x, \Pi_n)$ be the permutation distribution of the statistic $T_n(X_1, \ldots, X_n)$, and let r_n denote the critical value of the associated permutation test; let $J_n(x, F)$ be the unconditional distribution of this same statistic under F, and let Φ denote the standard normal distribution function.

Theorem A.8. *If F has mean zero and finite variance $\sigma^2 > 0$, and $T_n = n^{1/2}\bar{X}$, then as $n \to \infty$,*

$$\sup_x |R_n(x, \Pi_n) - J_n(x, F)| \longrightarrow 0 \text{ with probability } 1,$$

and

$$\sup_x |R_n(x, \Pi_n) - \Phi(x/\sigma)| \longrightarrow 0 \text{ with probability } 1.$$

Thus $r_n \to \sigma z_a$, with probability 1 and $E_F[\phi(R_n)] \to \alpha$.

A proof of this one-sample result is given in Romano [1990]; a similar one-sample result holds for a permutation test of the median subject to some mild continuity restrictions in the neighborhood of the median.

The two-sample case is quite different. Romano [1990] shows that if F_X and F_Y have common mean μ and finite variances σ_X^2 and σ_Y^2, respectively,

$T_{m,n} = n^{1/2}(\bar{X} - \bar{Y})$, and $m/n \to \lambda$ as $n \to \infty$, the unconditional distribution of $T_{m,n}$ is asymptotically Gaussian with mean zero and variance $\sigma_X^2 + (1-\lambda)\sigma_Y^2/\lambda$, while the permutation distribution of $T_{m,n}$ is asymptotically Gaussian with mean zero and variance $\sigma_Y^2 + (1-\lambda)\sigma_X^2/\lambda$. Thus, the two asymptotic distributions are the same only if either (a) the variances of the two populations are the same, or (b) the sizes of the two samples are equal (whence $\lambda = 1$).

Romano also shows that whatever the sample sizes, a permutation test for the difference of the medians of two populations will not be exact, even asymptotically (except in rare circumstances) unless the underlying distributions are the same.

A.7.5 Improved Bootstrap Confidence Intervals

As already noted, bootstrap confidence intervals are only asymptotically exact. The primitive bootstrap confidence interval is said to be first-order exact in that $\Pr\{\theta \le \theta_{LB}^*\} = \alpha + O(n^{-1/2})$ where θ_{LB}^* is the α lower confidence bound determined by the primitive bootstrap. Applying the Hall–Wilson corrections, we obtain a second-order accurate bootstrap $\Pr\{\theta \le \theta_{HW}^*\} = \alpha + O(n^{-1})$. The BC_α is also second-order accurate [Hall, 1988].

Beran [1987] and Hall [1992] describe iterative methods, known respectively as bootstrap pivoting and bootstrap inverting, that provide third-order accurate confidence intervals. Loh [1987, 1991] describes a bootstrap calibration method that yields confidence intervals that in some circumstances are fourth-order accurate.

A.8 Exercises

1. a) What is the complement of $\mathcal{L}$, that is, what events are not in $\mathcal{L}$?
   b) Prove that a σ-field is closed under countable intersections.
   c) Is $\mathcal{C} = \{\mathcal{L}, \emptyset\}$ a σ-field?
2. Consider the outcomes that could result from the flip of the coin, for example, "A force of 171 dynes is applied at a point 1 centimeter in from the edge, the coin rose outward at an angle of 70 degrees flipped over two and a half times, and fell to the ground heads upward." What is the smallest σ-field that would contain the event H ("lands heads upward").
3. Is the σ-field associated with the Poisson distribution countable or non-countable?
4. If $\mathcal{A} \subseteq \mathcal{A}$ and $\mathcal{B} \subseteq \mathcal{B}$ show that even if $P^Z(A' \times B') = P^X(A')XP^Y(B')$ for all A' in $\mathcal{A}'$ and B in $\mathcal{B}'$, there may exist a set $A \times B$, with A in $\mathcal{A}$ and B in $\mathcal{B}$ such that $P^Z(A \times B) \ne P^X(A)XP^Y(B)$.
5. Show that if G is function on the real line, nondecreasing and continuous on the right, such that $G(-\infty) = 0$ and $G(\infty) = 1$, then G uniquely determines a probability distribution over the Borel sets.

6. Prove that the positive and negative parts of any measurable function are also measurable.
7. Is a statistic a random variable?
8. Suppose β is the power of the most powerful level-α test for testing P_0 against P_1. Show that $\alpha < \beta$ unless $\alpha = 1$ or $P_0 = P_1$.
9. Suppose that X is normally distributed with zero mean. Is the information carried by the statistics $|X|$, X^2, and $\exp[-X^2]$ the same? (Hint: Consider the σ-fields induced by these statistics.)
10. Unbiased. The test $\varphi \equiv \alpha$ is a great timesaver: you don't have to analyze the data; you don't even have to gather data! All you have to do is flip a coin.

    a) Prove that this test is unbiased.
    b) Prove that a biased test cannot be uniformly most powerful.

11. Prove Lemma A.1. (Hint: Consider the test $\varphi[x] \equiv \alpha$ and apply Theorem A.1.)
12. Show that if $\psi_0(t) = 1$ when $a_1 \exp(b_1 t) + a_2 \exp(b_2 t) < 1$, with $b_1 < 0 < b_2$, and $b_1 = b_2 = 0$ otherwise, then ψ will satisfy condition A.2 only if $a_1 > 0$ and $a_2 > 0$.
13. If a critical function ψ satisfies the conditions of Lemma A.2, show that if $k_i \geq 0$ for $i = 1, \ldots, m$, then ψ maximizes $\int \psi f_{m+1} d\mu$.
14. Prove Lemma A.5
15. If T is a statistic and the functions $f, g, \ldots$ are integrable $(\mathcal{C}, P)$, show that the following properties hold almost everywhere with respect to $(\mathcal{B}, PT)$:

    a) $E(af(X) + bg(X)|t) = aE(f(X)|t) + bE(g(X)|t)$.
    b) $E(h(T)f(X)|t) = h(T)E(f(X)|t)$.
    c) $E(E(f(X)|t)) = E(f(X))$.

16. Use Basu's theorem to show that the mean and variance of a sample of n independent identically normally distributed observations are independent.
17. Sketch the power curves for the four critical functions defined in Section A.3. Prove that the curves must have the shapes you've sketched.
18. Let $X_1, \ldots, X_n$ be independent and identically distributed according to one of a continuous family of distributions $\mathcal{P}$. Suppose that the members of this family are symmetric with respect to the origin. Let $V_i = |X_i|$. Show that $\{V_{(1)} \leq \cdots \leq V_{(n)}\}$ is sufficient for $\mathcal{P}$.
19. Show that the set of summands $\{\sum X_i, \sum X_{i^2}, \ldots, \sum X_{i^n}\}$ is equivalent to the order statistics. [Hint: Consider also the set of symmetric functions $(\sum X_i, \sum_{i<j} X_i X_j, \sum_{i<j<k} X_i X_j X_k, \ldots, \sum_{i<j<\cdots<n} X_i \ldots X_n)$.]
20. Suppose that $\{X_i, i = 1, \ldots, n\}$ is $N(\mu, \sigma^2)$ and $\{Y_i, i = 1, \ldots, m\}$ is $N(\mu, \tau^2)$. Derive the most powerful unbiased permutation test for testing H: $\tau^2/\sigma^2 = 1$ against K: $\tau^2/\sigma^2 = 2$.
21. a) The times between successive decays of a radioactive isotope are said to follow the exponential distribution, that is, the probability that an atom will not decay until after an interval of length t is $1 - \exp[-t/\lambda]$. (A similar formula provides a first-order approximation to the time t you will spend waiting for the next bus.) Suppose you had two

potentially different isotopes with parameters λ_1 and λ_2, respectively. Derive a UMPU permutation test for testing $H\colon \lambda_1 = \lambda_2$, against $K\colon \lambda_1 > \lambda_2$.

b) More generally, suppose that an item is reliable for a fixed period b, after which its reliability decays at a constant rate λ. Then its lifetime has the exponential density $\lambda^{-1}\exp[(x-b)/\lambda]$. What statistic would you use for testing that $H\colon \lambda_1 = \lambda_2$, against $K\colon \lambda_1 > \lambda_2$? Is your answer the same as in part a? Why not? (Hint: Look for sufficient statistics. Note that the problem remains invariant under an arbitrary scale transformation applied to both sets of data. And see Section 3.7.)

22. Suppose you have taken n independent observations from a distribution with density function $f(x|a,b) = x^{a-1}(1-x)^{b-1}$, where $0 < x < 1$, $a > 0$ and $b > 0$. Can you describe a UMPU test of the hypothesis $1/8 < a < 3/8$ against the alternative $a < 1/8$ or $a > 3/8$?

23. a) Show that the permutation test based on the deviations about the sample medians described in Section 3.7.2 is asymptotically exact.

b) Show that the permutation test for interactions based on the residuals in a multifactor analysis (Section 7.4) is asymptotically exact.

Bibliography

Adderley EE. Nonparametric methods of analysis applied to large-scale seeding experiments. *J. Meteorology*. 1961; 18: 692–694.

Agresti A. A survey of exact inference for contingency tables. *Statist. Sci.* 1992; 7: 131–177.

Agresti A. *Categorical Data Analysis*. New York: John Wiley & Sons; 1990.

Agresti A; Wackerly D and Boyett JM. Exact conditional tests for cross-classifications: Approximations of attained significance levels. *Psychometrika*. 1979; 44: 75–83.

Agresti A and Wackerly D. Some exact conditional tests of independence for $R \times C$ cross-classification tables. *Psychometrika*. 1977; 42: 111–126.

Albers W; Bickel PJ and Van Zwet WR. Asymptotic expansions for the power of distribution-free tests in the one-sample problem. *Ann. Statist.* 1976; 4: 108–156.

Albert A; Chapelle JP, Huesghem C, Kulbertus GE and Harris EK. Evaluation of risk using serial laboratory data in acute myocardial infarction. In C Huesghem, A Albert and ES Benson (eds.), *Advanced Interpretation of Clinical Laboratory Data*. New York: Marcel-Dekker; 1982.

Alroy J. Permutation tests for the presence of phylogenetic structure: An editorial. *Systematics Biology*. 1994; 43: 430–437.

Aly E-E AA. Simple tests for dispersive ordering. *Stat. Prob. Ltr.* 1990; 9: 323–325.

Andersen PK; Borgan O; Gill RD and Keiding N. Linear nonparametric tests for comparison of counting processes with applications to censored survival data. *Int. Statist. Rev.* 1982; 50: 219–258.

Andersen PK; Borgan O; Gill RD and Keiding N. *Statistical Models Based on Counting Processes*. New York: Springer; 1993.

Arndt S; Cizadlo T; Andreasen NC; Heckel D; Gold S and Oleary DS. Tests for comparing images based on randomization and permutation methods. *J. Cerebral Blood Flow and Metabolism*. 1996; 16: 1271–1279.

Ascher S and Bailar J. Moments of the Mantel–Valand procedure. *J. Statist. Comput. Simul.* 1982; 14: 101–111.

Bailer AJ. Testing variance equality with randomization tests. *J. Statist. Comput. Simul.* 1989; 31: 1–8.

Baker RD. Two permutation tests of equality of variance. *Statist. Comput.* 1995; 5(4): 289–296.

Balakrishnan N and Ma CW. A comparative study of various tests for the equality of two population variances. *Statist. Comp. Simul.* 1990; 35: 41–89.

Baptista J and Pike MC. Exact two-sided confidence limits for the odds ratio in a 2×2 table. *J. Roy. Statist. Soc. C.* 1977; 26: 214–220.

Barnard GA. A new test for 2×2 tables (letter to the editor). *Nature.* 1945; 156: 177.

Barnard GA. Discussion of paper by MS Bartlett. *J. Roy. Statist. Soc. B.* 1963; 25: 294.

Barnard GA. In contradiction to J. Berkson's dispraise: Conditional tests can be more efficient. *J. Statist. Plan. Inference.* 1979; 3: 115–139.

Barnard GA. On alleged gains in power from lower p-values. *Statist. Med.* 1989; 8: 1469–1477.

Barnard GA. Statistical inference. *J. Roy. Statist. Soc. B.* 1949; 11: 115–139.

Barton DE and David FN. Randomization basis for multivariate tests. *Bull. Int. Statist. Inst.* 1961; 39(2): 455–467.

Bebbington AC. A simple method of drawing a sample without replacement. *Applied Statistics.* 1975; 24: 136.

Bell CB and Doksum KA. Distribution-free tests of independence. *Annals Math. Statist.* 1967; 38: 429–446.

Bell CB and Doksum KA. Some new distribution-free statistics. *Annals Math. Statist.* 1965; 36: 203–214.

Bell CB and Donoghue JF. Distribution-free tests of randomness. *Sankhya A.* 1969; 31: 157–176.

Beran R. Prepivoting to reduce error rate of confidence sets. *Biometrika*, 1987; 74: 151–173.

Berger JO and Wolpert RW. *The Likelihood Principle.* IMS Lecture Notes—Monograph Series. Hayward, CA: IMS; 1984.

Berger RL and Boos DD. p values maximized over a confidence set for a nuisance parameter. *JASA.* 1994; 89: 1012–1016.

Berry KJ. AS179 Enumeration of all permutations of multi-sets with fixed repetition numbers. *Applied Statistics.* 1982; 31.

Berry KJ; Kvamme KL and Mielke PW, Jr. Improvements in the permutation test for the spatial analysis of the distribution of artifacts into classes. *Amer. Antiquity.* 1983; 48: 547–553.

Berry KJ; Kvamme KL and Mielke PW, Jr. Permutation techniques for the spatial analysis of the distribution of artifacts into classes. *Amer. Antiquity.* 1980; 45: 55–59.

Berry KJ and Mielke PW, Jr. A generalization of Cohen's kappa agreement measure to interval measurement and multiple raters. *Educ. Psych. Measure.* 1992; 52: 97–101.

Berry KJ and Mielke PW, Jr. Computation of exact probability values for multi-response permutation procedures (MRPP). *Commun. Statist. B.* 1984; 13: 417–432.

Berry KJ and Mielke PW. Computation of exact and approximate probability values for a matched-pairs permutation test. *Commun. Statist. B.* 1985; 14: 229–248.

Besag J and Clifford P. Generalized Monte Carlo significance tests. *Biometrika.* 1989; 76: 633–642.

Besag J and Clifford P. Sequential Monte Carlo p-values. *Biometrika.* 1991; 78: 301–304.

Bickel PJ. A distribution-free version of the Smirnov two-sample test in the multivariate case. *Annals Math. Statist.* 1969; 40: 1–23.

Bickel PM and Van Zwet WR. Asymptotic expansion for the power of distribution-free tests in the two-sample problem. *Ann. Statist.* 1978; 6: 987–1004 (corr 1170–1171).

Birch MW. The detection of partial association. *J. Roy. Statist. Soc. B.* 1964 May; 26/27: I 313–324, II 1–124.

Birnbaum ZW. Combining independent tests of significance. *JASA,* 1954; 49: 559–574.

Bishop YMM; Fienberg SE and Holland PW. *Discrete Multivariate Analysis: Theory and Practice.* Cambridge, MA: MIT Press; 1975.

Blair C and Karinski W. Distribution-free statistical analyses of surface and volumetric maps. In RW Tatcher; M Hallett; T Zeffiro; ER John and M Huerta (eds.), *Functional Neuroimaging: Technical Foundations.* Academic Press: New York; 1994.

Blair RC; Sawilowsky SS and Higgins JJ. Limitations of the rank transform in factorial ANOVA. *Communications in Statistics: Computations and Simulations*, 1987; B16: 1133–1145.

Blair RC; Troendle JF and Beck RW. Control of familywise errors in multiple endpoint assessments via stepwise permutation tests. *Statistics in Medicine.* 1996; 15: 1107–1121.

Boik RJ. The Fisher–Pitman permutation test: A non-robust alternative to the normal theory F-test when variances are heterogeneous. *British J. Math. Stat. Psych.* 1987; 40: 26–42.

Boos DD and Browne C. Testing for a treatment effect in the presence of nonresponders. *Biometrics.* 1986; 42: 191–197.

Booth JG and Butler RW. Randomization distributions and saddlepoint approximations in general linear models. *Biometrika.* 1990; 77: 787–796.

Boothroyd J. Algorithm 29, permutation of the elements of a vector. *Computer J.* 1967; 60: 311.

Boschloo RD. Raised conditional level of significance for the 2×2 table when testing the equality of two probabilities. *Statist. Neer.* 1970; 24: 1–35.

Box GEP and Anderson SL. Permutation theory in the development of robust criteria and the study of departures from assumptions. *J. Roy. Statist. Soc. B.* 1955; 17: 1–34 (with discussion).

Box JF. *The Life of a Scientist*. New York: Wiley; 1978.

Boyett JM and Shuster JJ. Nonparametric one-sided tests in multivariate analysis with medical applications. *JASA*. 1977; 72: 665–668.

Bradbury IS. Analysis of variance vs randomization tests: A comparison (with discussion by White and Still). *Brit. J. Math. Stat. Psych.* 1987; 40: 177–195.

Bradbury IS. Approximations to permutation distributions in the completely randomized design. *Commun. Statist. T-M* A 1988; 17: 543–555.

Bradley JV. *Distribution-Free Statistical Tests*. Englewood Cliffs, NJ: Prentice Hall; 1968.

Bross, IDJ. Taking a covariable into account. *JASA*. 1964; 59: 725–736.

Brown BM. Robustness against inequality of variances. *Australian J. Statist.* 1982; 24: 283–295.

Bryant EH. Morphometric adaptation of the housefly, Musa domestica L, in the United States. *Evolution*. 1977; 31: 580–596.

Busby DG. Effects of aerial spraying of fenithrothion on breeding white-throated sparrows. *J. Appl. Ecol.* 1990; 27: 745–755.

Cade B. Comparison of tree basal area and canopy cover in habitat models: Subalpine forest. *J. Alpine Mgmt*. 1997; 61: 326–335.

Cade B and Richards L. Permutation tests for least absolute deviation regression. *Biometrics*. 1996; 52: 886–902.

Carpenter J and Bithell J. Bootstrap confidence intervals. *Statist. Med.* 2000; 19: 1141–1164.

Chan I. Exact tests of equivalence and efficacy with a nonzero lower bound for comparative studies. *Statist. Med.* 1998; 17: 1403–1413.

Chapelle JP; Albert A; Smeets JP; Heusghem C and Kulberts HE. Effect of the hyptoglobin phenotype on the size of a myocardial infarction. *NEJM*. 1982; 307: 457–463.

Chase PJ. Algorithm 382. Combinations of M out of N objects. *Commun. ACM*. 1970A; 13: 368.

Chatterjee SK and Sen PK. Nonparametric tests for the bivariate two-sample location problem. *Calcutta Statist. Ass. Bull.* 1964; 13: 18–58.

Chernick MR. *Bootstrap Methods: A Practitioner's Guide*. New York: Wiley; 1999.

Chernoff H and Savage IR. Asymptotic normality and efficiency of certain nonparametric test statistics. *Annals Math. Statist;* 1929.

Clark RM. A randomization test for the comparison of ordered sequences. *Math. Geology*. 1989; 21: 429–442.

Cliff AD and Ord JK. *Spatial Processes: Models and Applications*. London: Pion Ltd; 1981.

Coad DS and Rosenberger WF. A comparison of the randomized play-the-winner rule and the triangular test for clinical trials with binary responses. *Statist. Med.* 1999; 18: 761–769.

Cohen A and Sackrowitz HB. Methods of reducing loss of efficiency due to discreteness of distributions. *Statist. Methods Med. Res.* 2003; 12: 23–36.

Cohen A and Sackrowitz HB. An evaluation of some tests of trend in contingency tables. *JASA*. 1992; 87: 470–475.

Cohen A and Sackrowitz HB. Directional tests for one sided alternatives in multivariate models. *Annals Statist*. 1998; 26: 2321–2338.

Cohen A and Sackrowitz HB. Testing whether treatment is "better" than control with ordered categorical data: Definitions and complete class theorems. *Statistics and Decisions*. 2000; 18: 1–25.

Commenges D. Transformations which preserve exchangeability and application to permutation tests. *Nonparametric Statistics*. 2003; 15.

Conover WJ; Johnson ME and Johnson MM. Comparative study of tests for homogeneity of variances: With applications to the outer continental shelf bidding data. *Technometrics*. 1981; 23: 351–361.

Corcoran CD and Mehta CR. Exact level and power of permutation, bootstrap, and asymptotic tests of trend. *J. Modern Appl. Statist. Meth*. 2002; 1: 42–51.

Cornfeld J. A statistical problem arising from retrospective studies. In J Neyman (ed.), *Proc. 3rd Berkeley Symp. Math. Statist. Probab*. Berkeley: University of California Press; 1956; 4: 135–138.

Cory-Slechta DA; Weiss B and Cox C. Tissue distribution of Pb in adult vs old rats: A pilot study. *Toxicology*. 1989; 59: 139–150.

Cox DF and Kempthorne O. Randomization tests for comparing survival curves. *Biometrics*. 1963; 19: 307–317.

Cox DR and Snell EJ. *Applied Statististics. Principles and Examples*. London: Chapman and Hall; 1981.

Cramer H. *Mathematical Methods of Statistics*. Princeton, NJ: Princeton University Press; 1946.

Dardanoni V and Forcina A. A unified approach to likelihood inference of stochastic orderings in a non-parametric context. *JASA*. 1998; 93: 1112–1123.

David HA. *Order Statistics*. New York: Wiley; 1970.

Daw NC; Arnold JT; Abushullaih BA; Stenberg PE; White MM; Jayawardene D; Srivastava DK and Jackson CWA. Single intravenous dose murine megakaryocyte growth and development factor potently stimulates platelet production challenging the necessity for daily administration. *Blood*. 1998; 91: 466–474.

Dean AM and Verducci JS. Linear transformations that preserve majorization, Schur concavity, and exchangeability. *Linear Algebra Appl*. 1990; 127: 121–138.

DeCani J. An algorithm for bounding tail probabilities for two-variable exact tests. *Randomization*. 1979; 2: 23–24.

Dempster AP. A high dimensional two-sample significance test. *Ann. Math. Statist*. 1958; 29: 995–1010.

Dey DK; Rothenberg E; Sundh V; Bosaeus I and Steen B. Height and body weight in elderly adults: A 21-year population study on secular trends and related factors in 70-year-olds. *J. Gerontol A Biol. Sci. Med. Sci*. 2001; 56: M780–784.

Diaconis P and Efron B. Computer intensive methods in statistics. *Scientific American.* 1983; 48: 116–130.

Diaconis P and Sturmfels B. Algebraic algorithms for sampling from conditional distributions. *Annals Statist.* 1998; 26: 363–397.

Dodge Y. (ed.), *Statistical Data Analysis Based on the L1 Norm and Related Methods.* Amsterdam: North Holland; 1987.

Donegani, M. An adaptive and powerful test. *Biometrika.* 1991; 78: 930–933.

Draper D; Hodges JS; Mallows CL and Pregibon D. Exchangeability and data analysis (with discussion). *J. Roy. Statist. Soc. A.* 1993; 156: 9–28.

Draper NR and Stoneman DM. Testing for the inclusion of variables in linear regression by a randomization technique. *Technometrics.* 1966; 8: 695–699.

Dunnett CW and Tamhane AC. A step-up multiple test procedure. *JASA.* 1992; 87: 162–170.

Dupont WD. Sensitivity of Fisher's exact test to minor perturbations in 2×2 contingency tables. *Statist. Med.* 1986; 5: 629–635.

Durstenfield R. Random permutations. *Commun. ACM.* 1964; 7: 420.

Dwass M. On the asymptotic normality of some statistics used in nonparametric tests. *Ann. Math. Statist.* 1955; 26: 334–339.

Eden T and Yates F. On the validity of Fisher's z test when applied to an actual sample of nonnormal data. *J. Agricultural Sci.* 1933; 23: 6–16.

Edgington ES. Overcoming obstacles to single-subject experimentation. *J. Educ. Statist.* 1980b; 5: 261–267.

Edgington ES. Randomized single-subject experimental designs. *Behav. Res. Therapy.* 1996; 34: 567–574.

Edgington ES. The role of permutation groups in randomization tests. *J. Educ. Statist.* 1983; 8: 121–145.

Edgington ES. Validity of randomization tests for one-subject experiments. *J. Educ. Statist.* 1980a; 5: 235–251.

Edgington ES and Bland BH. Randomization tests: Application to single-cell and other single-unit neuroscience experiments. *J. NeuroSci. Meth.* 1993; 47: 169–177.

Efron B. Better bootstrap confidence intervals. (with disc.) *JASA.* 1987; 82: 171–200.

Efron B. Bootstrap methods: Another look at the jackknife. *Annals Statist.* 1979; 7: 1–26.

Efron B. Nonparametric standard errors and confidence intervals. *Canadian J. Statist.* 1981; 9: 139–172.

Efron B. Three examples of computer intensive statistical inference. *Sankhya A.* 1988; 50: 338–362.

Efron B and Tibshirani R. Bootstrap measures for standard errors, confidence intervals, and other measures of statistical accuracy. *Statist. Sci.* 1986; 1: 54–77.

El Barmi H and Dykstra R. Testing for and against linear inequality constraints in a multinomial setting. *Canadian J. Statist.* 1995; 23: 131–143.

Entsuah AR. Randomization procedures for analyzing clinical trend data with treatment related withdrawls. *Commun. Statist. A.* 1990; 19: 3859–3880.

Erdos P and Renyi A. On a central limit theorem for samples from a finite population. *Publ. Math. Inst. Hung. Acad. Sci.* 1959; 4: 49–61.

Faris PD and Sainsbury RS. The role of the Pontis oralis in the generation of RSA activity in the hippocampus of the guinea pig. *Psych. and Behav.* 1990; 47: 1193–1199.

Farrar DA and Crump KS. Exact statistical tests for any carcinogenic effect in animal assays. *Fund. Appl. Toxicol.* 1988; 11: 652–663.

Farrar DA and Crump KS. Exact statistical tests for any carcinogenic effect in animal assays. II. Age adjusted tests. *Fund. Appl. Toxicol.* 1991; 15: 710–721.

Fears TR; Tarone RE and Chu KC. False-positive and false-negative rates for carcinogenicity screens. *Cancer Res.* 1977; 37: 1941–1945.

Feinstein AR. Clinical biostatistics XXIII. The role of randomization in sampling, testing, allocation, and credulous idolatry (part 2). *Clinical Pharm.* 1973; 14: 989–1019.

Feldman SE and Kluger E. Shortcut calculations to Fisher–Yates "exact tests." *Psychometrika.* 1963; 2: 289–291.

Finney DJ. Fisher–Yates test of significance in 2×2 contingency table. *Biometrika.* 1948; 35: 145–156.

Fisher RA. Coefficient of racial likeness and the future of craniometry. *J. Royal Anthrop Soc.* 1936; 66: 57–63.

Fisher RA. *Design of Experiments.* New York: Hafner; 1935.

Fisher RA. The logic of inductive inference (with discussion). *J. Roy. Statist. Soc. A.* 1934; 98: 39–54.

Fisher RA. The use of multiple measurements in taxonomic problems. *Ann. Eugenics.* 1936; 7: 179–188.

Fisher RA. *Statistical Methods for Research Workers.* Edinburgh: Oliver and Boyd; 1st ed 1925.

Fix E; Hodges JL, Jr and Lehmann EL. The restricted chi-square test. In *Studies in Probability and Statistics Dedicated to Harold Cramer.* Stockholm: Almquist and Wiksell; 1959.

Flehinger BJ and Louis TA. Sequential treatment allocation in clinical trials. *Biometrika.* 1971; 58: 419–426.

Forster JJ; McDonald JW and Smith PWF. Markov chain Monte Carlo exact inference for binomial and multinomial logistic regression models. *Statistics and Computing.* 2003; 13: 169–177.

Forster JJ; McDonald JW and Smith PWF. Monte Carlo exact conditional tests for log-linear and logistic models. *J. Roy. Statist. Soc. B.* 1996; 58: 445–453.

Forsythe AB; Engleman L and Jennrich R. A stopping rule for variable selection in multivariate regression. *JASA.* 1973; 68: 75–77.

Foutz RN; Jensen DR and Anderson GW. Multiple comparisons in the randomization analysis of designed experiments with growth curve responses. *Biometrics.* 1985; 41: 29–37.

Frank, D; Trzos RJ and Good P. Evaluating drug-induced chromosome alterations. *Mutation Res.* 1978; 56: 311–317.

Freedman DA. A note on screening regression equations. *Amer. Statist.* 1983; 37: 152–155.

Freeman GH and Halton JH. Note on an exact treatment of contingency, goodness of fit, and other problems of significance. *Biometrika.* 1951; 38: 141–149.

Friedman JH and Rafsky LC. Multivariate generalizations of the Wald–Wolfowitz and Smirnov two-sample test. *Annals Statist.* 1979; 7: 697–717.

Gabriel KR and Feder P. On the distribution of statistics suitable for evaluating rainfall simulation experiments. *Technometrics.* 1969; 11: 149–160.

Gabriel KR and Hsu CF. Evaluation of the power of rerandomization tests, with application to weather modification experiments. *JASA.* 1983; 78: 766–775.

Gabriel KR and Sokal RR. A new statistical approach to geographical variation analysis. *Systematic Zoology.* 1969; 18: 259–270.

Gail MH and Mantel N. Counting the number of r × c contingency tables with fixed marginals. *JASA.* 1977; 72: 859–862.

Gail MH; Tan WY and Piantadosi S. Tests for no treatment effect in randomized clinical trials. *Biometrika.* 1988; 75: 57–64.

Galambos J. Exchangeability. In S Kotz and NL Johnson (eds.), *Encyclopedia of Statistical Sciences.* New York: Wiley; 1986; 7: 573–577.

Garthwaite PH. Confidence intervals from randomization tests. *Biometrics.* 1996; 52: 1387–1393.

Gastwirht JL. Statistical reasoning in the legal setting. *Amer. Statist.* 1992; 46: 55–69.

Gastwirht JL and Rubin H. Effects of dependence on the level of some one-sample tests. *JASA.* 1971; 66: 816–820.

Gelbaum BR and Olmsted JMH. *Counterexamples in Analysis.* Dover; 2003.

Gerig TM. A multivariate extension of Friedman's chi-square test with random covariates. *JASA.* 1975; 70: 443–447.

Gerig TM. A multivariate extension of Friedman's chi-square test. *JASA.* 1969; 64: 1595–1608.

Gill DS and Siotani M. On randomization in the multivariate analysis of variance. *J. Statist. Plan. Infer.* 1987; 17: 217–226.

Gine E and Zinn J. Necessary conditions for a bootstrap of the mean. *Annals Statist.* 1989; 17: 684–691.

Glass AG; Mantel N; Gunz FW and Spears GFS. Time-space clustering of childhood leukemia in New Zealand. *J. Nat. Cancer Inst.* 1971; 47: 329–336.

Gliddentracey CE and Parraga MI. Assessing the structure of vocational interests among Bolivian university students. *J. Vocational Beh.* 1996; 48: 96–106.

Goldberg P; Leffert F; Gonzales M; Gorgenola I and Zerbe GO. Intraveneous aminophylline in asthma: A comparison of two methods of administration in children. *Amer. J. of Diseases of Children.* 1980; 134: 12–18.

Gong G. Cross-validation, the jackknife and the bootstrap: Excess error in forward logistic regression. *JASA.* 1986: 81: 108–113.

Gonzalez Jose R; Garcia-Moro C; Dahinten S and Hernandez M. Origin of Fueguian-Patagonians: An approach to population history and structure using R matrix and matrix permutation methods. *Amer. J. Human Biol.* 2002; 14: 308–320.

Good PI. Extensions of the concept of exchangeability and their applications, *J. Modern Appl. Statist. Methods.* 2002; 1: 243–247.

Good PI. *Applying Statistics in the Courtroom.* London: Chapman and Hall; 2001.

Good PI. Shave and a haircut. *Echoes.* 1994; 1(3): 43–61.

Good PI. Globally almost powerful tests for censored data. *Nonpar. Statist.* 1992; 1: 253–262.

Good PI. Most powerful tests for use in matched pair experiments when data may be censored. *J. Statist. Comp. Simul.* 1991; 38: 57–63.

Good PI. Almost most powerful tests for composite alternatives. *Comm. Statist.—Theory and Methods.* 1989; 18: 1913–1925.

Good PI. Detection of a treatment effect when not all experimental subjects respond to treatment. *Biometrics.* 1979; 35(2): 483–489.

Good PI and Hardin J. *Common Errors in Statistics.* New York: Wiley; 2003.

Good PI and Kemp P. Almost most powerful test for censored data. *Randomization.* 1969; 2: 25–33.

Goodman LA and Kruskal WH. *Measures of Association for Cross-Classifications.* New York: Springer-Verlag; 1979.

Goodman SN. Of p-values and Bayes: A modest proposal. *Epidemiology.* 2001; 12: 295–297.

Graubard BI and Korn EL. Choice of column scores for testing independence in ordered 2 by K contingency tables. *Biometrics.* 1987; 43: 471–476.

Green BF. A practical interactive program for randomization tests of location. *Amer. Statist.* 1977; 31: 37–39.

Greenland S. On the logical justification of conditional tests for 2 × 2 contingency tables. *Amer. Statist.* 1991; 45: 248–251.

Grossman DC; Cummings P and Koepsell TD et al. Firearm safety counseling in primary care pediatrics: A randomized controlled trial. *Pediatrics.* 2000; 106: 22–26.

Gupta PC et al. Community dentistry and oral epidemiology. *Community Dentistry and Oral Epidemiology.* 1980; 8: 287–333.

Haber M. A comparison of some conditional and unconditional exact tests for 2 × 2 contingency tables. *Commun. Statist. A.* 1987; 18: 147–156.

Haberman SJ. Log-linear models for frequency tables with ordered classifications. *Biometrics*. 1974; 30: 589–600.

Hajek J. Limiting distributions in simple random sampling from a finite population. *Publ. Math. Inst. Hung. Acad. Sci.* 1960; 5: 361–374.

Hajek J. Some extensions of the Wald–Wolfowitz–Noether theorem. *Ann. Math. Statist.* 1961; 32: 506–523.

Hajek J and Sidak Z. *Theory of Rank Tests*. Academic Press: New York; 1967.

Hall P. *The Bootstrap and Edgeworth Expansion*. New York: Springer Verlag; 1992.

Hall P. On efficient bootstrap simulation. *Biometrika*. 1989; 76: 613–617.

Hall P. Theoretical comparisons of bootstrap confidence intervals (with discussions). *Ann. Statist.* 1988; 16: 927–953.

Hall P and Wilson SR. Two guidelines for bootstrap hypothesis testing. *Biometrics*. 1991; 47: 757–762.

Halmos P. *Measure Theory*. New York: Springer; 1997.

Halter JH. A rigorous derivation of the exact contingency formula. *Proc. Cambridge Phil. Soc.* 1969; 65: 527–530.

Hampel FR; Ronchetti EM; Rousseuw PJ and Stahel WA. *Robust Statistics: The Approach Based on Influence Functions*. New York: Wiley; 1986.

Hardy GH. *A Course of Pure Mathematics*, 10th ed. Cambridge, England: Cambridge University Press; 1992.

Hartigan JA. Using subsample values as typical values. *JASA*. 1969; 64: 1303–1317.

Hasegawa M; Kishino H and Yano T. Phylogentic inference from DNA sequence data. In K Matusita (ed.), *Statistical Theory and Data Analysis*. Amsterdam: North Holland; 1988.

Hastings WK. Monte Carlo sampling methods using Markov chains and their applications. *Biometrika*. 1970; 57: 97–109.

Henze N. A multivariate two-sample test based on the number of nearest neighbor coincidence. *Annals Statist.* 1988; 16: 772–783.

Hettmansperger TP. *Statistical Inference Based on Ranks*. New York: Wiley; 1984.

Hiatt WR; Fradl DC; Zerbe GO; Byyny RL and Niels AS. Comparative efforts of selective and nonselective beta-blockers on the peripheral circulation. *Clinical Pharmacology and Therapeutics*. 1983; 35: 12–18.

Higgins JJ and Noble W. A permutation test for a repeated measures design. *Applied Statistics Agriculture*. 1993; 5: 240–255.

Hilton JF and Mehta CR. Exact power of conditional and unconditional tests. Going beyond the 2×2 contingency table. *JASA*. 1993; 47: 91–98.

Hilton JF and Mehta CR. Power and sample size for exact conditional tests with ordered categorical data. *Biometrics*. 1993; 49: 609–616.

Hirji KF; Mehta CR and Patel NR. Computing distributions for exact logistic regression. *JASA*. 1987; 82: 1110–1117.

Hisdal H; Stahl K and Tallaksen LM et al. Have streamflow droughts in Europe become more severe or frequent? *Int. J. Climatol.* 2001; 21: 317–321.

Hodges JL and Lehmann EL. Estimates of location based on rank tests. *Ann. Math. Statist.* 1963; 34: 598–611.

Hodges JS. Uncertainy, policy analysis, and statistics. *Statist. Sci.* 1987; 2: 259–291.

Hoeffding W. The large-sample power of tests based on permutations of observations. *Ann. Math. Statist.* 1952; 23: 169–192.

Hoeffding W. Combinatorial central limit theorem. *Annals Math. Statist.* 1951; 22: 556–558.

Hoel DG and Walburg HE. Statistical analysis of survival experiments. *J. Nat. Cancer Inst.* 1972; 49: 361–372.

Hogg RV and Lenth RV. A review of some adaptive statistical techniques. *Commun. Statists.* 1984; 13: 1551–1579.

Holm S. A simple sequentially rejective multiple test procedure. *Scand. J. Statist.* 1979; 6: 65–70.

Holmes MC and Williams REO. The distribution of carriers of streptococcus pyrogenes among 2413 healthy children. *J. Hyg. Camd.* 1954; 52: 165–179.

Hossein-Zadeh GA; Ardekani BA and Soltanian-Zadeh H. Activation detection in fMRI using a maximum energy ratio statistic obtained by adaptive spatial filtering. *IEEE Trans. Med. Imaging.* 2003; 22: 795–805.

Howard M (pseud. for Good P). Randomization in the analysis of experiments and clinical trials. *American Laboratory.* 1981; 13: 98–102.

Huber PJ. *Robust Statistics.* New York: Wiley; 1981.

Hubert LJ. Combinatorial data analysis: Association and partial association. *Psychometrika.* 1985; 50: 449–467.

Hubert LJ. Nonparametric tests for patterns in geographic variation: Possible generalizations. *Geographical Analysis.* 1978; 10: 86–88.

Hubert LJ and Baker FB. Evaluating the conformity of sociometric measurements. *Psychometrika.* 1978; 43: 31–42.

Hubert LJ; Golledge RG; Costanzo CM; Gale N and Halperin WC. Nonparametric tests for directional data. In G Bahrenberg, M Fischer and P Nijkamp (eds.), *Recent Developments in Spatial Analysis: Methodology, Measurement, Models.* Aldershot UK: Gower; 1984: 171–190.

Hubert LJ and Schultz J. Quadratic assignment as a general data analysis strategy. *Brit. J. Math. Stat. Psych.* 1976; 29: 190–241.

Irony TZ and Pereira CAB. Exact tests for equality of two proportions: Fisher vs Bayes. *J. Statist. Comput. Simul.* 1986; 25: 83–114.

Izenman AJ. Recent developments in nonparametric density estimation. *JASA.* 1991; 86(413): 205–224.

Jagers P. Invariance in the linear model—An argument for chi-square and F in nonnormal situations. *Mathematische Operationsforschung und Statistik*, 1980; 11: 455–464.

James GS. The comparison of several groups of observations when the ratios of the population variances are unknown. *Biometrika.* 1950; 38: 324–329.

Janssen A. Conditional rank tests for randomly censored data. *Annals Statist.* 1991; 19: 1434–1456.

Jennings JM; Mcintosh AR; Kapur S; Tulving E and Houle S. Cognitive subtractions may not add up—The interaction between semantic processing and response-mode. *Neuroimage.* 1997; 5: 229–239.

Jennison C and Turnbull BW. *Group Sequential Methods with Applications to Clinical Trials.* Chapman and Hall/CRC: Boca Raton; 2000.

Jockel KH. Finite sample properties and asymptotic efficiency of Monte Carlo tests. *Annals Statist.* 1986; 14: 336–347.

Joe H. Extreme probabilities for contingency tables under row and column independence with applications to Fisher's exact test. *Commun. Statist: Theory Methods.* 1988; 17: 3677–3685.

Jogdeo K. Asymptotic normality in nonparametric methods. *Annals Math. Statist.* 1968; 39: 905–922.

John RD and Robinson J. Edgeworth expansions for the power of permutation tests. *Annals Statist.* 1983; 11: 625–631.

John RD and Robinson J. Significance levels and confidence intervals for randomization tests. *J. Statist. Comput. Simul.* 1983; 16: 161–173.

Jorde LB; Rogers AR; Bamshad M; Watkins WS; Krakowiak P; Sung S; Kere J and Harpending HC. Microsatellite diversity and the demographic history of modern humans. *Proc. Nat. Acad. Sci. USA.* 1997; 94: 3100–3103.

Kalbfleisch JD and Prentice RL. *The Statistical Analysis of Failure Time Data.* New York: John Wiley and Sons; 1980.

Karlin S; Ghandour G; Ost F; Tauare S and Korph K. New approaches for computer analysis of DNA sequences. *Proc. Nat. Acad. Sci. USA.* 1983; 80: 5660–5664.

Karlin S and Williams PT. Permutation methods for the structured exploratory data analysis (SEDA) of familial trait values. *Amer. J. Human Genetics.* 1984; 36: 873–898.

Kazdin AE. Obstacles in using randomization tests in single-case experiments. *J. Educ. Statist.* 1980; 5: 253–260.

Kazdin AE. Statistical analysis for single-case experimental designs. In M Hersen and DH Barlow (eds.), *Strategies for Studying Behavioral Change.* New York: Pergammon; 1976.

Kempthorne O. In dispraise of the exact test: Reactions. *J. Statist. Plan. Infer.* 1979; 3: 199–213.

Kempthorne O. Why randomize? *J. Statist. Prob. Infer.* 1977; 1: 1–26.

Kempthorne O. Inference from experiments and randomization. In JN Srivastava (ed.), *A Survey of Statistical Design and Linear Models.* Amsterdam: North Holland; 1975: 303–332.

Kempthorne O. The randomization theory of experimental inference. *JASA.* 1955; 50: 946–967.

Kempthorne O. *Design and Analysis of Experiments.* New York: Wiley; 1952.

Kempthorne O; Zyskind G; Addelman S; Throckmorton T and White R. Analysis of variance procedures. Publication of Aeronautical Research Laboratory, USAF; 1961.

Kendall MG; Stuart A and Ord JK. *Advanced Theory of Statistics.* London: Charles Griffin and Co; 1977.

Kendall MG and Stuart A. *The Advanced Theory of Statistics*, Vol 2. 4th ed. New York: Macmillan; 1979.

Kennedy PE. Randomization tests in econometrics. *J. Business and Economic Statist.* 1995; 13: 85–95

Kim MJ; Nelson CR and Startz R. Mean revision in stock prices? A reappraisal of the empirical evidence. *Rev. Econ. Stud.* 1991; 58: 515–528.

Kolassa JE and Tanner MS. Approximate conditional inference in exponential families via the Gibbs sampler. *JASA.* 1994; 89: 697–702.

Kolchin VF and Christyakov VP. On a combinatorial central limit theorem. *Theor. Prob. Appl.* 1973; 18: 728–739.

Konigsberg LW. Comments on matrix permutation tests in the evaluation competing models for modern human origins. *J. Human Evolution.* 1997; 32: 479–488.

Koziol JA; Maxwell DA; Fukushima M; Colmer A and Pilch YH. A distribution-free test for tumor-growth curve analyses with applications to an animal tumor immunotherapy experiment. *Biometrics.* 1981; 37: 383–390.

Krewski D; Brennan J and M Bickis. The power of the Fisher permutation test in 2 by k tables. *Commun. Statist. B.* 1984; 13: 433–448.

Laitenberger O; Atkinson C and Schlich M et al. An experimental comparison of reading techniques for defect detection in UML design documents. *J. Syst. Software.* 2000; 53: 183–204.

Lambert D. Robust two-sample permutation tests. *Ann. Statist.* 1985; 13: 606–625.

Lambert, D. Influence functions for testing. *JASA.* 1981; 76: 649–657.

Lancaster HO. Significance tests in discrete distributions. *JASA.* 1961; 56: 223–234.

Latscha R. Tests of significance in a r × r contingency table: Extension of Finney's table. *Biometrika.* 1953; 40: 74–86.

Lee D. Analysis of phase-locked oscillations in multi-channel single-unit spike activity with wavelet cross-spectrum. *J. Neurosci. Methods.* 2002; 115: 67–75.

Leemis LM. Relationships among common univariate distributions. *Amer. Statist.* 1986; 40: 143–146.

Lefebvre M. Une application des methodes sequentielles aux tests de permutations. *Canad. J. Statist.* 1982; 10: 173–180.

Lehmann EL. *Elements of Large Sample Theory.* New York: Springer; 1999.

Lehmann EL. *Testing Statistical Hypotheses.* 2nd ed. New York: John Wiley and Sons; 1986.

Lehmann EL. *Non-Parametrics: Statistical Methods Based on Ranks.* San Francisco: Holden-Day; 1975.

Lehmann EL and Stein C. On the theory of some nonparametric hypotheses. *Ann. Math. Statist.* 1949; 20: 28–45.

Leslie PH. A method of calculating the exact probabilities in 2×2 contingency tables with small marginal totals. *Biometrika*. 1955; 42: 522–523.

Levin DA. The organization of genetic variability in Phlox drummondi. *Evolution*. 1977; 31: 477–494.

Liebtrau AM. *Measures of Association*. Newhall CA: Sage Publications; 1983.

Lin DY; Wei IJ and DeMets DL. Exact statistical inference for group sequential trials. *Biometrics*. 1991; 47: 1399–1408.

Liptak I. On the combination of independent tests. *Magyar Tudomanmyos Akademia Matematikai Kutato Intezenek Kozlomenyei*, 1958; 3: 17–141.

Liu RY. Bootstrap procedures under some non i.i.d. models. *Annals Statist*. 1988; 16: 1696–1788.

Lock RH. A sequential approximation to a permutation test. *Commun. Statist.—Simul.* 1991; 20: 341–363.

Loh W-Y. Bootstrap calibration for confidence interval construction and selection. *Statist. Sinica*. 1991; 1: 479–495.

Loh W-Y. Calibrating confidence coefficients. *JASA*. 1987; 82: 155–162.

Mackay DA and Jones RE. Leaf-shape and the host-finding behavior of two ovipositing monophagous butterfly species. *Ecol. Entom.* 1989; 14: 423–431.

Mackert JR Jr; Twiggs SW; Russell CM and Williams AL. Evidence of a critical leucite particle size for microcracking in dental porcelains. *J. Dent Res.* 2001; 80: 1574–1579.

Madow, WG. On the limiting distribution of estimates based on samples from finite universes. *Ann. Math. Statist.* 1948; 19: 534–545.

Majeed AW; Troy G; Nicholl JP; Smythe A; Reed MWR; Stoddard CJ; Peacock J and Johnson AG. Randomized prospective single-blind comparison laparoscopic versus small-incision cholecystectomy. *Lancet*. 1996; 347: 989–994.

Makinodan T; Albright JW; Peter CP; Good PI and Hedrick ML. Reduced humoral activity in long-lived mice. *Immunology*. 1976; 31: 400–408.

Manly B and Francis C. Analysis of variance by randomization when variances are unequal. *Aust. New Zeal. J. Statist.* 1999; 41: 411–430.

Manly BFJ. Analysis of polymorphic variation in different types of habitat. *Biometrics*. 1983; 39.

Manly BFJ. *Randomization, Bootstrap and Monte Carlo Methods in Biology*, 2nd ed. London: Chapman and Hall; 1997.

Manly BFJ. The comparison and scaling of student assessment marks in several subjects. *Applied Statististics*. 1988; 37: 385–395.

Mantel N. The detection of disease clustering and a generalized regression approach. *Cancer Res.* 1967; 27: 209–220.

Mantel N and Hankey BJ. Programmed analysis of a 2×2 contingency table. *Amer. Statist.* 1971; 25: 40–44.

Marascuilo LA and McSweeny M. *Nonparametric and Distribution-Free Methods for the Social Sciences*. Monterey, CA: Brooks/Cole; 1977.

March DL. Exact probabilities for R × C contingency tables. *Commun. ACM*. 1972; 15: 991–992.

Marcus LF. Measurement of selection using distance statistics in prehistoric orangutan pongo pygamous palaeosumativens. *Evolution.* 1969; 23: 301.

Maritz JS. *Distribution Free Statistical Methods*, 2nd ed. London: Chapman and Hall; 1996.

Marriott FHC. Barnard's Monte Carlo tests: How many simulations? *Applied Statist.* 1979; 28: 75–77.

Maxwell SE and Cole DA. A comparison of methods for increasing power in randomized between-subjects designs. *Psych. Bull.* 1991; 110: 328–337.

McCarthy PJ. Psuedo-replication: Half samples. *Review Int. Statist. Inst.* 1969; 37: 239–264.

McDonald JW; DeRoure DC and Michaelides DT. Exact tests for two-way symmetric contingency tables. *Statist. Comp.* 1998; 8: 391–399.

McDonald JW and Smith PWF. Exact conditional tests of quasi-independence for triangular contingency tables: Estimating attained significance levels. *Applied Statist.* 1995; 44: 131–151.

McDonald JW; Smith PWF and Forster JJ. Exact tests of goodness of fit of log-linear models for rates. *Biometrics* 1999; 55: 620–624.

McDonald LL; Davis BM and Miliken GA. A nonrandomized unconditional test for comparing two proportions in 2×2 contingency tables. *Technometrics.* 1977; 19: 145–158.

McKinney PW; Young MJ; Hartz A and Bi-Fong Lee M. The inexact use of Fisher's exact test in six major medical journals. *J. Amer. Medical Assoc.* 1989; 261: 3430–3433.

McLachlan G. *Finite Mixture Models.* New York: Wiley; 2000.

McQueen G. Long-horizon mean-reverting stock prices revisited. *J. Financial Quant. Anal.* 1992; 27: 1–17.

Mead R. A test for spatial pattern at several scales using data from a grid of contiguous quadrants. *Biometrics.* 1974; 30: 295–307.

Mehta CR; Patel NR and Gray R. On computing an exact confidence interval for the common odds ratio in several 2×2 contingency tables. *JASA.* 1985; 80: 969–973.

Mehta CR; Patel NR and Senchaudhuri P. Exact power and sample size computations for the Cochran–Armitage trend test. *Biometrics.* 1998; 54: 1615–1621.

Mehta CR; Patel NR and Senchaudhuri P. Importance sampling for estimating exact probabilities in permutational inference. *JASA.* 1988; 83: 999–1005.

Mehta CR; Patel NR and Wei LJ. Computing exact permutational distributions with restricted randomization designs. *Biometrika.* 1988; 75: 295–302.

Mehta CR and Patel NR. FEXACT: A Fortran subroutine for Fisher's exact test on unordered $r \times c$ contingency tables. *ACM Trans. on Math. Software.* 1986a; 12: 154–161.

Mehta CR and Patel NR. A hybrid algorithm for Fisher's exact test in unordered $r \times c$ contingency tables. *Commun. Statist.* 1986b; 15: 387–403.

Mehta CR and Patel NR. A network algorithm for performing Fisher's exact test in $r \times c$ contingency tables. *JASA.* 1983; 78: 427–434.

Mehta CR and Patel NR. A network algorithm for the exact treatment of the 2 × K contingency table. *Commun. Statist. B.* 1980; 9: 649–664.

Micceri, T. The unicorn, the normal curve, and other improbable creatures. *Psychol. Bull.* 1989; 105: 156–166.

Michelat D and Giraudoux P. The feeding behaviour of breeding short-eared owls (Asio flammeus) and relationships with communities of small mammal prey. *Rev. Ecol-Terre Vie.* 2000; 55: 77–91.

Mielke PW. Non-metric statistical analysis: Some metric alternatives. *J. Statist. Plan. Infer.* 1986; 13: 377–387.

Mielke PW. On asymptotic nonnormality of null distributions of MRPP statistics. *Commun. Statist. A.* 1979; 8: 1541–1550.

Mielke PW. Clarification and appropriate inferences for Mantel and Valand's nonparametric multivariate analysis technique. *Biometrics.* 1978; 34: 277–282.

Mielke PW, Jr and Berry KJ. *Permutation Methods—A Distance Function Approach.* New York: Springer; 2001.

Mieike PW and Berry KJ. Permutation covariate analyses of residuals based on Euclidean distance. *Psychological Reports.* 1997a; 81: 795–802.

Mieike PW and Berry KJ. Permutation-based multivariate regression analysis: The case for least sum of absolute deviations regression. *Annals Operations Research.* 1997b; 74: 259–268.

Mielke PW, Jr and Berry KJ. Permutation tests for common locations among samples with unequal variances. *J. Educ. Behav. Statist.* 1994; 19: 217–236.

Mielke PW and Berry KJ. An extended class of permutation techniques for matched pairs. *Commun. Statist—Theory Meth.* 1982; 11: 1197–1207.

Mielke PW; Berry KJ and Brier GW. Application of multiresponse permutation procedures for examining seasonal changes in monthly mean sea-level pressure patterns. *Monthly Weather Rev.* 1981; 109: 120–126.

Mielke PW; Berry KJ and Johnson ES. Multiresponse permutation procedures for a priori classifications. *Commun. Statist.* 1976; A5(14): 1409–1424.

Mielke PW; Berry KJ; Brockwell PJ and Williams JS. A class of nonparametric tests based on multiresponse permutation procedures. *Biometrika.* 1981; 68: 720–724.

Milano F; Maggi E and del Turco MR. Evaluation of the effect of a quality control programme in mammography on technical and exposure parameters. *Radiat. Prot. Dosim.* 2000; 90: 263–266.

Miller RA; Bookstein F; Vandermeulen J; Engle S; Kim J; Mullins L and Faulkner J. Candidate biomarkers aging—Age-sensitive indexes of immune and muscle function covary in genetically heterogeneous mice. *J. Gerontology A—Biol. Sci. Med. Sci.* 1997; 52: B39–B47.

Miller RG. Jackknifing variances. *Annals. Math. Statist.* 1968; 39: 567–582.

Mitani JC; Sanders WJ and Lwanga JS et al. Predatory behavior of crowned hawk-eagles (Stephanoaetus coronatus) in Kibale National Park, Uganda. *Behav. Ecol. Sociobiol.* 2001; 49: 187–195.

Mitchell-Olds T. Analysis of local variation in plant size. *Ecology*. 1987; 68: 82–87.

Mitra SK. On the F-test in the intrablock analysis of a balanced incomplete block design. *Sankhya*. 1961; 22: 279–284.

Motoo, M. On the Hoeffding's combinatorial central limit theorem. *Ann. Inst. Statist. Math.* 1957; 8: 145–154.

Mudholkar GS and Hutson AD. Continuity corrected approximations for "exact" inference with Pearson's chi-square. *J. Statist. Plan. Infer.* 1997; 23: 61–78.

Mueller PS. Plasma free fatty acid response to insulin in schizophrenia. *Arch. Gen. Psych.* 1962; 7: 140–146.

Nelson DE and Zerbe GO. A SAS/IML program to execute randomization of response curves with multiple comparisons. *Amer. Statistician*. 1988; 42: 231–232.

Nelson LS. A randomization test for ordered alternatives. *J. Quality Technology*. 1992; 24: 51–53.

Neyman J and Pearson ES. On the use and interpretation of certain test criteria for purposes of statistical inference. Part I. *Biometrika*. 1928; 20A: 175–240.

Noreen E. *Computer Intensive Methods for Testing Hypotheses*. New York: John Wiley and Sons; 1989.

North BV; Curtis D; Cassell PG; Hitman GA and Sham PC. Assessing optimal neural network architecture for identifying disease-associated multi-marker genotypes using a permutation test, and application to calpain 10 polymorphisms associated with diabetes. *Ann. Hum. Genet.* 2003; 67: 348–356.

O'Brien P. Comparing two samples: Extension of the t, rank-sum, and log-rank tests. *JASA*. 1988; 83: 52–61.

Oden NL. Allocation of effort in Monte Carlo simulations for power of permutation tests. *JASA*. 1991; 86: 1074–1076.

Ogawa J. On the null distribution of the F-statistic in a randomized block under the Neyman model. *Annals Math. Statist.* 1963; 34: 1558.

Oliva A; Farina J and Llabres M. Comparison of shelf-life estimates for a human insulin pharmaceutical preparation using the matrix and full-testing approaches. *Drug. Dev. Ind. Pharm.* 2003; 29: 513–521.

O'Reilly FJ and Mielke PW. Asymptotic normality of MRPP statistics from invariance principles of U-statistics. *Commun. Statist. A*. 1980; 9: 629–637.

Orlowski LA; Grundy WD; Mielke PW and Schumm S. Geological applications of multi-response permutation procedures. *Math. Geol.* 1993; 25: 483–500.

O'Sullivan F; Whitney P; Hinshelwood MM and Hauser ER. Analysis of repeated measurement experiments in endocrinology. *J. Animal Science*. 1989; 59: 1070–1079.

Pagano M and Halvorsen K. An algorithm for finding the exact significance levels of r $\times$ C contingency tables. *JASA*. 1981; 76.

Pagano M and Tritchler D. On obtaining permutation distributions in polynomial time. *JASA*. 1983; 78: 435–441.

Pampoulie C and Morand S. Nonrandom association patterns in parasite infections caused by the host life cycle: Empirical evidence from Kudoa caguensis (Myxosporea) and Aphalloides coelomicola (Trematoda). *J. Parasitol.* 2002; 88: 817–819.

Patefield WM. Exact tests for trends in ordered contingency tables. *Appl. Statist.* 1982; 31: 32–43.

Pattison P; Wasserman S and Robins G et al. Statistical evaluation of algebraic constraints for social networks. *J. Math. Psychol.* 2000; 44: 536–568.

Penninckx W; Hartmann C; Massart DL and Smeyersverbeke J. Validation of the calibration procedure in atomic absorption spectrometric methods. *J. Analytical Atomic Spectrometry.* 1996; 11: 237–246.

Perneger TV. What is wrong with Bonferroni adjustments. *British Medical Journal.* 1998; 136: 1236–1238.

Pesarin F. *Multivariate Permutation Tests.* New York: Wiley & Sons; 2001.

Pesarin F. A resampling procedure for a nonparametric combination method of several dependent permutation tests. *J. Ital. Statist. Soc.* 1992; 1: 87–101.

Pesarin F. On a nonparametric combination method for dependent permutation tests with applications. *Psychotherapy and Psychosomatics.* 1990; 54: 172–179.

Pitman EJG. Significance tests which may be applied to samples from any population. *Roy. Statist. Soc. Suppl.* 1937; 4: 119–130, 225–232.

Plackett RL and Hewlett PS. A unified theory of quantal responses to mixtures of drugs. The fitting to data of certain models for two non-interactive drugs with complete positive correlation of tolerances. *Biometrics.* 1963; 19: 517–531.

Ponton D and Copp GH. Early dry-season community structure and habitat use of young fish in tributaries of the river Sinnamary (French-Guyana South America) before and after hydrodam operation. *Environ. Biol. Fishes.* 1997; 50: 235–256.

Posten HO; Yeh HC and Owen DB. Robustness of the two-sample t-test under violation of the homeogeneity of variance assumption. *Commun. Statist.* 1982; 11: 109–126.

Potthoff RF; Peterson BL and George SL. Detecting treatment-by-centre interaction in multi-centre clinical trials. *Statist. Med.* 2001; 20: 193–213.

Priesendorfer RW and Barnett TP. Numerical model/reality intercomparison tests using small-sample statistics. *J. Atmospheric Sciences.* 1983; 40: 1884–1896.

Puri ML and Sen PK. *Nonparametric Techniques in Multivariate Analysis.* New York: Wiley; 1971.

Puri ML and Sen PK. A class of rank order tests for a general linear hypothesis. *Ann. Math. Statist.* 1969; 40: 1325–1343.

Puri ML and Sen PK. On a class of multivariate, multisample rank-order tests. *Sankyha Ser A.* 1966; 28: 353–376.

Puri ML and Shane HD. Statistical inference in incomplete blocks design. In ML Puri (ed.), *Nonparametric Techniques in Statistical Inference*. Cambridge: University Press. 1970: 131–155.

Quinn JF. On the statistical detection of cycles in extinctions in the marine fossil record. *Paleobiology*. 1987; 13: 465–478.

Ramsey PH. Exact Type I error rates for robustness of Student's t-test with unequal variances. *J. Edu. Statistis*. 1980; 5: 337–349.

Raz J; Zheng H; Ombao H and Turetsky B. Statistical tests for fMRI based on experimental randomization. *Neuroimage*. 2003; 19: 226–232.

Ripley BD. *Spatial Statistics*. New York: Wiley; 1981.

Ritland C and Ritland K. Variation of sex allocation among eight taxa of the Minimuls guttatus species complex (Scrophulariaceae). *Amer. J. Botany*. 1989; 76.

Robbins H. Some aspects of the sequential design of experiments. *Bull. Amer. Math. Soc.* 1952; 58: 527–535.

Robbins H and Siegmund D. Sequential tests involving two populations. *JASA*. 1974; 69: 132–139.

Roberson P and Fisher L. Lack of robustness in time-space disease clustering. *Commun. Statist. B: Simulation and Computing*. 1986; 12: 11–22.

Roberts GO. Markov chain concepts related to sampling algorithms. In WR Gilks, S Richardson and DJ Spiegelhalter (eds.), *Markov Chain Monte Carlo in Practice*. 45–57. London: Chapman and Hall; 1996.

Robertson T; Wright FT and Dykstra RL. *Order Restricted Statistical Inference*. Wiley: Chichester; 1988.

Robinson J. Approximations to some test statistics for permutation tests in a completely randomized design. *Austr. J. Statist*. 1983; 25: 358–369.

Robinson J. Saddlepoint approximations to permutation tests and confidence intervals. *JRSS, B*. 1982; 44: 91–101.

Robinson, J. An asymptotic expansion for samples from a finite population. *Annals Statist*. 1978; 6: 1005–1011.

Robson AJ; Jones TK; Reed DW and Bayliss AC. A study of national trend and variation in UK floods. *Int. J. Climatology*. 1998; 18: 165–182.

Romano JP. On the behavior of randomization tests without a group invariance assumption. *JASA*. 1990; 85(411): 686–692.

Romney AK; Moore CC and Batchelder WH et al. Statistical methods for characterizing similarities and differences between semantic structures. *P. Natl. Acad. Sci. USA*. 2000; 97: 518–523.

Roper RJ; Doerge RW; Call SB; Tung KSK; Hickey WF and Teuscher C. Autoimmune orchitis epididymitis and vasitis are immunogenetically distinct lesions. *Amer. J. Path*. 1998; 152: 1337–1345.

Rosenbaum PR. Permutation tests for matched pairs with adjustments for covariates. *Appl. Statist*. 1988; 37(3): 401–411.

Rosenbaum PR. Conditional permutation tests and the propensity score in observational studies. *JASA*. 1984; 79: 565–574.

Rosenberg J. A methodology for evaluating predictive metrics. *Adv. Comput.* 2000; 53: 285–318.

Royaltey HH; Astrachen E and Sokal RR. Tests for patterns in geographic variation. *Geographic Analysis.* 1975; 7: 369–395.

Ryan JM; Tracey TJG and Rounds J. Generalizability of Holland's structure of vocational interests across ethnicity, gender, and socioeconomic status. *J. Counseling Psych.* 1996; 43: 330–337.

Salmaso L. Synchronized permutation tests in 2^k factorial designs. *Communications in Statistics—Theory and Methods.* 2003; 32: 1419–1438.

Salsburg DS. *The Use of Restricted Significance Tests in Clinical Trials.* New York: Springer-Verlag; 1992.

Santner TJ and Snell MK. Small-sample confidence intervals for $p_1 - p_2$ and p_1/p_2 in 2×2 contingency tables. *JASA.* 1980; 75: 386–394.

Schemper M. A survey of permutation tests for censored survival data. *Commun. Statist. A.* 1984; 13: 433–448.

Sen PK. Nonparametric tests for multivariate interchangeability. Part 2. The problem of MANOVA in two-way layouts. *Sankhya A.* 1969; 31.

Sen PK. Nonparametric tests for multivariate interchangeability. Part 1: Problems of location and scale in bivariate distributions. *Sankhya A.* 1967; 29: 351–372.

Sen PK. On some multisample permutation tests based on a class of U-statistic. *JASA.* 1967; 62: 1201–1213.

Sen PK. On some permutation tests based on U-statistics. *Bull. Calcutta Stat. Assoc.* 1965; 14: 106–126.

Sen PK and Puri ML. On the theory of rank order tests for location in the multivariate one sample problem. *Annals Math. Statist.* 1967; 38: 1216–1228.

Servy EC and Sen PK. Missing variables in multi-sample rank permutation tests for MANOVA and MANCOVA. *Sankhya A.* 1987; 49: 78–95.

Shane HD and Puri ML. Rank order tests for multivariate paired comparisons. *Annals Math. Statist.* 1969; 40: 2101–2117.

Shapiro CP and Hubert L. Asymptotic normality of permutation probabilities derived from the weighted sums of bivariate functions. *Annals Statist.* 1979; 7: 788–794.

Shen CD and Quade D. A randomization test for a three-period three-treatment crossover experiment. *Commun. Statist. B.* 1986; 12: 183–199.

Shorack GR. Testing against ordered alternatives in model I analysis of variance: Normal theory and nonparametric. *Annals Math. Statist.* 1967; 38: 1740–1753.

Siegel S. *Practical Nonparametric Statistics.* New York: Wiley; 1956.

Siegmund H. *Sequential Analysis: Tests and Confidence Intervals.* New York: Springer-Verlag; 1985.

Siemiatycki J and McDonald AD. Neural tube defects in Quebec: A search for evidence of 'clustering' in time and space. *Brit. J. Prev. Soc. Med.* 1972; 26: 10–14.

Silvey SD. Asymptotic distributions of statistics arising in certain nonparametric tests. *Proc. Glasgow Math. Assoc.* 1956; 2: 47–51.

Silvey SD. The equivalence of asymptotic distributions arising under randomization and normal theories. *Proc. Glasgow Math. Assoc.* 1954; 1: 139–147.

Simmons RB and Weller SJ. What kind of signals do mimetic tiger moths send? A phylogenetic test of wasp mimicry systems (Lepidoptera: Arctiidae: Euchromiini). *Proc. Roy. Soc. Lond. B—Biol. Sci.* 2002; 26: 983–990.

Smith AFM and Roberts GO. Bayesian computation via the Gibbs sampler and related Markov chain Monte Carlo methods (with discussion). *JRSS B.* 1993; 55: 3–23.

Solomon H. Confidence intervals in legal settings. In MH DeGroot, SE Fienberg and JB Kadane (eds.), *Statistics and the Law.* New York: John Wiley and Sons; 1986: 455–473.

Soms AP. Permutation tests for k-sample binomial data with comparisons of exact and approximate P-levels. *Commun. Statist. A.* 1985; 14: 217–233.

Spitz MR; Shi HH; Yang F; Hudmon KS; Jiang H; Chamberlain RM; Amos CI; Wan Y; Cinciripini P; Hong WK and Wu XF. Case-control study of the d2-dopamine-receptor gene and smoking status in lung-cancer patients. *J. Nat. Cancer Inst.* 1998; 90: 358–363.

Srivastava MS and Awan HM. On the robustness of Hotelling's T^2-test and distribution of linear and quadratic forms in sampling from a mixture of two multivariate distributions. *Commun. Statist. Theor. Meth.* 1982; 11: 81–107.

Stallard N and Rosenberger WF. Exact group-sequential designs for clinical trials with randomized play-the-winner allocation. *Statist. Med.* 2002; 21: 467–480.

Still AW and White AP. The approximate randomization test as an alternative to the F-test in the analysis of variance. *Brit. J. Math. Stat. Psych.* 1981; 34: 243–252.

Storer BE and Kim C. Exact properties of some exact test statistics for comparing two binomial populations. *JASA.* 1990; 85: 146–155.

Streitberg B and Roehmel J. On tests that are uniformly more powerful than the Wilcoxon–Mann–Whitney test. *Biometrics* 1990; 46: 481–484.

Streitberg B and Rohmed R. Exact distributions for permutation and rank tests: An introduction to some recently published algorithms. *Stat. Software Newsletter.* 1986; 12: 10–17.

Stuart GW; Maruff P and Currie J. Object-based visual-attention in luminance increment detection. *Neuropsychologia.* 1997; 35: 843–853.

Suissa S and Shuster J. Exact unconditional sample sizes for the 2×2 binomial trial. *J. Roy. Statist. Soc A.* 1985; 148: 317–327.

Suissa S and Shuster JJ. Are uniformly most powerful unbiased tests really best? *Amer. Statistician.* 1984; 38: 204–206.

Sukhatme BV. A two sample distribution free test for comparing variances: *Biometrika.* 1958; 45: 544–548.

Syrjala SE. A statistical test for a difference between the spatial distributions of two populations. *Ecology.* 1996; 77: 75–80.

Tallis GM. Goodness of fit. In *Encyclopedia of Statistical Science*. V3. New York: Wiley; 1983.

Tardif S. On the almost sure convergence of the permutation distribution for aligned rank test statistics in randomized block designs. *Annals Statist.* 1981; 9: 190–193.

ter Braak CJF. Permutation versus bootstrap significance tests in multiple regression and ANOVA. In KH Jockel, G Rothe and W Sendler (eds.), *Bootstrapping and Related Techniques*. 79–86. Berlin: Springer-Verlag; 1992.

Thomas DG. Exact confidence limits for the odds ratio in a 2×2 table. *J. Roy. Statist. Soc. C.* 1971; 20: 105–110.

Tiku ML and Singh M. Robust test for means when population variances are unequal. *Commun. Statist.* A10: 2057–2071.

Tocher KD. Extension of the Neyman–Pearson theory of tests to discontinuous variates. *Biometrika.* 1950; 37: 1301–1444.

Tracy DS and Khan KA. Fourth exact moment results for MRPB and related power performance. *Commun. Statist. A.* 1991; 20: 2701–2718.

Tracy DS and Khan KA. Comparison of some MRPP and standard rank tests for three equal sized samples. *Commun. Statist. B.* 1990; 19: 315–333.

Tracy DS and Tajuddin IH. Empirical power comparisons of two MRPP rank tests. *Commun. Statist. A.* 1986; 15: 551–570.

Tracy DS and Tajuddin IH. Extended moment results for improved inferences based on MRPP. *Commun. Statist. A.* 1985; 14: 1485–1496.

Trieschmann JS and Pinches GE. A multivariate model for predicting financially distressed P-L insurers. *J. Risk Insurance.* 1973; 40: 327–338.

Tritchler D. On inverting permutation tests. *JASA.* 1984; 79: 200–207.

Troendle JF. A stepwise resampling method of multiple hypothesis testing. *JASA.* 1995; 90: 370–378.

Tu D and Zhang L. Jackknife approximations for some nonparametreic confidence intervals of functional parameters based on normalizing transformations. *Compu. Statist.* 1992; 7: 3–5.

Tukey JW; Brillinger DR and Jones LV. *Management of Weather Resources*, Vol II: The role of statistics in weather resources management. Washington DC: Department of Commerce, US Government Printing Office; 1978.

Upton GJG. On Mead's test for pattern. *Biometrics.* 1984; 40: 759–766.

Vadiveloo, J. On the theory of modified randomization tests for nonparametric hypothesis. *Commun. Statist.—Theory and Methods.* 1983; 12: 1581–1596.

Valdes-Perez RE. Some recent human-computer studies in science and what accounts for them. *AI Magazine.* 1995; 16: 37–44.

vanKeerberghen P; Vandenbosch C; Smeyers-Verbeke J and Massart DL. Some robust statistical procedures applied to the analysis of chemical data. *Chemometrics Intelligent Lab. Sys.* 1991; 12: 3–13.

Vanlier JB. Limitations of thermophilic anaerobic waste-water treatment and the consequences for process design. *Antonie Van Leeuwenhoek Int. J. General Molecular Microbiol.* 1996; 69: 1–14.

van-Putten B. On the construction of multivariate permutation tests in the multivariate two-sample case. *Statist. Neerlandica.* 1987; 41: 191–201.

Varga J and Toth B. Genetic variability and reproductive mode of Aspergillus fumigatus. *Infect. Genet. Evol.* 2003; 3: 3–17.

Vecchia DF and Iyer HK. Exact distribution-free tests for equality of several linear models. *Commun. Statist. A.* 1989; 18: 2467–2488.

Vecchia DF and Iyer HK. Moments of the quartic assignment statistic with an application to multiple regression. *Commun. Statist.—Theor. and Meth.* 1991; 20: 3253–3269.

Vollset SE; Hirji KF and Afifi AA. Evaluation of exact and asymptotic interval estimators in logistic analysis of matched case-control studies. *Biometrics.* 1991; 47: 1311–1325.

Wald A. *Statistical Decision Functions.* New York: Wiley; 1950.

Wald A and Wolfowitz J. Optimum character of the sequential probability ratio test. *Ann. Math. Statist.* 1948; 19: 326–339.

Wald A and Wolfowitz J. Statistical tests based on permutations of the observations. *Ann. Math. Statist.* 1944; 15: 358–372.

Wallis WA. Compounding probabilities from independent significance tests. *Econometrica.* 1942; 10: 229–248.

Wang FT and Scott DW. The L_1 method for robust nonparametric regression. *JASA.* 1994; 89: 65–76.

Weerahandi S. *Exact Statistical Methods for Data Analysis.* Berlin: Springer Verlag; 1995.

Wei LJ and Durham S. The randomized play-the-winner rule in medical trials. *JASA.* 1978; 73: 840–843.

Wei LJ and Lachin JM. Properties of urn-randomization in clinical trials. *Controlled Clinical Trials.* 1988; 9: 345–364.

Welch BL. On the z-test in randomized blocks and Latin squares. *Biometrika.* 1937; 29: 21–52.

Welch WJ and Guitierrez LG. Robust permutation tests for matched pairs designs. *JASA.* 1988; 83: 450–461.

Werner M; Tolls R; Hultin J and Mellecker J. In *Fifth International Congress on Automation, Advances in Automated Analysis.* Sex and age dependence of serum calcium, inorganic phosphorous, total protein, and albumin in a large ambulatory population. 1970; 2: 59–65.

Westfall DH and Young SS. *Resampling-Based Multiple Testing: Examples and Methods for p-value Adjustment.* New York: John Wiley; 1993.

Weth F; Nadler W and Korsching S. Nested expression domains for odorant receptors in zebrafish olfactory epithelium. *Proc. Nat. Acad. Sci. USA.* 1996; 93: 1321–1326.

Whaley FS. The equivalence of three individually derived permutation procedures for testing the homogeneity of multidimensional samples. *Biometrics*. 1983; 39: 741–745.

Wilcoxon F. Individual comparisons by ranking methods. *Biometrics*. 1945; 1: 80–83.

Wilson HG. Least squares versus minimum absolute deviations estimates in linear models. *Decision Sci.* 1978; 9: 322–335.

Witztum D; Rips E and Rosenberg Y. Equidistant letter sequences in the Book of Genesis. *Statist. Science.* 1994; 89: 768–776.

Wright T. A note on Pascal's triangle and simple random sampling. *College Math J.* 1984; 20: 59–66.

Wu JC; Bell K; Najafi A; Widmark C; Keator D; Tang C; Klein E; Bunney BG; Fallon J and Bunney WE. Decreasing striatal 6-fdopa uptake with increasing duration cocaine withdrawal. *Neuropsychopharmacology*. 1997; 17: 402–409.

Wu XF; Amos CI; Kemp BL; Shi HH; Jiang H; Wan Y and Spitz MR. Cytochrome-p450 2E1 Drai polymorphisms in lung-cancer in minority populations. *Cancer Epidemiology Biomarkers and Prevention.* 1998; 7: 13–18.

Xu R and Li X. A comparison of parametric versus permutation methods with applications to general and temporal microarray gene expression data. *Bioinformatics.* 2003; 19: 1284–1289.

Yale PB. *Geometry and Symmetry.* Holden-Day: San Francisco; 1968.

Yucesan E. Randomization tests for initialization bias in simulation output. *Naval Res. Logistics.* 1993; 40: 643–663.

Zelen M. Play the winner rule and the controlled clinical trial. *JASA.* 1969; 80: 974–984.

Zelen M. The analysis of several 2×2 contingency tables. *Biometrika.* 1971; 58: 129–137.

Zempo N; Kayama N; Kenagy RD; Lea HJ and Clowes AW. Regulation of vascular smooth-muscle-cell migration and proliferation in vitro and in injured rat arteries by a synthetic matrix metalloprotinase inhibitor. *Art. Throm. V.* 1996; 16: 28–33.

Zerbe GO and Murphy JR. On multiple comparisons in the randomization analysis of growth and response curves. *Biometrics.* 1986; 42: 795–804.

Zerbe GO and Walker SH. A randomization test for comparison of groups of growth curves with different polynomial design matrices. *Biometrics.* 1977; 33: 653–657.

Zhang Ji and Boos DD. Adjusted power estimates in Monte Carlo experiments. *Comm. Statist. B.* 1994; 23: 165–173.

Zimmerman DL. A bivariate Cramer-von Mises type of test for spatial randomness. *Appl. Statist.* 1993; 42: 43–54.

Zucker S and Mazeh T. On the statistical significance of the Hipparcos astronomic orbit of ρ Coronae Borealis. *Mon. Not. Roy. Astron. Soc.* 2003.

Zumbo BD and Hubley AM. A note on misconceptions concerning prospective and retrospective power. *Statistician.* 1998; 47: 385–388.

Author Index

Adderley, 5
Agresti, 156, 161, 163, 164, 166, 237, 239
Alber, 55, 63, 247, 273
Albert, 184
Alroy, 7
Aly, 60
Andersen, 57, 58, 107, 204
Ardekani, 7
Arndt, 6
Ascher, 247
Astrachen, 6
Awan, 176

Baglivo, 235
Bailar, 58, 235, 247
Baker, 60, 61, 236
Balakrishnan, 57
Baptista, 151
Barker, 6
Barnard, 150, 233
Barnett, 222, 223
Barton, 176
Bayliss, 6
Bebbington, 253
Beck, 12, 80, 81
Bell, 11, 200
Beran, 276
Berger, 150
Berry, 5, 12, 17, 156, 191, 201, 212, 217, 247, 253
Besag, 234, 235, 250
Bickel, 11, 47, 52, 55, 63, 247, 273, 275

Bickis, 237
Birch, 166
Birnbaum, 82
Bishop, 155
Bithell, 44
Blair, 12, 80, 81, 176
Bland, 6
Boik, 57
Boos, 150, 199, 216, 231
Booth, 247
Boothroyd, 253
Boschloo, 147
Box, 57, 58, 107, 149
Boyett, 177, 237
Bradbury, 246
Bradley, 199
Brennan, 237
Brier, 191
Brillinger, 5
Brockwell, 12
Bross, 6, 164
Brown, 57
Browne, 199, 216
Bryant, 6
Busby, 6
Butler, 247

Cade, 6, 201
Carpenter, 44
Casella, 244
Chan, 150
Chapelle, 6, 184
Chase, 253
Chatterjee, 176

Chernick, 47, 214
Chernoff, 200
Christyakov, 246
Chung, 177
Clark, 6
Clarke, 157–159
Cliff, 12, 191
Clifford, 234, 235, 250
Coad, 115
Cohen, 161, 162, 242, 243
Cole, 104
Commenges, 214, 270, 272
Conover, 57
Copp, 5
Cornfield, 149, 151, 163
Cory-Slechta, 7
Cox, 7, 11, 198
Cramer, 39
Crump, 7
Currie, 6

Dahinten, 5
Daniels, 247
Dardanoni, 173
David, 176, 200
Davis, 147, 176
Daw, 5
Dayhoff, 6
De Cani, 97, 238
de Finetti, 268
Dean, 215, 272
del Turco, 7
DeMets, 12
Dempster, 175
DeRoure, 160
Dey, 6
Diaconis, 12, 245
Dodge, 201, 212
Doksum, 11, 200
Donegani, 207
Donoghue, 11
Draper, 102, 232, 268
Dubins, 24
Dunnett, 80
Dupont, 148
Durham, 114
Durstenfeld, 253
Dwass, 233, 246, 274
Dykstra, 173

Eden, 5
Edgington, 6, 12, 102, 235
Efron, 12, 44, 51, 61, 207, 226
El Barmi, 173
Elashoff, 246
Entsuah, 196
Erdos, 246

Farina, 6
Faris, 6
Farrar, 7
Feder, 5
Feinstein, 6
Feldman, 237, 239
Fienberg, 155
Finney, 236
Fisher, 5, 7, 11, 57, 82, 149, 190
Fix, 211
Flehinger, 113
Forcina, 173
Forster, 244, 245
Forsythe, 102
Foutz, 184
Francis, 55
Frank, 93, 94
Fraser, 177
Freedman, 24
Freeman, 79, 154
Friedman, 47, 177, 178, 192

Gabriel, 5, 7, 42, 249
Gail, 101, 237, 239
Galambos, 268
Garcia-Moro, 5
Garthwaite, 247
Gastwirht, 232
Gelbaum, 256
George, 6, 244
Gerig, 177, 178
Gill, 176
Giraudoux, 6
Girshick, 110
Glass, 6
Gliddentracey, 7
Goldberg, 6, 184
Gong, 221
Gonzalez, 5

Good, 6, 7, 47, 63, 93, 101, 116, 128, 148, 177, 199, 202, 204, 205, 232, 239, 271
Goodman, 83, 156
Gossett, 11
Graubard, 160–162
Gray, 165, 239
Gree, 238
Greenland, 150
Grossman, 6
Guiterrez, 202
Gupta, 153

Haber, 149, 150
Haberman, 163
Hajek, 200, 246
Hall, 44, 57, 125, 224, 276
Halmos, 256
Halter, 154
Halton, 154
Halvorsen, 237, 239
Hampel, 199
Hankey, 151
Hardin, 63, 101, 177, 232
Hartigan, 12
Hasegawa, 218
Henze, 181
Hernandez, 5
Hettmansperger, 232
Hewlett, 6
Hiatt, 184
Higgins, 5, 176, 181
Hill, 110
Hilton, 249
Hirji, 239, 246
Hisdal, 6
Hochberg, 80
Hodges, 11, 83, 211
Hoeffding, 11, 107, 246, 273, 274
Hoel, 6
Hogg, 207
Holland, 155
Holm, 79
Holmes, 168
Hossein-Zadeh, 7
Howard, 5
Hsu, 42, 249
Huber, 201
Hubert, 6, 12, 190, 192, 247

Hubley, 248
Hutson, 154

Irony, 149
Iyer, 192, 194
Izenman, 12

Jagers, 92
James, 125
Janssen, 204
Jennings, 7
Jennison, 113
Jockel, 235
Joe, 242
Jogdeo, 246
John, 42, 247
Johnson, 12, 57
Jones, 5, 6
Jorde, 6

Kalbfleisch, 7, 204
Karlin, 6, 217, 218
Karniski, 80
Kazdin, 12
Kelly, 7
Kempthorne, 5, 11, 39, 101, 107, 246
Kendall, 149, 190
Kennedy, 6
Khan, 191
Kim, 6, 150
Klauber, 5
Kluger, 237, 239
Koch, 24
Kolassa, 244
Kolchin, 246
Konigsberg, 5
Korn, 160–162
Korsching, 6
Koziol, 183, 184
Krewski, 237
Krishino, 218
Kruskal, 156
Kvamme, 5

Lachin, 6
Laitenberger, 6
Lambert, 199, 201
Latscha, 236
Lawley–Hotelling, 173

Lee, 6, 231
Leemis, 75
Lefebvre, 12
Lehmann, 11, 39, 42, 49, 50, 58, 88, 90, 111, 121, 148, 166, 183, 211, 246, 259, 265, 267
Lenth, 207
Leslie, 237, 239
Levin, 6
Li, 5
Liebetrau, 155
Lin, 12, 115
Liptak, 82
Liu, 47, 57
Llabres, 6
Lock, 234
Loh, 276
Louis, 113
Lwanga, 6

Ma, 57
Mackay, 6
Mackert, 6
Madow, 246
Maggi, 7
Majeed, 7
Makinodan, 6
Manly, 6, 55, 157, 219
Mantel, 12, 151, 189, 190, 194, 237, 239
Marascuilo, 7
March, 237
Marcus, 6
Maritz, 42, 59, 201
Marriott, 234
Maruff, 6
Massart, 6
Maxwell, 104
Mazeh, 5
McCarthy, 12
McDonald, 147, 160, 189, 244, 245
McKinney, 6, 145, 229
McLachlan, 75
McQueen, 6
McSweeny, 7
Mead, 193
Mehta, 96, 149, 152, 165, 212, 235, 237, 239–242, 249, 250
Micceri, 11
Michaelides, 160

Michelat, 6
Mielke, 5, 12, 17, 157, 190, 191, 201, 212, 217, 247
Milano, 7
Miliken, 147
Miller, 6, 57, 58
Mitani, 6
Mitchell-Olds, 5
Mitra, 246
Morand, 6
Motoo, 246
Mudholkar, 154
Murphy, 184

Nadler, 6
Nelson, 6, 7, 184
Neyman, 11, 34, 70, 83, 176, 209
Noble, 5, 181
Noreen, 235
North, 6

O'Brien, 55
O'Reilly, 247
O'Sullivan, 6
Oden, 249
Oliva, 6
Olmsted, 256
Onghena, 19
Ord, 12, 191
Orlowski, 6
Owen, 55

Pagano, 237, 239, 245
Pampoulie, 6
Park, 5
Parraga, 7
Patefield, 163, 237, 239
Patel, 96, 149, 152, 165, 212, 235, 237, 239–242, 249, 250
Pattison, 7
Pearson, 11, 34, 83, 190, 209
Penninckx, 6
Pereira, 149
Perneger, 79
Pesarin, 12, 82, 128, 169, 183
Peterson, 6
Piantadosi, 101
Pike, 151
Pinches, 186

Pitman, 11, 94, 190, 246
Plackett, 6
Ponton, 5
Posten, 55
Potthoff, 6
Prentice, 7, 204
Priesendorfer, 222, 223
Puri, 11, 182, 183

Quade, 101
Quinn, 5, 6

Rafesky, 178
Rafsky, 192
Ramsey, 55
Raz, 6, 7, 96
Reed, 6
Renis, 199
Renyi, 246
Richards, 201
Ripley, 193
Rips, 7
Ritland, 5
Robbins, 113, 114
Roberson, 190
Roberts, 245
Robertson, 173
Robinson, 42, 47, 246, 247, 275
Robson, 6
Roehmel, 162
Rohmed, 237
Romano, 39, 55, 63, 202, 275
Romney, 6
Roper, 6
Rosenbaum, 102, 103, 197
Rosenberg, 6, 7, 115
Rosenberger, 115
Rounds, 7
Royaltey, 6
Rubin, 232
Russell, 6
Ryan, 7

Sackrowitz, 161, 162, 242, 243
Sainsbury, 6
Salmaso, 128
Salsburg, 6
Sampford, 204
Sanders, 6

Santner, 167
Savage, 200
Sawilowsky, 176
Schemper, 204
Schultz, 6, 192
Scott, 70, 201
Sen, 11, 176, 182, 183
Senchaudhuri, 96, 152, 235, 239, 242, 249, 250
Servy, 176
Shane, 11, 182
Shao, 263
Shapiro, 247
Shell, 198
Shen, 101
Shorack, 173
Shuster, 150, 177
Sidak, 200
Siegel, 11
Siegmund, 111, 113
Siemiatycki, 189
Silvey, 246
Simmons, 6
Singh, 55
Siotani, 176
Smeyers-Verbeke, 6
Smirnov, 178
Smith, 160, 244, 245
Snell, 167
Sokal, 6, 7
Solomon, 6
Soltanian-Zadeh, 7
Soms, 163
Spearman, 190
Spitz, 6
Srivastava, 176
Stallard, 115
Startz, 6
Stein, 11
Stoneman, 102
Storer, 150
Streitberg, 162, 237
Stuart, 6, 149
Sturmfels, 245
Suissa, 150
Sukhatme, 57
Syrjala, 193

Tajuddin, 191

Tallis, 118
Tan, 101
Tanner, 244
Tardif, 246
Tarnhane, 80
Taylor, 204
ter Braak, 63
Thomas, 151
Tibshirani, 44, 51
Tiku, 55
Tilbury, 236
Tippett, 177
Tocher, 144
Toth, 6
Tracey, 7
Tracy, 191
Trieschman, 186
Tritchler, 42, 245
Troendle, 12, 80, 81
Trzos, 93
Tu, 45
Tukey, 5
Turnbull, 113
Twiggs, 6

Upton, 193

Vadiveloo, 235
Valdesperez, 6
van Keerberghen, 6
van Putten, 181
Van Zwet, 11, 47, 52, 55, 63, 247, 273, 275
Vandenbosch, 6
Vanlier, 5
Varga, 6
Vecchia, 192, 194
Verducci, 215, 272
Vollset, 246

Wackerly, 164, 237
Walburg, 6
Wald, 11, 97, 111, 174, 178, 273, 274
Walker, 184
Wallis, 82
Wang, 201
Weerahandi, 55
Wei, 6, 12, 114, 239, 242
Weiss, 7
Welch, 58, 91, 202, 246
Weller, 6
Werner, 185
Westfall, 12, 80
Weth, 6
Whaley, 190, 192
Wilcoxon, 11
Williams, 6, 12, 168, 218
Wilson, 44, 201, 224
Witztum, 7
Wolfowitz, 11, 97, 111, 174, 178, 273, 274
Wolpert, 46
Wright, 173, 236
Wu, 6

Xu, 5, 231

Yano, 218
Yates, 5
Yeh, 55
Young, 12, 80
Yucesan, 6

Zelen, 114, 151
Zempo, 6
Zerbe, 184
Zhang, 45, 231
Zimmerman, 193
Zucker, 5
Zumbo, 248

Subject Index

a *prior* probability density, 26
acceptance region, 17, 40, 41, 272
accuracy, 57
acyclic network, 240
adaptive rule, 114
adaptive test, 207
admissible test, 14, 20, 50
advertising, 5
after-the-fact hypothesis, 146
aging, 7, 138
agriculture, 5, 99, 119, 125, 127, 196
AIDS, 114
algorithm, 106, 152, 160, 234, 237, 242, 253
alternative hypothesis, 8, 20, 49, 86, 87, 148, 155, 248, 250, 274
Aly's statistic, 60
analog, 69
analysis of covariance, 102
analytic function, 261, 266
ancillary, 262
anencephalic infants, 190
animal, 193, 204
ANOVA, 89, 120, 122, 123, 140, 182
anthropology, 5
antibody, 167
aquatic science, 5
archaeology, 5, 191
ASN, 112
assessment, 219
association, 163, 164, 166
assumptions, 14, 158, 198, 232
astronomy, 5, 70

asymptotic approximation, 107, 247, 250, 276
asymptotic consistency, 39
asymptotic expansions, 274
asymptotic properties, 193
asymptotically exact, 47, 57, 61, 141
asymptotically normal, 115
atmospheric science, 5, 222

back-up statistic, 147, 162, 242
bacterial infection, 167
balanced design, 134, 139
baseline, 99
Bayes' factor, 83
Bayes' risk, 26, 27
Bayes' rule, 83
Bayes' solution, 27, 31
BC_a interval, 44, 48, 225
Behrens–Fisher, 55
beta distribution, 27
between-group sum of squares, 91
biases, 103
binomial, 75, 82, 84, 90, 110, 116, 167, 235
binomial distribution, 67, 71, 76, 112
binomial trials, 148
bioequivalence, 184
biological significance, 218, 232
biology, 5, 138, 217
biotechnology, 5
birth control, 186
birth defects, 230
birth weight, 164

bivariate correlation, 62
bivariate normal distribution, 62
block, 98, 99, 165, 177, 182, 197
blocking variables, 229
blood chemistries, 170
blood pressure, 17
bootstrap, 12, 21, 44, 46, 125, 139, 140, 206, 218, 221, 224, 231, 247, 251, 262, 275
bootstrap confidence interval, 61, 223, 276
Borel field, 256
Borel measurable functions, 256
botany, 5
boundedly complete, 262
Boyle's law, 1
branch and bound method, 235–250
Breslow and Day statistic, 153

C code, 72, 174
cancer, 146
carcinogenicity, 15
cardiology, 6
categories, 143, 156, 211
Cauchy, 34
cause-and-effect model, 177
cell cultures, 7–8, 69, 76
censored, 202–204, 236
central limit theorem, 39
characteristic function, 245, 266
chemistry, 6
chi-square, 75, 154
chi-square distribution, 82, 121, 152, 155, 176
chi-square statistic, 146, 153, 157, 161, 166, 247
chi-square test, 149
cholera, 189
choosing
    statistic, 180
    test, 157
circadian series, 268
climatology, 6
clinical study, 197, 202
clinical trials, 4, 6, 26, 27, 79, 121, 169, 239, 250
clone, 70
cluster analysis, 12
Cochran–Armitage statistic, 249

Cochran–Armitage test, 96
Cohen's kappa, 156
coin, 5
collinear, 214
comparing two treatments, 114
complement of a set, 255
composite hypothesis, 11, 21, 22, 49
computational efficiency, 242
computer science, 6
conditional independence, 166
conditional probability, 263
confidence interval, 40–47, 64, 139, 150, 151, 224, 272
confounded, 196
conservative test, 22, 47, 60, 147
contingency table, 143, 146, 149, 154, 155, 158, 239, 243, 244
continuous distributions, 52, 53, 89
control group, 98, 100, 101
control variables, 232
correlation, 44, 62, 190
corticosteroids, 104
countably additive, 255
covariance matrix, 175, 185, 270
covariances, 44
covariate, 101, 102, 103, 164, 197, 232
Cramer's V, 155
critical function, 267
crop yield, 119, 125, 127, 196
culture, 69
cumulative distribution function, 3, 84
curtailed inspection, 112
cut-off values, 38

decision making, 3, 4, 10, 11, 91, 204, 212
decision rule, 25
degrees of freedom, 22, 149, 172
demographics, 6
density function, 131
dentistry, 6, 153
design matrices, 102
deterministic, 1, 136
deviations, 89
deviations from normality, 140
diagnostic imaging, 6
digital, 69
directed chi-square, 162
directional data, 12

discrepancy measure, 240
discrete observations, 49, 90
discrimination, 150
disease, 108, 151
distribution-free, 97
DNA sequencing, 217
dot product, 97
double blind, 70, 230
double exponential, 34
doubly ordered tables, 163
dynamic programming, 241

ecology, 6, 217
econometrics, 6
Edgeworth expansions, 246
education, 6
empirical distribution function, 3
endocrinology, 6
entomology, 6
environment, 189
epidemiological, 167
epidemiology, 6
ergonomics, 6
error rate, 79
estimation, 29, 45
Euclidean distance, 191
Euclidean spaces, 257, 264
event, 3
exact test, 14, 22, 47, 51, 59, 128, 231
Excel, 72
exchangeable, 24, 40, 63, 100, 101,
    127, 128, 158, 213, 215, 230,
    268, 272, 275
exchanges, 130
exchanging labels, 229
expectation, 273
experimental design, 99, 104, 119, 124,
    127, 129, 132, 195, 201
exponential distribution, 2, 71, 75
exponential family, 72, 110, 262, 263

F-distribution, 58, 88, 89, 90, 121, 172
F-ratio test, 57–58, 88–93, 107, 125
factorial design, 254
false negative, 16
false positive, 15
familial traits, 218
fertilizer, 104
finite measures, 256

Fisher's exact test, 143, 148, 154, 166,
    229, 236, 237
Fisher's omnibus statistic, 170
forensics, 6
Freeman–Halton statistic, 153, 154, 237
Friedman's test, 47
fundamental lemma,
    (see Neyman–Pearson lemma.)

gamma distribution, 22, 96
GAMP test, 204, 205, 206
gas mileage, 187
Gaussian distribution,
    (see Normal distribution)
general linear models, 247
genetics, 6
geography, 6
geology, 6
gerontology, 6
Gibbs sampler, 244
global protection, 155
goodness-of-fit, 222
gradient, 104
group of transformations, 85, 88

habitat, 157
Hall–Wilson criteria, 224
health hazard, 113
histogram, 2
history, 10
Hodges–Lehmann estimator, 223
Hotelling's T^2, 172, 174, 176, 180,
    252, 274
hospital, 76
Householder rotations, 102
hypergeometric distribution, 145, 154
hypertension, 4
hypothesis, 5, 68, 80, 138
hypothesis versus alternative, 28

identically distributed, 63, 67
immunology, 6, 138
impartiality, 22, 82, 85
importance sampling, 235, 241, 250
independence, 23, 63, 67, 78, 155
independent pairs, 100
independent tests, 121
indicator function, 81
indifference region, 203

insurance company, 27
interaction, 105, 120, 122, 126, 127, 136, 196
interdependent variables, 67, 158
interval estimate, 44, 224
invariance, 85, 213, 268
invariant, 23, 87, 88, 92
iterative methods, 276

jury, 2

k-sample permutation tests, 125
Kaplan-Meir estimates, 207
kernel estimation, 96

L_1 test, 198, 201
laboratory techniques, 7
Latin square, 104–106
lattice, 129, 132
law suits, 17, 113
least favorable distribution, 64
least squares methods, 101
leukemia, 193
likelihood ratio, 49, 161
likelihood ratio test, 155
limitations, 14
linear correlation, 94
linear forms, 273
linear models, 120, 192
linear rank statistic, 166
linear transformation, 215, 272
linguistics, 6
Liptak combining function, 82
local regression, 96
log-likelihood ratio, 153, 164
log-linear model, 160, 163, 244
log-normal distribution, 77
logarithms, 48
logistic model, 198
logistic regression, 239
loss function, 16, 25, 28, 87, 91, 124, 212, 226

M-factor design, 134, 136
main effect, 119, 120, 124, 126, 135, 196
Mann–Whitney test, 47, 200
Mantel's U, 163, 189, 192
Mantel–Haenszel trend test, 239
manufacturing process, 57

marginal, 143, 144, 163
Markov chain, 244, 245
matched pair, 86, 100, 204
maximal invariant, 85, 86, 87, 269
maximum likelihood, 212
maximum t-test, 177
measurable function, 257, 263, 266
measurable set, 256, 267
measurable transformation, 257
median, 1, 59, 74
medicine, 4, 6, 17, 99, 108, 146, 151, 167, 170, 229
metric, 222
microarray data, 231
micronuclei, 94
mid-p-value, 147
midrank scores, 161–162
mini-max, 27
minimal spanning tree, 178
missing data, 183, 195, 219, 232
mixture of distributions, 11, 78, 123
molecular biology, 6
monotone function, 94, 97, 135, 174, 239
monotone increasing transformation, 45, 225
monotone likelihood ratio, 37, 65, 72, 110, 259, 260
Monte Carlo, 54, 57, 62, 96, 211, 231, 233, 234, 241, 250, 274
most powerful test (see Uniformly most powerful tests)
most stringent test, 85, 269
MRPP, 190
multinomial, 68, 149, 160
multiparameter exponential family, 73
multiple comparison procedure, 81
multiple regression, 79, 177
multivariate data, 158, 169
multivariate distributions, 171, 214, 244
multivariate statistics, 189
mutagenicity, 93
mutually exclusive, 256
myocardial infarction, 197

natural measure, 212
nearest-neighbor, 193
network, 241, 250
network algorithm, 115, 242, 249

network representation, 239
neurobiology, 6
neurology, 6
neuropsychology, 6
neuropsychopharmacology, 6
Neyman structure, 50, 65, 70, 73, 267
Neyman–Pearson lemma, 9, 34, 37, 49, 209, 257, 262
node, 241, 249
nonparametric bootstrap, 47, 57
nonparametric combination, 12, 169, 176
nonparametric test, 34
nonresponders, 216
nonsymmetric distribution, 44
normal alternatives, 55, 201
normal approximation, 180
normal distribution, 2, 36, 42, 45, 58, 63, 74, 75, 88, 90, 123, 138, 210, 225, 246, 270, 273
normal scores, 200
nuisance parameters, 21
null hypothesis, 8, 27, 106, 145, 147, 206, 250, 273

observation vectors, 170
oceanographers, 222
odds ratio, 148, 150, 151, 163, 165, 239
omnibus alternative, 172
omnibus statistics, 82
oncology, 6
one-parameter exponential family, 260
one-sample tests, 48
one-sided alternative, 161
one-sided stopping rules, 112
one-sided test, 22, 145, 232
operating characteristic, 111
optimal algorithm, 254
optimal scores, 200
oral lesions, 153
order statistics, 60, 209, 269
ordered contingency tables, 160
ordered dose response, 95
ordinal scale, 160
organisms, 217
ornithology, 6
orthogonal transformation, 87, 171
outliers, 198, 200

p-value, 40, 120, 152
paired comparisons, 182
pairs of observations, 62
pairwise exchange, 133, 134, 137
paleontology, 6
parameter, 45, 48, 58, 63, 71
parameter space, 261
parametric bootstrap, 45, 46
parametric test, 21, 36, 47, 57, 63, 89, 212, 231, 275
parasitology, 6
partial correlation, 102
path length, 241
pathologist, 231
Pearson type III curve, 191, 247
Pearson's chi-square, 170
pediatrics, 6
permutation, 46, 63, 72, 86, 146, 215, 262, 269, 275
permutation distribution, 34, 94, 96, 107, 153, 173, 191, 194, 218, 219, 237, 245–247
permutation test, 11, 21, 39, 41, 47, 54, 55, 57, 58, 60, 101, 140, 154, 231, 268, 273, 274
pharmacology, 4, 6, 216
physics, 6
physiology, 6, 7
Pitman correlation, 96, 97, 106, 124, 170, 194, 274
Pitman statistic, 95
pivotal quantity, 46
placebo effect, 98
plant growth, 104
play-the-winner, 114
Poisson distribution, 68–76, 90, 91, 107, 116, 149, 167, 210
Polya urn models, 24
power, 14, 18, 20, 23, 33, 49, 52, 55, 97, 107, 109, 111, 123, 148, 149, 199, 200, 207, 216, 231, 248, 258, 274
power function, 22, 110, 261
precision, 57
predictors, 215
premature infants, 164
preranking, 206
primary hypothesis, 128, 148
probability density, 3, 37, 62
probability distribution, 25, 212, 265

probability ratio, 109
probability space, 256
psychology, 7
publication, 231

quality control, 111, 219
quasi-independence, 160

R, 78, 251
radioactive isotopes, 70, 76
radioimmune assay, 202
radiology, 7
random number generator, 72, 104
random variable, 3, 10, 72, 256
random vector, 265
randomization, 39, 104, 229
randomize on the boundary, 147
randomized block design, 124, 246
ranks, 54, 99–100, 170–178, 198–199, 206
rank test, 47, 239
ratings, 14
ratio, 44
rearrangement, 128, 130, 132
recursive relationship, 237
region of indifference, 17, 202, 205
regression, 185, 221, 271
regression coefficients, 101, 184
rejection region, 17, 54, 203
reliability, 7
resampling methods, 61, 212
residuals, 119, 214
response profiles, 183
restricted alternatives, 210
restricted randomizations, 103
robust transformation, 201
runs test, 169

S-plus, 78, 251
sample size, 90, 104, 106, 109, 196, 216, 231, 232, 248
scale parameter, 48, 50, 247
schizophrenics, 188
semiparametric tests, 34, 48
sensitivity analysis, 107, 148
sequential probability ratio test, 109–111
sequential sampling plan, 110–112
sequential test, 11, 234

servicemen, 108
shift algorithm, 236, 238
shift alternatives, 51, 175, 274
side effects, 169
sigma fields, 256, 264
significance level, 17, 18, 33, 40, 95, 108, 111, 123, 126, 150, 153, 173, 176, 235, 248, 249, 257
silicon implants, 113
similarity, 133
simple hypothesis, 33
simulated distributions, 122
single-case designs, 12
singly ordered tables, 163
sociology, 7, 192
software, 235, 251
spatial distributions, 193
standard deviation, 45, 48, 74, 235, 273
standard error, 67
Stata, 78, 252
step-down procedure, 80
step-length constant, 247
step-up method, 80–81
stepwise regression, 221
Still–White test, 141
stochastic, 11, 52–53, 136
stopping rule, 234
strata, 99
Student's t, (see t-test)
Student's t-distribution, 138
studentized, 224, 246
subdesign, 137
subpopulations, 98
subspace, 87, 121, 171, 182, 214
sufficiency, 34, 35, 48, 230, 268
sufficient statistics, 48, 49, 69, 78, 209, 262, 267
surgery, 7
surrogate variables, 232
survival rates, 145, 165
symmetric distribution, 35, 224
symmetric table, 160
symmetry, 42, 51, 82, 213, 232
synchronized exchange, 129, 130
synchronous rearrangement, 133–137, 140

t-statistics, 11, 138, 170
t-test, 18, 38, 51–64, 216

$tau(\tau)$, 156
taxonomy, 7
tensile strength, 27, 116
test of independence, 217
test statistic, 8, 9, 92, 99, 106, 120, 125, 132, 177, 183, 191, 210, 211, 234, 242, 247
theology, 7
ties, 59, 95
time-to-event data, 90
Tippett combining function, 82
tobacco, 197
toxicology, 7
tranquilizer, 188
transformably exchangeable, 270
transformation, 23, 24, 85
triple blinding, 98
tumor, 181
two-sample comparison, 97, 178, 215
two-sided test, 42, 232
Type I censoring, 203
Type I error, 16, 17, 21, 48, 58, 112, 124, 147
Type II error, 16, 17, 19, 224

UMP, 20, 37, 48, 49, 72, 86
UMP invariant test, 91, 115, 183
UMP unbiased test, 41, 50–51, 73, 107, 166, 259–270
unbalanced designs, 122, 137, 140

unbiased, 22, 45, 116
unbiased test, 22
uncertainty coefficient, 156
uniform distribution, 27, 34, 71–77
union, 255
univariate linear hypothesis, 87

vaccine, 70, 107
validation, 222
validation methods, 232
variance, 76, 97, 98, 125, 216, 248
variance comparisons, 61
variation, 1, 4, 14, 16
vasectomy, 197
vector space, 133
vector-valued function, 198
vertex peeling, 242
virology, 7, 116, 217
vitamin E, 8
vocational guidance, 7

Wald's equation, 111
weak compactness theorem, 259
weak exchangeability, 128, 269
weakly mixing, 269
within-group sum of squares, 91

z-test, 58
Zelen statistic, 153
zero point, 86, 88

Springer Series in Statistics *(continued from p. ii)*

Huet/Bouvier/Poursat/Jolivet: Statistical Tools for Nonlinear Regression: A Practical Guide with S-PLUS and R Examples, 2nd edition.
Ibrahim/Chen/Sinha: Bayesian Survival Analysis.
Jolliffe: Principal Component Analysis, 2nd edition.
Knottnerus: Sample Survey Theory: Some Pythagorean Perspectives.
Kolen/Brennan: Test Equating: Methods and Practices.
Kotz/Johnson (Eds.): Breakthroughs in Statistics Volume I.
Kotz/Johnson (Eds.): Breakthroughs in Statistics Volume II.
Kotz/Johnson (Eds.): Breakthroughs in Statistics Volume III.
Küchler/Sørensen: Exponential Families of Stochastic Processes.
Kutoyants: Statistical Influence for Ergodic Diffusion Processes.
Lahiri: Resampling Methods for Dependent Data.
Le Cam: Asymptotic Methods in Statistical Decision Theory.
Le Cam/Yang: Asymptotics in Statistics: Some Basic Concepts, 2nd edition.
Liu: Monte Carlo Strategies in Scientific Computing.
Longford: Models for Uncertainty in Educational Testing.
Manski: Partial Identification of Probability Distributions.
Mielke/Berry: Permutation Methods: A Distance Function Approach.
Molenberghs/Verbeke: Models for Discrete Longitudinal Data.
Pan/Fang: Growth Curve Models and Statistical Diagnostics.
Parzen/Tanabe/Kitagawa: Selected Papers of Hirotugu Akaike.
Politis/Romano/Wolf: Subsampling.
Ramsay/Silverman: Applied Functional Data Analysis: Methods and Case Studies.
Ramsay/Silverman: Functional Data Analysis, 2nd edition.
Rao/Toutenburg: Linear Models: Least Squares and Alternatives.
Reinsel: Elements of Multivariate Time Series Analysis, 2nd edition.
Rosenbaum: Observational Studies, 2nd edition.
Rosenblatt: Gaussian and Non-Gaussian Linear Time Series and Random Fields.
Särndal/Swensson/Wretman: Model Assisted Survey Sampling.
Santner/Williams/Notz: The Design and Analysis of Computer Experiments.
Schervish: Theory of Statistics.
Shao/Tu: The Jackknife and Bootstrap.
Simonoff: Smoothing Methods in Statistics.
Singpurwalla and Wilson: Statistical Methods in Software Engineering: Reliability and Risk.
Small: The Statistical Theory of Shape.
Sprott: Statistical Inference in Science.
Stein: Interpolation of Spatial Data: Some Theory for Kriging.
Taniguchi/Kakizawa: Asymptotic Theory of Statistical Inference for Time Series.
Tanner: Tools for Statistical Inference: Methods for the Exploration of Posterior Distributions and Likelihood Functions, 3rd edition.
van der Laan: Unified Methods for Censored Longitudinal Data and Causality.
van der Vaart/Wellner: Weak Convergence and Empirical Processes: With Applications to Statistics.
Verbeke/Molenberghs: Linear Mixed Models for Longitudinal Data.
Weerahandi: Exact Statistical Methods for Data Analysis.
West/Harrison: Bayesian Forecasting and Dynamic Models, 2nd edition.

ALSO AVAILABLE FROM SPRINGER!

STATISTICAL TOOLS FOR NONLINEAR REGRESSION
Second Edition
S. HUET, A. BOUVIER, M.-A. POURSAT, E. JOLIVET

This book presents methods for analyzing data using parametric nonlinear regression models. The new edition has been expanded to include binomial, multinomial and Poisson non-linear models. It concentrates on presenting the methods in an intuitive way rather than developing the theoretical backgrounds. The examples are analyzed with the free software nls2 updated to deal with the new models included in the second edition. The nls2 package is implemented in S-PLUS and R.

2003/200 PP./HARDCOVER/ISBN 0-387-40081-8
SPRINGER SERIES IN STATISTICS

MONTE CARLO STATISTICAL METHODS
Second Edition
CHRISTIAN P. ROBERT and **GEORGE CASELLA**

The second edition has been revised towards a coherent and flowing coverage of these simulation techniques. This is a textbook intended for a second year graduate course, but someone who either wants to apply simulation techniques for the resolution of practical problems or wishes to grasp the fundamental principles behind those methods can also use it. Chapters 1–5 cover non-Markov Monte Carlo techniques for integration and optimization, while Chapters 7–12 provide a complete coverage of Markov chain Monte Carlo (MCMC) methods. Chapters 13 and 14 provide a path to more recent developments.

2004/680 PP./ HARDCOVER/ISBN 0-387-21239-6
SPRINGER TEXTS IN STATISTICS

RESAMPLING METHODS FOR DEPENDENT DATA
S.N. LAHIRI

This book gives a detailed account of bootstrap methods and their properties for dependent data, covering a wide range of topics such as block bootstrap methods, bootstrap methods in the frequency domain, resampling methods for long range dependent data, and resampling methods for spatial data. The first five chapters of the book treat the theory and applications of block bootstrap methods at the level of a graduate text. The rest of the book is written as a research monograph, with frequent references to the literature, but mostly at a level accessible to graduate students familiar with basic concepts in statistics.

2003/374 PP./HARDCOVER/ISBN 0-387-00928-0
SPRINGER SERIES IN STATISTICS

Free journal sample copies!

Stay current with the latest research; visit springerlink.com to view complimentary electronic sample copies of Springer journals.

To Order or for Information:

In the Americas: **CALL:** 1-800-SPRINGER or **FAX:** (201) 348-4505 • **WRITE:** Springer, Dept. S7870, PO Box 2485, Secaucus, NJ 07096-2485
• **VISIT:** Your local technical bookstore
• **E-MAIL:** orders-ny@springer-sbm.com

Outside the Americas: **CALL:** +49 (0) 6221 345-217/8 • **FAX:** + 49 (0) 6221 345-229 • **WRITE:** Springer Customer Service, Haberstrasse 7, 69126 Heidelberg, Germany • **E-MAIL:** bookorder@springer-sbm.com

PROMOTION: S7870

Springer
the language of science

springeronline.com